NASA Conference Publication 2190

Aircraft Engine Diagnostics

A conference held at
Lewis Research Center
Cleveland, Ohio
May 6-7, 1981

National Aeronautics
and Space Administration

**Scientific and Technical
Information Branch**

1981

PREFACE

During the past four decades, the Lewis Research Center has been providing advances in aeronautical propulsion from the research activities of its staff and its university and industrial grantees and contractors. These advances have helped create the preeminence in aeronautics that has contributed to our national defense, has provided swift and reliable transportation for our people and their goods, and has greatly aided our position in international trade. In recent years substantial resources have also been directed at improving our nation's utilization of energy.

NASA is well aware that the aviation industry is an important segment of our national economy. In 1979 aircraft sales led all U.S. manufacturing industries with a trade surplus of over $10 billion - without which the country would have experienced a one-third greater trade deficit. This favorable balance attributable to the aircraft industry is largely a result of being able to provide a superior product and to continue to upgrade the product. Efforts at improving the performance retention of today's and future engines which will power commercial and military aircraft represent a positive step toward this end.

To provide to representatives from government, industry, and universities the latest findings directly related to improved aircraft engine performance retention, a two-day conference was held in May 1981. This publication contains the papers presented at that conference.

John F. McCarthy, Jr.
Director

CONTENTS

Page

PREFACE iii

ENGINE DIAGNOSTICS FOR PERFORMANCE RETENTION

CF6 JET ENGINE DIAGNOSTICS
    Ron Stricklin, General Electric Company 1

CF6 HIGH PRESSURE COMPRESSOR AND TURBINE CLEARANCE EVALUATIONS
    M. A. Radomski and L. D. Cline, General Electric Company 19

JT9D ENGINE DIAGNOSTICS PROGRAM
    W. J. Olsson and W. J. Stromberg, Pratt & Whitney
    Aircraft Group 43

JT8D ENGINE PERFORMANCE RETENTION
    Albert D. James and David R. Weisel, Pratt & Whitney
    Aircraft Group 63

PERFORMANCE RETENTION OF THE RB211 POWER PLANT IN SERVICE
    B. L. Astridge, Rolls-Royce Ltd., and J. T. Pinder,
    Rolls-Royce Inc. 83

PERFORMANCE DETERIORATION - AIRLINE PERSPECTIVE
    Niels B. Andersen, Pan American World Airways, Inc. 103

ENGINE DURABILITY AND PERFORMANCE RETENTION CONCEPTS

IMPROVING TURBINE ENGINE COMPRESSOR PERFORMANCE RETENTION THROUGH
  AIRFOIL COATINGS
    L. A. Friedrich, Pratt & Whitney Aircraft Group 109

ADVANCED OXIDE DISPERSION STRENGTHENED SHEET ALLOYS FOR IMPROVED
  COMBUSTOR DURABILITY
    R. J. Henricks, Pratt & Whitney Aircraft Group 119

ADVANCED TURBINE BLADE TIP SEAL SYSTEM
    J. W. Zelahy, General Electric Company 137

AN INTRODUCTION TO NASA'S TURBINE ENGINE HOT SECTION TECHNOLOGY
  (HOST) PROJECT
    Daniel J. Gauntner and C. Robert Ensign, NASA Lewis
    Research Center 153

THE NATURE OF OPERATING FLIGHT LOADS AND THEIR EFFECT ON
  PROPULSION SYSTEM STRUCTURES
    Kenneth H. Dickenson and Richard L. Martin, Boeing
    Commercial Airplane Company 175

CONSERVATION OF STRATEGIC AEROSPACE MATERIALS (COSAM)
    Joseph R. Stephens, NASA Lewis Research Center. 189

ENGINE CONDITION MONITORING SYSTEMS

WHO NEEDS ENGINE MONITORING?
    James L. Pettigrew, Wright-Patterson Air Force Base 209

F100 ENGINE DIAGNOSTICS SYSTEM STATUS TO DATE
    James A. Boyless, Wright-Patterson Air Force Base 225

TURBINE ENGINE PERFORMANCE ESTIMATION AND ITS ROLE IN FUTURE
  SYSTEMS
    Ronald L. DeHoff, Systems Control, Inc., and
    Charles A. Skira, Wright-Patterson Air Force Base 243

IMPACT OF AUTOMATED ENGINE MONITORING ON RELIABILITY CENTERED
  MAINTENANCE AND LOGISTICS SUPPORT
    Laura E. Baker and W. Earl Hall, Jr., Systems Control, Inc. 263

A-10/TF34 TURBINE ENGINE MONITORING SYSTEM (TEMS)
    Robert G. Christophel, San Antonio Air Logistics Center 271

REVIEW OF AIDS DEVELOPMENT
    Henk C. Vermeulen, KLM Royal Dutch Airlines, and
    Sven G. Danielsson, SAS Scandinavian Airlines System. 285

HELICOPTER PROPULSION SYSTEM RELIABILITY AND ENGINE
  MONITORING ASSESSMENTS
    John A. Murphy, Bell Helicopter Textron 311

ENGINE HEALTH MONTIORING SYSTEMS - TOOLS FOR IMPROVED
  MAINTENANCE MANAGEMENT IN THE 1980's
    Jonathan C. Kimball, Pratt & Whitney Aircraft Group 323

ENGINE "ON CONDITION" MONITORING - CF6 FAMILY
  60's THROUGH THE 80's
    Harry J. Kent, General Electric Company, and
    Gerwin Dienger, Lufthansa German Airlines 341

ENGINE HEALTH MONITORING - AN ADVANCED SYSTEM
    R. J. E. Dyson, General Electric Company. 357

AN OVERVIEW OF SAE ARP 1587 "AIRCRAFT GAS TURBINE ENGINE
  MONITORING SYSTEM GUIDE"
    John A. Murphy, Bell Helicopter Textron 377

CF6 ENGINE DIAGNOSTICS

Ron Stricklin
General Electric Company

INTRODUCTION

The energy demands of the United States far exceed domestic fuel supplies which creates a severe dependence on foreign oil. This dependence was accentuated by the OPEC embargo in the winter of 1973/1974 which triggered a rapid rise in fuel prices. This price rise (Figure 1) further compounded by other inflation factors has brought about a set of changing economic circumstances with regard to the use of energy. As a result, our government, with the support of the Aviation Industry, initiated programs aimed at reducing fuel demands. One such program sponsored by NASA is the Aircraft Energy Efficiency Program which is directed toward reducing fuel consumption for commercial air transports. An integral portion of this program is the Engine Component Improvement (ECI) Program aimed at improving fuel efficiency of current engines. This ECI Program consists of two parts, 1) Performance Improvement and 2) Engine Diagnostics.

General Electric is participating in both parts of the Engine Component Improvement Program. As part of the program, performance deterioration studies for the CF6-6D and the CF6-50 Engine Models have been conducted. The basic objectives of the latter effort were: 1) to determine the specific causes for engine deterioration which increase engine fuel consumption rates, 2) to isolate short term losses from longer term losses and 3) identify potential means to minimize the deterioration effects. The deterioration studies have been completed and final NASA reports published.

To quantify the effect of engine performance deterioration, the fleet statistics for the CF6 family of engines in 1981 were projected. It is anticipated that the CF6-50 family of engines will amass approximately 3.4 million flight hours and the CF6-6 family over one million flight hours in 1981.

An average deterioration in cruise specific fuel consumption of 1 percent over new engine levels will result in excess fuel consumption of approximately 36 million gallons for the CF6 fleet alone. The effects of small amounts of deterioration throughout the fleet are obviously substantial.

This paper presents a summary of the activities which led to defining deterioration rates of the CF6 family of engines, a description of what was learned and an identification of means of conserving fuel based upon the program findings.

HOW DID WE DO THE JOB

The program to define the deterioration levels and modes for the CF6 family of engines involved four distinct phases: analysis of inbound engine test results, analysis of airline cruise data, analysis of airline test cell data resulting from testing of refurbished engines and inspection of engine hardware.

INBOUND ENGINE TESTS

Testing of engines, removed from aircraft after extensive revenue service, was conducted in order to define, on a specific engine basis, how much specific fuel consumption had increased and provide some insight into which components were the prime contributors to the observed deterioration.

Through the CF6-6 and CF6-50 phases of the program, 15 inbound engine tests were conducted. One of these tests conducted as part of the CF6-6 Program, was specifically accomplished to identify short term losses.

For each of the inbound engine tests, sufficient instrumentation was installed to measure overall engine deterioration and to indicate the magnitude of deterioration of each major component. After the inbound tests had been conducted, three of the engines were subjected to a detailed teardown inspection by design engineers to relate hardware condition to inbound test results.

CRUISE DATA ANALYSIS

Inbound engine tests, however, which are conducted on specific engines, yield only limited information concerning the degradation in performance of the average fleet. Recognizing that the intent of the Diagnostics Program was defined to determine the deterioration characteristics of the typical CF6 engine in revenue service, it was concluded that analysis of fleet performance data accumulated during flight was the best means of accomplishing this objective.

Data from many airlines are supplied to General Electric on a periodic basis. These data are supplied in many forms from logs recorded by flight engineers in the cockpit to data recorded via automatic data acquisition systems. These data, which are in general recorded during every flight of an aircraft, were used to define the deterioration characteristics of individual engines during the life of the engines during a given installation period. The process to define the performance trend was to compare the performance indicating parameters (fuel flow level and exhaust gas temperature level) to a reference engine parameter level at the flight condition and power setting.

Data from five airlines using CF6-6D engines and from 9 airlines using CF6-50 engines were reviewed. In all, data from 239 CF6-6D engines and 263

CF6-50 engines were analyzed in defining deterioration rates of initial installation and multiple installation engines in revenue service.

General Electric obtained and analyzed data recorded at cruise during initial aircraft checkout flights conducted by the aircraft manufacturer to determine if performance degradation occurred within an engine prior to initial revenue service. Data from 82 CF6-6D engines and data from 111 CF6-50 engines were analyzed in order to determine the magnitude of any "Short-Term" deterioration of engine performance prior to airline receipt of the aircraft and engines. As will be discussed in more detail later, it was concluded after this analysis that significant deterioration did occur during these aircraft checkout activities.

The CF6-6D engine removed from a DC-10 aircraft and subjected to an inbound performance run verified that the indicated loss based upon cruise data analysis was indeed real and non-reversible. As mentioned, this engine was disassembled and critically inspected by a team of General Electric engineers to define the area of performance degradation. Another engine, removed early after entrance into revenue service due to vibration problems, was also tested inbound and similarly confirmed that the short-term loss of performance was real.

AIRLINE CELL DATA ANALYSIS

An important part of the analysis effort to understand airline fleet engine performance levels centered around the definition of basic engine performance levels after overhaul in the airline shops.

Performance levels were reviewed for engines outbound after overhaul at a major airline overhaul facility during the CF6-6D Program and at one consortium central agency and 5 other overhaul facilities during the CF6-50 Program.

Performance levels from these facilities were compared to new engine performance levels from the General Electric Production Facilities in order to define the effectiveness of typical engine workscopes in restoring performance by refurbishment to new engine levels.

COMPONENT DETERIORATION MECHANISMS

The actual modes of deterioration were identified by hardware observation by General Electric teams. Teams of Mechanical and Aerodynamic Design personnel visited various maintenance facilities and conducted detailed inspections of the various engine modules in the disassembled stage to assess the condition of component parts relative to the condition of new hardware. Observations of rotor clearances, surface finishes of the airfoils, cleanliness and smoothness of various static structures and potential air leakage paths were reviewed and yielded estimates of component performance relative to a non-deteriorated component.

Hardware from each major module at various stages of engine life was observed, thus allowing estimation of the deterioration associated with any module degradation mechanism as a function of time and cycles.

Combination of the trends established for each module degradation mechanism yielded module performance deterioration trends. Combination of the module deterioration characteristics using appropriate knowledge of the engine cycle then led to establishing overall engine deterioration characteristics. In all cases throughout both the CF6-6 and CF6-50 Programs, the estimates of engine deterioration established based upon hardware examinations showed excellent agreement with the overall deterioration rates established by cruise and cell data analysis.

WHAT WAS LEARNED

Figure 2 shows the resulting assessment of CF6-6D performance deterioration characteristics for the typical engine thru its initial installation and experience in review service and for the same typical engine after several multiple installations. Each of the elements of deterioration is presented in Figure 2 for the CF6-6D engine. This Figure shows equivalent cruise specific fuel consumption increases relative to a production new engine. The initial installation is shown on the left. Engines incur an average Short-Term loss of 0.9 percent prior to revenue service. During their initial installation, SFC increases an average of 1.7 percent based on the 4000-hour family of engines. The total increased SFC of the deteriorated engine is thus 2.6 percent from production new. Insufficient data is available to determine the amount of performance restoration during the first shop visit.

During the "nth" installation, the serviceable engine re-enters revenue service after a shop visit with an average unrestored cruise SFC loss of 2.1 percent. During revenue service, the cruise SFC of this multiple-build engine increases 0.9 percent for the 3000-hour engine. The total increased SFC of this deteriorated engine at 3000 hours was 3.0 percent from new. On the average, 0.9 percent cruise SFC is restored during the shop visit. During the next installation, an average revenue service deterioration of 0.9 percent is incurred. This amount is restored on the average during maintenance and the cycle is repeated.

Though engine-to-engine variations within this cycle are significant, the data presented reflects the typical or average engine deterioration characteristic for the CF6-6D engine.

Figure 3 shows the deterioration characteristics resulting from cruise data analysis of data obtained for the CF6-50 engine on various aircraft. General findings of the program were that the Short-Term losses which occurred during the airframer checkout of the aircraft tended to be the same for operation on all three aircraft. Also, the unrestored performance level of the multiple-build engine as refurbished by the various airlines was essentially the same. It can be noted from Figure 3 that the deterioration rate shown during typical 747 operation was lower than observed with DC-10 and A300 operations. The same relationship holds for both the initial and multiple installations. The unrestored SFC of the typical engine re-entering revenue service after airline shop visits is 1.8 percent poorer than the new engine baseline for the CF6-50 engine as compared with the 2.1 percent determined in CF6-6D analysis.

The deterioration rates shown on Figures 2 and 3 are presented as a function of flight hours since installation. An analysis was conducted as part of the CF6-50 program, to understand the variability in deterioration rates which resulted from analysis of DC-10, 747 and A300B data. The conclusion was that deterioration rates for the data surveyed was most strongly influenced by average flight length per cycle and the amount of derate or reduced power being used by the individual operators. Table 1 shows the data from Figure 3 translated into the deterioration rate per 1000 cycles basis. The conclusion is that while the DC-10 and 747 data are reasonably consistent and show approximately the same deterioration rate per 1000 cycles, the A300B data shows a much lower deterioration rate per 1000 cycles. Since the A300B data studied as part of this program were consistent with flight cycle lengths of approximately 1.9 hours, the lower deterioration rate per 1000 cycles suggests that deterioration rates are not only influenced by numbers of cycles but also time at temperature.

Figure 4 illustrates the results of the hardware inspection analyses and the resulting deterioration model compared to the performance-data-derived deterioration level for the CF6-6D initial installation. It shows that the largest portion of the 0.9 percent Short-Term SFC loss resulted from High Pressure Turbine (HPT) performance losses. This loss was due largely to HPT clearance increases during the initial checkout phases of the airplane. During initial operation of the aircraft by the aircraft manufacturer, there is little attendant loss in fan, high pressure compressor and low pressure turbine. It is also to be noted that the combined performance deterioration level created by the stackup of the individual component deterioration losses at 4000 hours shows 2.3 percent total performance degradation from the "as new" condition compared to the 2.6 percent level which resulted from performance data analysis.

Figure 5 shows the deterioration mechanisms as assessed by hardware inspection for the CF6-6D multiple-build engines. The major deterioration of a multiple-build engine is within the HPT module. Typically, HPT performance is restored during every shop visit while fan, HP compressor and LPT performance levels are not. Therefore, each engine as it re-enters revenue service after an overhaul shop visit has new HPT hardware and somewhat deteriorated fan, HPC, and LPT performance levels. It can be noted again from Figure 5 that the results of the hardware inspection show 3.3 percent performance loss at 3000 hours on multiple-build engines compared to the performance data analysis level which indicated 3.0 percent. Again, agreement is good. Similar findings for losses associated with the initial installation and the multiple installations of CF6-50 engines resulted.

Of prime importance to the program was the finding that the unrestored loss for the typical engine out of the overhaul shop, based on hardware inspections, was 2.08 percent in terms of cruise SFC compared to the 2.1 percent unrestored performance level as identified by performance data analysis.

Figure 6 describes the component breakdown for both CF6-6D and CF6-50 engine models as shipped from the airline overhaul facilities compared to the new engine performance levels. It shows that, of the 2.1 percent unrestored performance for the CF6-6D engine and 1.8 percent unrestored performance for the CF6-50 engine, a large portion of these performance losses are due to lack of performance restoration in the fan area with lesser amounts of the performance loss associated with the high pressure compressor and the LP turbines. Note that there is very little performance left to restore in the HPT area for typical outbound engines, again, this is due to the fact that HP turbines are typically completely refurbished during shop visit. Again, the hardware inspection data and the performance data show excellent agreement.

The unrestored performance identified in Figure 6 represents a potential gold mine in terms of fuel and dollars savings to the airlines, if it can be reduced on a cost effective basis. The presence of large amounts of unrestored performance associated with performance degradation of the fan module, HPC module and the LPT module, relative to new modular performance levels, is due to early workscope definitions. These airline shop overhaul work scope definitions were primarily aimed at maintaining reduced EGT levels and at restoring the condition of the hardware primarily from a reliability standpoint. The engine modules, which have the most direct impact on EGT margin and direct impact on reliability, are primarily associated with the hot section of the engine, the combustor and high pressure turbine area. Larger efforts (dollars and manhours) are required to achieve the same amount of EGT margin restoration in the LP system components than in the HP system components. In the early 1970's, it was concluded that it was not cost effective to do significant performance restoration in the fan and LPT areas with fuel prices at a 30 cents per gallon level. With current and projected fuel prices, the cost effectiveness of doing performance restoration work in all of the engines' major components must be re-examined.

HARDWARE INSPECTION DETAILS

The prime modes of deterioration within each module were established primarily by design team inspections at two major CF6-50 overhaul facilities and at one major CF6-6 overhaul facility. The details of the findings of these inspections are identified in references 1 and 2, including identification of the amounts of cruise SFC increase associated with each deterioration mechanism. However, some general statements concerning the more significant deterioration mechanisms are in order.

FAN DETERIORATION

The major areas of performance degradation within the fan section for both engine models were: 1) increases in tip clearance due to shroud erosion and the current maintenance philosophy which requires controlling only minimum clearance; this can result in local grinding and, in turn, results in increased shroud out-of-roundness and increased average clearance, 2) fan blade leading edge bluntness due to erosion and 3) fan bypass OGV erosion and leading edge bluntness due to loss of the polyurethene protective coating. During typical shop visits, the leading edge contours of the stage one blades are typically restored (with approximately 75 percent frequency). However, the "on-condition" maintenance philosophy, requiring only durability repairs, generally results in very little refurbishment to restore to new engine average clearance and to restore the OGV surfaces to the as new condition.

HPC DETERIORATION

The major deterioration modes of the CF6 engine high pressure compressors are: 1) increases in airfoil tip clearances, 2) degradation of airfoil surface finishes and leading edges and 3) creation of airflow leakage paths primarily through the variable stator vane bushings. With increased time in revenue service, the assembly of engine compressor stator cases (as engine parts are interchanged during shop visits) develop significant tendencies toward out-of-roundness. The maintenance philosophy in matching rotors and stators is to establish a minimum clearance. Thus, any tendency of the stator case to distort inward creates the requirement for short rotor blades (with resulting increased average clearance) and locally short stator vanes. The eventual result is increased airfoil tip clearances and associated deteriorated performance. Casing distortion and design changes which will result in a reduction in casing distortion is the subject of a separate paper at this conference.

HP TURBINE DETERIORATION

The primary mode of deterioration noted in the high pressure turbine during revenue service is the increase in blade tip-to-shroud clearances, resulting from rubs with some losses in performance due to increased airflow leakage and airfoil surface finish degradation.

It has been found that tip clearances for both stages of the CF6-50 high pressure turbine typically increase during the first 1500 hours of operation and continue to increase, but a a lower rate, thereafter. At 4000 hours, the average increase in tip clearances is 0.013 inch on stage 1 and 0.011 inch on stage 2, which accounts for 0.55 percent increase in cruise specific fuel consumption. These rubs and resulting clearance changes are primarily due to shroud support distortion, shroud swelling and bowing, shrinkage of the shroud supports and thermal mismatch between rotating and static structures during engine transients. Again, shroud distortion is the subject of another paper at this conference.

LP TURBINE DETERIORATION

As in the case of the high pressure turbine, increases in blade tip clearances and interstage seal clearances result in the major portion of deterioration occurring within the low pressure turbine in service. Degradation of airfoil surface finish is another contributor but results in very little performance loss. The increase in clearances was found to be primarily due to wear of the stationary surfaces which result from engine axial mismatches during different phases of engine operation. While there is little loss of material from the rotating components, the wear of the tip shrouds and interstage seals results in approximately 0.4 percent loss in cruise SFC after 4000 hours of operation with both the CF6-6 and the CF6-50 turbines.

SUMMARY OF DETERIORATED ENGINE

As is evident, a large part of degradation of engine performance in revenue service results from rubs and subsequent increases in clearance in the high pressure compressor, the high pressure turbine and the low pressure turbine.

Considerable effort is being expended by General Electric and the other engine manufacturers to create functional clearance control systems designed to eliminate rubs in these components and to maintain optimum clearances at the required cruise condition to maintain peak engine performance.

USE OF WHAT HAS BEEN LEARNED

The objectives of this part of the Diagnostic Program were to: 1) determine the specific causes for engine deterioration, 2) to isolate Short-Term losses from the longer term losses and 3) to identify potential ways to minimize the deterioration effects. Two potential means are available for minimizing deterioration effects on the current fleet. First is identification of product improvements which will provide better performance retention characteristics in the current engine, and the second is to identify improved engine work scopes which can be used by the airlines to improve performance restoration and, therefore, absolute performance levels of the engines coming out of the airline overhaul shops.

PERFORMANCE RETENTION

As a result of knowledge gained from the Diagnostic Programs, a Performance Improvement and Performance Retention Improvement Program has been identified for the CF6 family of engines. The complications of introducing new performance retention features into an existing engine arises from limitations on changes to aircraft power management and functional interchangeability. However, some features are currently planned by General Electric for introduction into the CF6-50 engine production models and will be retrofitable within the current fleet. Items being considered which can be included into the current fleet of engines include: smooth solid shrouds in booster stages 1, 2, and 3, which result in reduction in shroud erosion and better clearance control; a modified front engine mount, and steel front compressor casing which reduce bending deflections and locally reduce rub potential; improved surface finishes on high pressure compressor blades and vanes; replacement of three stages of titanium stator vanes and four stages of HPC rotor blades with steel which increases erosion resistance; and incorporation of new VSV bushings in the compressor stator case to increase durability and reduce leakage.

Other performance retention features are also being incorporated into the production configuration of the CF6-80 family of engines in addition to the performance retention items just mentioned. The HPC casing is a stiffer, two piece case with insulated rear stages which provides reduced deflections and better roundness thereby reducing rubs. The HPC rotor is cooled by introducing fan air into the bore, resulting in better matching of rotors and stators which again reduces rubs during transients. There will be an improved HPT shroud support system and improved HPT shrouds which reduce distortion and blade tip rubs. A passive cooling system for the HPT stator is being utilized which will provide a better match with the rotor and reduce blade rubs. Also to be included is an active clearance control system in the LPT which will produce close clearances at cruise and larger clearances at takeoff to reduce shroud rubs and prevent deterioration. The deterioration portion of the Diagnostics Program also verified that the performance retention features being designed into the Energy Efficient Engine (E^3) Program will have a definite payoff. These performance retention features include: a low tip speed, wide-chord, rugged fan blade; a short stiff compressor case; ruggedized fan OGV's; and active clearance controls on the high pressure compressor, the high pressure turbine and the low pressure turbine. Current estimates are that deterioration rate on the E^3 engine should be reduced by 51 percent relative to deterioration rates established for the CF6-50 engine as part of the NASA Diagnostic Program.

IMPROVED ENGINE WORK SCOPES

The most immediate reduction in fuel usage by todays CF6 fleet, which can be achieved as a result of information gained during the NASA Engine Diagnostics Program, lies in the definition of improved engine work scopes during engine shop visits by individual airlines. An integral part of the Engine Diagnostics Program with both the CF6-6D and CF6-50 engines were studies conducted to define how much of the unrestored performance losses associated with the typical engine as currently shipped from the overhaul test cells could be restored on a cost effective basis. The results of these studies were intended to be used as guidelines for improved definition of modular work scopes at the overhaul facilities.

As part of these studies, assumptions were made which included: material cost as defined by either repair cost or replacement hardware cost established in the General Electric catalogues; estimates of the cost of doing work based upon General Electric experience; performance gains and the life of the gain consistent with the deterioration rates established as part of the Engine Diagnostics Program; and typical missions assumed consistent with DC-10-10 and DC-10-30 operation. Fuel price for these studies was assumed to be a dollar a gallon.

It was concluded, based upon these studies, that approximately 60 percent of the unrestored performance currently existing on engines being shipped from the various overhaul test sites could be restored on a cost effective basis for the typical engine. Table 2 shows the results of the cost effectiveness feasibility study conducted by General Electric for the CF6-50 engine based upon typical overhaul test cell performance levels. It is noted that the greatest potential for cost effective refurbishment exists in restoring fan performance. This restoration includes surface finishes, leading edges, and maintaining clearances. Cost effective performance restoration is also achievable on the HP compressor, and slight additional cost effective gains are achievable on the HPT. General Electric has concluded to date based upon the studies for both engine models that performance restoration resulting from tearing down the LPT module and restoring performance is not cost effective.

A word of caution, however. These studies are based upon a typical or average engine as it is shipped from the various overhaul facilities. Some of the restoration work used in these cost effective studies is currently being done by some airlines on a part-time basis. Not all engines that are shipped from the overhaul facilities are equivalent (low) in performance as the typical engine identified and used as part of this study. The cost effectiveness study for an average engine can be misleading on an individual engine basis. The key point to emphasize is that each airline should conduct its own cost effectiveness studies based on individual practices, labor rates and work scopes to define the actual fuel and dollars savings available. General Electric's conclusions concerning the actual deterioration mechanisms within each module which contribute to the overall module deterioration are established and documented in extreme detail within the referenced NASA reports. These deterioration mechanisms can be used as a basis for each airline to conduct its own cost effectiveness refurbishment study. The implications of these studies are overwhelming. General Electric believes that potential savings of between 50 and 60 millions gallons of fuel could be realized in one-year's time period based upon the current CF6-6 and CF6-50 fleet of engines.

WHAT ELSE CAN BE DONE TO SAVE FUEL?

Discussions to this point have dealt with what is known about engine deterioration and refurbishment practices in todays operation and what can and s being done to further fuel conservation. There are other factors which must be considered in order not to use excess fuel. Careful attention to operational practices and use of derate power ratings are two such areas.

ENGINE ABUSE

Any turbofan engine can be operated in a manner which could produce excessive deterioration. For example, an engine which has been stabilized at high power, then subjected to a reduction in power and subsequently subjected to another accel is exposed to a condition where engine static cases have cooled faster than the rotor during the down time and could interfere with the hot rotor blades as they stretch during the accel thereby resulting in rubs and performance losses. This is known as "hot rotor reburst".

Every engine manufacturer publishes guidelines for engine operation which, if heeded, should result in avoiding the "hot rotor reburst" situation and any other similar situation. Proper discipline by all personnel responsible for any phase of engine operation from line maintenance personnel through flight crews is required in order not to abuse the engine.

USE DERATE POWER

It is common knowledge throughout the industry that use of reduced power settings has a strong influence on parts life and maintenance cost.

CF6-50 data analyzed as part of the Engine Diagnostic Program substantiated the fact that a larger amount of derate (reduced power) results in a lower deterioration rate. Figure 7 shows the average deterioration rates of the data from the 9 airlines studied. Shown are the average deterioration rates expressed in terms of EGT (at fan speed) and a percent fuel flow increase (at fan speed) for 1000 hours of operation as a function of average flight cycle length (hours/cycle) for each airline studied. The average of the A300B data, the average of the DC-10-30 data and the average of the 747 data are used to define a "composite characteristic". The numbers enclosed in parentheses indicate the average percentage thrust derate typically used by the indicated airline. While this summary is not sufficiently accurate to define an exact relationship between deterioration rate and average percentage derate, it does show a correlation between derate usage and reduced deterioration rates.

Although not implicitly suggested by these data, it is most probably a fact that continued usage of a given percentage of derate power will result in lower deterioration rates than alternately operating above and below that same percentage of derate. The same is true for maintenance cost. Maximum derate usage is encouraged.

SUMMARY

To summarize, the portion of the NASA Engine Diagnostics Program aimed at defining CF6 deterioration characteristics was highly successful. Deterioration rates and modes were identified as were areas of design improvement which can and will result in improved performance retention characteristics.

Also defined were potential means of fuel conservation today with improved cost effective engine performance restoration practices during engine shop visits.

The potential for additional fuel conservation is there if we make maximum use of this information. The engine manufacturer must design more performance retention into his product; the airlines must analyze and modify engine (and aircraft) maintenance practices.

REFERENCES

1. NASA CR-159786, "CF6-6D Engine Performance Deterioration", RH Wulf, January 1980.

2. NASA CR-159330, "CF6-6D Engine Short Term Performance Deterioration", WH Kramer, JE Paas, JJ Smith, RH Wulf, April 1980.

3. NASA CR-159367, "CF6-50 Engine Performance Deterioration", RH Wulf, November 1980.

4. "CF6 High Pressure Compressor and Turbine Clearance Evaluations", M. Radomski, and L. Cline, Paper Presented at NASA Aircraft Engine Diagnostics Conference, May 6, 1981.

TABLE 1. CF6-50 DETERIORATION IN 1000 CYCLES

	A/C TYPE		
INSTALLATION	DC10	B747	A300
INITIAL INSTALLATION ΔSFC	1.71%	2.07%	.83%
MULTIPLE BUILD INSTALLATION ΔSFC	1.03%	1.36%	.51%

TABLE 2. COST EFFECTIVE PERFORMANCE REFURBISHMENT
CF6-50 ENGINE

	% CRUISE SFC	
	UNRESTORED PERFORMANCE	COST EFFECTIVE REFURBISHMENT
FAN SECTION		
FAN BLADE TIP CLEARANCE	.38	.38
FAN BLADE L.E. CONTOUR	.12	.12
FAN BLADE SURFACE FINISH	.01	.01
SPLITTER LEADING EDGE	.07	.07
BYPASS OGV - L.E.	.06	.06
BYPASS OGV - SURFACE FINISH	.18	.18
BOOSTER TIP CLEARANCE	.03	0
BOOSTER AIRFOIL ROUGHNESS	.01	0
HP COMPRESSOR		
BLADE & VANE TIP CLEARANCE	.16	.16
AIRFOIL LEADING EDGE BLUNTNESS	.05	0
AIRFOIL SURFACE FINISH	.03	0
CASING/SPOOL SURFACE FINISH	.01	0
HP TURBINE		
STAGE 1 NOZZLE DISTORTION	.05	0
AIRFOIL SURFACE FINISH	.10	.10
LP TURBINE		
BLADE TIP CLEARANCE	.30	0
INTERSTAGE SEAL CLEARANCE	.22	0
AIRFOIL SURFACE FINISH	.04	0
TOTAL	1.82	1.08

60% OF UNRESTORED PERFORMANCE CAN BE RESTORED ON A COST EFFECTIVE BASIS

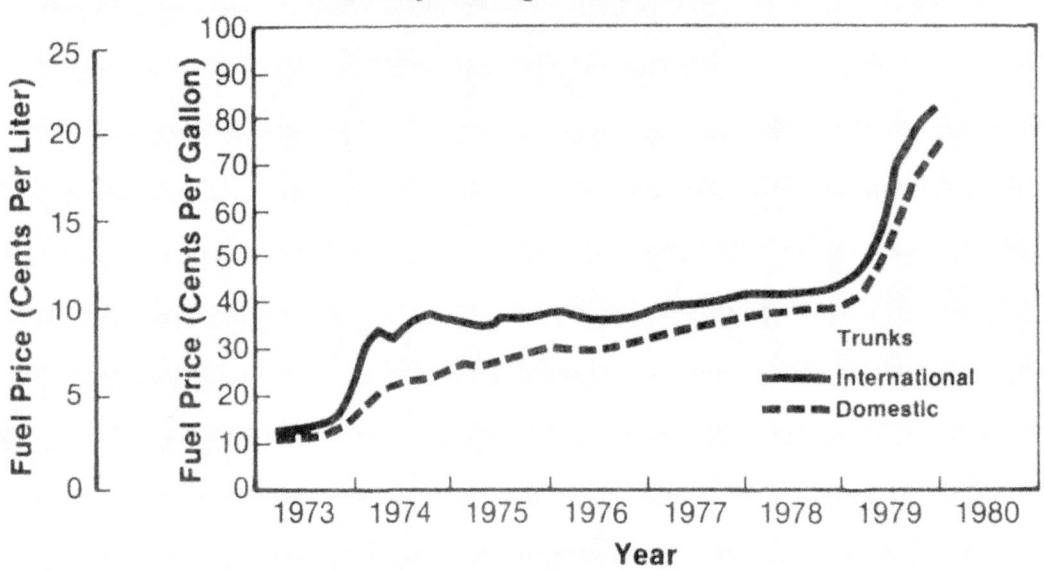

FIGURE 1

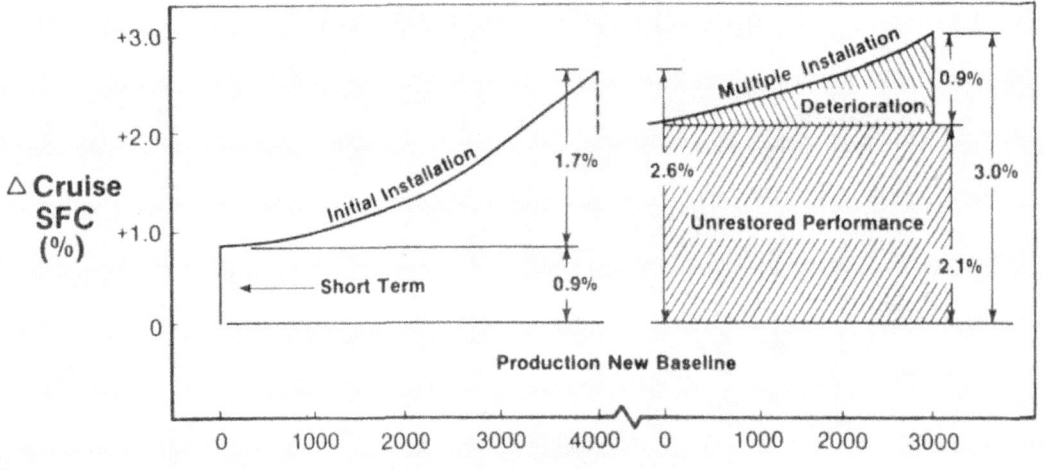

FIGURE 2

CF6-50 Deterioration Characteristics

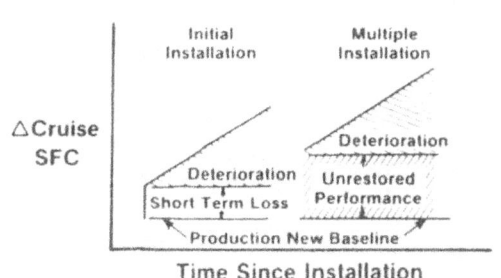

	DC-10	B747	A300
Short-Term SFC Loss (%)	0.7	0.7	0.7
Typical Time to Removal Initial Installation (Hrs)	3000	4000	2000
Initial Installation SFC Loss (%)	1.5	1.6	1.1
Typical Time to Removal Multiple Installations (HRS)	3000	3850	2000
Multiple Installation SFC Loss (%)	0.79	0.95	0.79
Unrestored SFC (%)	1.8	1.8	1.8

FIGURE 3

CF6-6D Initial Installation Performance Deterioration

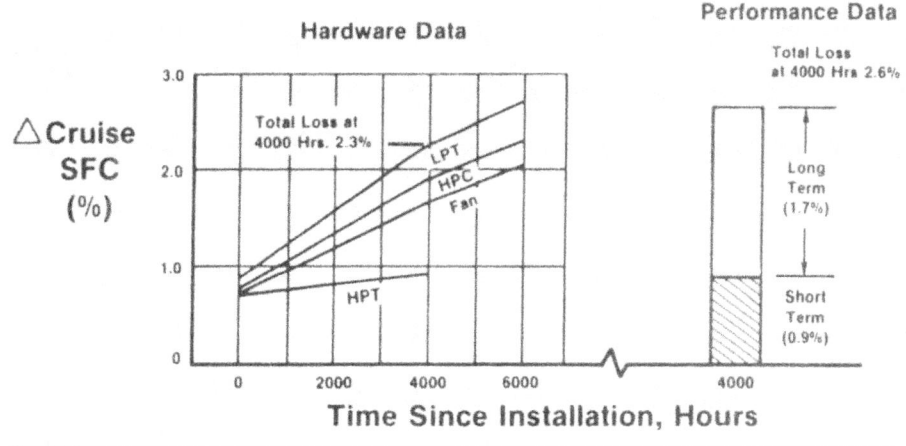

FIGURE 4

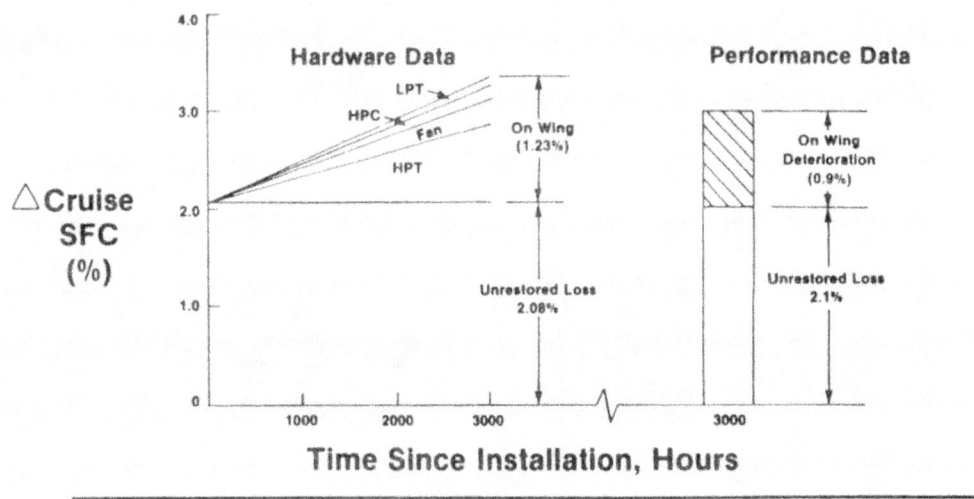

FIGURE 5

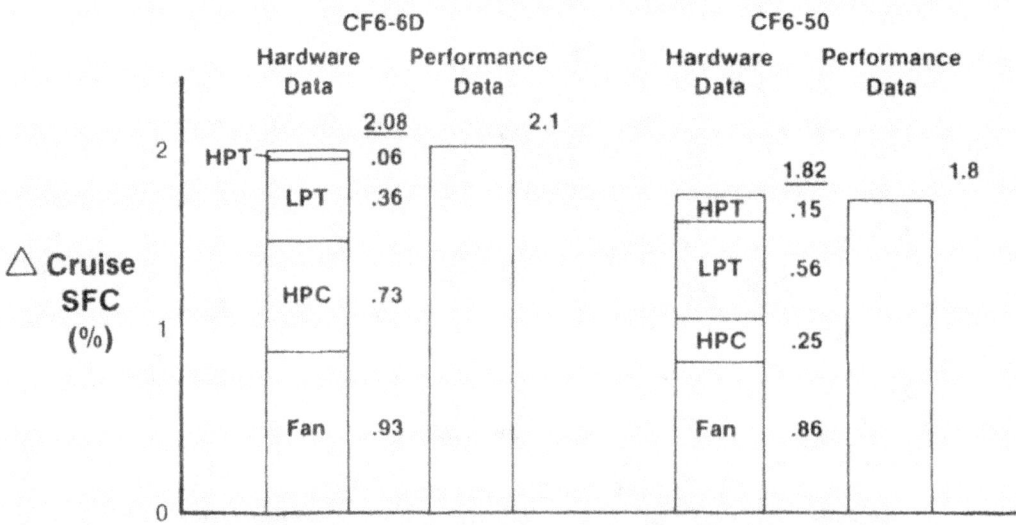

FIGURE 6

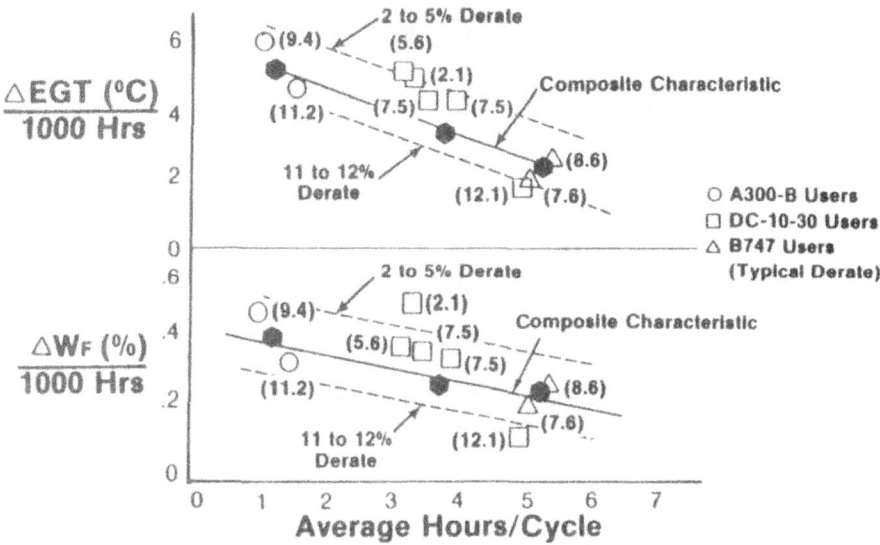

FIGURE 7

CF6 HIGH PRESSURE COMPRESSOR AND TURBINE CLEARANCE EVALUATIONS

M.A. Radomski and L.D. Cline
General Electric Company

SUMMARY

In the CF6 Jet Engine Diagnostics Program the causes of performance degradation were determined for each component of revenue service engines. It was found that a significant contribution to performance degradation was caused by increased airfoil tip radial clearances in the high pressure (HP) compressor and turbine areas.

Since the influence of these clearances on engine performance and fuel consumption is significant, it is important to accurately establish these relationships, especially now when fuel prices are rapidly escalating. It is equally important to understand the causes of clearance deterioration so that they can be reduced or eliminated.

This paper describes the results of factory engine tests run to enhance the understanding of the high pressure compressor and turbine clearance effects on performance. It also indicates the causes of clearance deterioration and discusses potential improvements in clearance control.

INTRODUCTION

The CF6 Jet Engine Diagnostics Program showed measurable degradation of compressor and turbine airfoil tip clearances in revenue service engines.

The degradations of the compressor tip clearances were caused by spalling of abradable coatings from the stator casings and rotor spools, by blade and vane tip rubs and by field assembly procedures involving out-of-round stator casings. Degradation of clearances in the CF6 compressor was estimated to produce, on the average, a 0.4 percent increase in the specific fuel consumption at cruise. This is estimated to amount to 15 million gallons per year for the 1981 CF6 engine fleet.

The effect of compressor clearances on compressor efficiency has been studied at General Electric for many years using a low speed research compressor. The data obtained from this research vehicle have been applied to the compressors of the CF6 engine family. Attempts have been made to verify these data by power calibrations of the CF6 production engines, but they have not been entirely successful because, in these tests, compressor efficiency was affected by several factors, and the effects of the airfoil tip clearances alone on efficiency could not be isolated with accuracy. There was, therefore, a need to conduct instrumented CF6 factory engine tests to accurately verify the influence of compressor clearances on engine performance.

The tests were conducted as part of the CF6 Jet Engine Diagnostics Program using an instrumented core engine.

The high pressure turbine tip clearance degradation was caused by blade tip rubs on the stator casing. The major cause of rubs was out-of-roundness. The principal causes of HP turbine out-of-roundness are the thermal gradients and transient responses of adjacent structures, such as the compressor rear frame, turbine mid-frame, and the low pressure turbine case. The increase in HP turbine clearances was estimated to contribute, on the average, a 0.6 percent increase in cruise specific fuel consumption. This equates to approximately 22 million gallons per year for the 1981 CF6 engine fleet.

Accurate measurements of HP turbine clearances and out-of-roundness have been difficult to achieve until now. Rub pins and High Energy X-Ray (HEX) tests have yielded reasonable approximations to date. The current performance sensitivity factors are based on turbine efficiency sensitivities established during air turbine testing and SFC effects determined from the engine cycle deck. Direct, on-engine analysis of the sensitivity factors related to turbine clearance were required to determine the turbine's contribution to overall engine performance deterioration. A test program was conducted as part of the CF6 Jet Engine Diagnostics Program to evaluate the effects of stator out-of-roundness and stage 1 blade tip clearance upon performance.

This paper outlines the scope of the HP compressor and HP turbine clearance evaluation programs, describes the unique instrumentation and the test procedures used, discusses the data reduction and presents some preliminary results.

HIGH PRESSURE COMPRESSOR CLEARANCE EVALUATION

Program Scope

The program was designed to determine the influence of compressor clearances on engine performance and also to evaluate potential improvements in clearance control. The approach used was to run instrumented core engine tests in which airfoil tip clearances were varied by varying quantities of rotor bore cooling air. The greater the cooling air flow, at any power setting, the lower was the bulk temperature of the rotor structure and, hence, the greater were the tip clearances. The running clearances were calculated from measured temperatures of the stator and rotor structures. The calculations were verified using measured blade tip clearances in stages 10, 12 and 13. The clearance changes and the corresponding measured performance changes were then correlated.

Rotor Bore Cooling - Externally supplied shop air was used to cool the rotor and thus vary the clearances. The total cooling air flow was measured by an instrumented orifice and remotely controlled by a valve upstream of the orifice shown in Figure 1. From the orifice, the air flowed to the manifold around the slave front frame, from there through flexible hoses into the frame struts and then into the HP compressor inlet inner cavity shown in Figure 2. From this cavity, some of the air leaked out through the air/oil and air seals bounding the cavity (only the rotating seals are shown in the figure) and the remainder, the net cooling air flow, entered the rotor main cavity through the holes in the forward shaft. The cooling air exited the rotor cavity through holes in the rear shaft and from there was discharged through the compressor rear frame struts into the test facility exhaust. The net cooling air flow was calculated for each test point by subtracting the seal leakages from the total measured flow.

Instrumentation - In addition to the standard factory test engine instrumentation, there were air temperature and pressure rakes at the compressor inlet and discharge to measure compressor efficiency, and also temperature and pressure probes at the rotor main cavity inlet and discharge to monitor the cooling air flow.

The compressor mechanical instrumentation shown in Figure 2 included the stator casing and rotor structure thermocouples, the clearanceometers and the touch probes. The touch probes were used to doublecheck the clearanceometers at the steady-state engine running conditions.

The clearanceometers used were electrical capacitance probes whose output voltage varies with the distance between the clearanceometer and the passing blade tips. A touch probe is a traverse probe with an open electrical circuit which closes when the probe touches the passing blade tip. When contact is made, the probe automatically backs off. The distance traversed by the probe to touch the blade tip is measured by a linear potentiometer.

Engine Tests

The tests were run in the General Electric Altitude Test Facility. They included steady-state power calibration tests, rapid accels and decels and a simulated typical flight cycle. The steady-state power calibration tests were made with three different engine inlet conditions, which were the core engine ambient, the simulated fan engine sea level static and cruise inlet conditions, and at a number of different engine speeds. At each speed point, at least three different sets of clearances were produced and their effects on engine performance were measured.

The test procedure was as follows. After the engine inlet condition and speed were stabilized, the rotor bore cooling air flow was set at the desired level, and three minutes later all instrumentation sensors were scanned and their outputs recorded. Instrumentation scanning and recording of data was repeated approximately every three minutes until the rotor disk temperatures became stable, which took fifteen to twenty minutes. At this point, the cooling air flow was changed and scanning of instrumentation sensors and data recording was repeated.

The simulated flight cycle and the rapid accels and decels, which included hot rotor rebursts, were made to evaluate potential improvements in clearance control. The tests were run with ambient engine inlet conditions and two different sets of clearances. The rapid accels were made from ground idle to take-off power setting which was then held constant until rotor disk temperatures became stable after which the engine was rapidly deceled to ground idle.

A hot rotor reburst is said to occur when an engine is rapidly deceled from a high to a low power setting and after a short time at the low power is reburst back to the high power setting. This type of power throttle maneuver may result in the most adverse tip clearances in the aft end of the compressor, particularly if the engine dwells at the low power setting for such a time period so as to produce the maximum temperature difference between the rotor and stator structures. In the hot rotor reburst tests, the metal temperatures were stabilized at take-off, then the engine was rapidly deceled to ground idle and a short time later, it was reburst back to take-off where the metal temperatures were again permitted to stabilize. Time at ground idle varied from approximately one minute to thirty minutes. In these tests, both the transient and the steady-state data were recorded.

Results and Discussion

Results discussed here are the steady-state differences in rotor temperature, compressor clearances, compressor efficiency and engine fuel flow produced by different quantities of the rotor bore cooling air flow. Typical results are shown for one particular speed point, the simulated sea level static take-off point. The compressor efficiency changes and the engine fuel flow changes are then shown as functions of normalized average compressor airfoil tip clearance changes. Analysis of the transient test data has not yet been completed, and, therefore, these data could not be included. Finally, the most significant cause of compressor clearance degradation in the CF6-50 field engines is briefly discussed and some data are presented to underscore the relevance of this compressor clearance evaluation program.

Rotor Temperatures - The axial temperature profiles in the rotor main cavity are shown in Figure 3 for different rates of cooling air flow. The data indicate a greater rate of change in the aft end than in the forward end of the rotor, which is primarily due to the geometrical differences of these two areas of the rotor structure and, to some small extent, due to the higher conductivity of Inco 718 as compared to titanium. The data also show, as would be expected, that the higher the cooling air flow the lower the air temperature of the rotor cavity and of the disk bores as shown in Figure 4. At stage 14, the maximum cooling air flow reduced the air and metal temperatures by at least 300° F, which is a very significant reduction and somewhat greater than the pre-test predictions indicated.

Radial temperature profiles in the stage 14 disk are shown in Figure 5. The cooling air was most effective at the disk bore. The rim of the disk was much less affected where the maximum temperature reduction was about one quarter of that at the bore. In other stages, this was even a smaller fraction. Similar temperature profiles were generated for all other power calibration points, and they were used to calculate the airfoil tip clearance changes which are discussed in the next paragraph. As will be noted, the pre-test predictions significantly underestimated the effectiveness of rotor bore cooling.

Airfoil Tip Clearance Changes - At the power calibration points, the stator casing temperatures remained constant, and only the rotor temperatures were affected by the cooling air flow. Therefore, to calculate the clearance changes, only the rotor temperature changes needed to be considered. Clearance changes calculated in this manner, for the simulated sea level static take-off power calibration point, are shown in Figure 6 for different cooling air flow rates. Similar calculations were made for all other power calibration points. These data were then used to calculate the normalized average clearance changes which were later correlated with the corresponding measured compressor efficiency and engine fuel flow changes. The correlations are discussed in a later paragraph where the normalized average clearance is also defined.

The measured and calculated clearances for stages 10, 12 and 13 are shown in Figures 7-9. There is an excellent agreement for stage 10. For the other two stages, there are small discrepancies between measured and calculated data. The causes of these discrepancies have not yet been determined.

Efficiency and Fuel Flow Changes - Compressor efficiency changes as a function of the cooling air flow are shown in Figure 10. The data were measured at the simulated sea level static take-off conditions. Indicated and corrected values are shown in the figure. The corrections were made to obtain the net effect of clearance changes by allowing for

1) leakage of cooling air into the compressor inlet and

2) heat removed from the gas path by the cooling air.

The magnitude of the latter correction was about ten times as large as that of the former. Colder cooling air, leaking into the compressor inlet, slightly reduced the actual air temperature downstream of the inlet temperature rakes and, therefore, made the actual efficiency reduction, caused by increased clearances, slightly larger than the indicated reduction calculated from the measured temperatures and pressures at compressor inlet and discharge. Heat removed from the gas path by the cooling air reduced the indicated compressor discharge temperature and, hence, made the indicated efficiency reduction, due to the increased clearances, somewhat smaller than the actual reduction.

Engine fuel flow changes as a function of the cooling air flow are shown in Figure 11. The corrections made to the fuel flow were in the opposite direction to that of the efficiency corrections, i.e., the fuel flow corrections resulted in a smaller actual fuel flow change than that indicated, because the measured fuel flow included the energy removed from the cycle by the cooling air which was vented overboard. If the clearance changes were produced mechanically, there would be no heat loss from the cycle, and therefore, the total fuel required would be less.

The compressor efficiency and fuel flow changes were corrected in this manner for all power calibration speed points and then correlated with the average normalized clearance changes which are discussed in the next paragraph.

<u>Correlation of Efficiency and Fuel Flow Changes with Clearances</u> - Compressor efficiency changes are shown as a function of the normalized average clearance changes in Figure 12 for four different speed points with three different engine inlet conditions. There is a good correlation of measured efficiency changes with the calculated normalized average clearance changes. The normalized average clearance change is defined as follows:

$\Sigma \Delta CL/L$, where:

ΔCL = clearance change in a given stage and

L = airfoil length in the same stage.

The line shown in the figure is the best line drawn through the data points.

The engine fuel flow changes versus the normalized average clearance changes are shown in Figure 13. Only the data obtained at the simulated sea level static take-off and cruise conditions are presented. The data for the other two speed points were inaccurate because of a fuel flow meter malfunction. The line shown in the figure is derived from the efficiency line in Figure 12 through the derivative of fuel flow as a function of efficiency for the test engine. Because the line correlates well with the fuel flow data, it, therefore, indicates consistency of the efficiency and fuel flow changes.

Compressor Clearance Degradation

Compressor stator casing out-of-roundness has been found to be the most significant cause of airfoil tip clearance degradation in revenue service engines. The effect of casing out-of-roundness on compressor blade and vane clearances is shown in Figures 14 and 15, respectively. To meet the minimum clearance at build-up, all blade tips have to be machined shorter by the amount equal to at least the inward distortion. Vanes, on the other hand, only need to be machined shorter in the area of the inward casing distortion.

The effect of out-of-roundness on clearances is magnified for two reasons. First, the permitted interchangeability of modules and, second, because of the field shop practices at engine build-up. At every shop visit, the distorted casing may be installed in a different compressor module, thus causing short blades and hence increasing the clearances in all of them.

The field shop practices involving out-of-round casings at engine build-up require that the individual compressor rotor and stator casing modules are machined to meet a target minimum clearance. However, verification of the actual minimum clearance is required, and this is accomplished by applying wax strips of known thicknesses to the stator casing and rotor spool lands and then installing the casing halves around the rotor. The rotor is then rotated through 360° after which the stator casing halves, upper and lower, are removed and the wax strip thicknesses are measured. If the wax strips were rubbed, indicating below minimum clearance, then the airfoil tips are hand ground to correct this condition. How the magnifying effect on clearances is produced will be illustrated by a specific field engine incident. An engine that failed to meet the minimum performance standards was disassembled and inspected. Inspection of the compressor, which had been refurbished prior to the test cell run, indicated out-of-roundness in the stator casing which is shown in Figure 16. The rotor blade tips in the aft end of this compressor were rounded off by hand grinding at assembly, because of below minimum clearance due to the out-of-round stator casing. A rounded-off blade tip from this compressor is compared to a machine ground blade tip from another compressor in Figure 17. The casing out-of-roundness in the aft end was a maximum of 20 mils, but the blade tips were up to 40 mils shorter at the leading and trailing edges. This is a good example of how the effects of an out-of-round casing on the average clearances are magnified. Hand grinding is not acceptable for this purpose. A procedural change has been specified to require remachining rather than hand grinding in similar cases.

Out-of-roundness data from a survey of twenty-six stator casings are summarized in Figure 18, where three sets of values are shown, i.e., the average of all measured and the largest measured in modules B and C. Although the average values were only about 10 mils, the maximum values were as much as 30 mils. Out-of-roundness of the module B stator casing, in particular, had a very significant impact on performance, since the clearances in the aft end of this module had to be increased by at least 20 to 30 mils. To avoid this large performance penalty, field procedures have been specified for the repair of casing out-of-roundness.

Concluding Remarks

The data presented are preliminary and are still being analyzed, but they indicate that the test technique was effective. By means of rotor bore cooling, appreciable changes were produced in rotor temperatures, compressor airfoil tip clearances, compressor efficiency and core engine fuel flow. A good correlation was obtained of measured and calculated clearances. Compressor efficiency changes correlated well with the normalized average clearance changes, and, furthermore, they were consistent with the corresponding fuel flow changes. The clearance changes produced in this engine test were comparable to the maximum clearance degradation observed in the revenue service engines. Significant fuel savings can be achieved if clearance degradations in the revenue service engines are reduced or eliminated. The results of this test are being applied to the development of new General Electric commercial engines.

HIGH PRESSURE TURBINE ROUNDNESS/CLEARANCE EVALUATION

Program Scope

As an engine accumulates operating time in revenue service, its performance deteriorates as a function of time and operating cycles. A significant part of the CF6 engine performance deterioration is chargeable to the high pressure turbine. This deterioration is primarily due to increased blade tip-to-shroud clearances caused by rubbing of the blade tips on the shrouds. The objective of the HP turbine clearance and roundness diagnostics program was to provide test data to improve the understanding of the effect of blade tip clearance on performance as affected by transient engine operating conditions and of stator out-of-roundness.

Engine Tests

The tests were conducted in Test Cell 2 at the General Electric Company Plant in Evendale, Ohio. The test vehicle was a CF6-50C engine. The engine was mounted in an overhead frame as shown in Figure 19.

Blade tip clearance data were obtained from clearanceometer probes especially designed for this purpose. Eight of these clearanceometer probes were installed in the engine which had been modified to accept the probes as shown in Figures 20 and 21. The probes were located circumferentially around the engine over the stage 1 blade tips as identified in Figure 22. Clearance data was then recorded for the type of engine operations shown typically in Figures 23 and 24. These operations included numerous steady-state and transient operating conditions.

Test Results

Measured data from each of the eight clearanceometer probes were averaged to obtain the "round engine" clearance. This clearance, plotted against time, defines the round engine clearance response. The "round engine" data were then used for tuning axisymmetric analytical models of the modified test engine configuration. This knowledge was then utilized to upgrade the production configuration engine analytical model.

Representative plots of a throttle burst (steady-state idle to steady-state take-off) and throttle chop (steady-state take-off to steady-state idle) are presented in Figures 25 and 26.

Evaluating the results of each of the individual clearanceometer probes relative to the averaged data for any given time yields a measured "out-of-roundness." These data were then used in conjunction with calculated mechanical loads and calculated thermal distortions to identify deficiencies, and they formed the basis for correction of the calculated results. Out-of-roundness results are shown in Figures 27 and 28 for throttle bursts and chops respectively.

Discussion of Results

The "round engine" clearance transient response results matched pre-test calculated predictions closely as did clearance at steady-state operating points. This clearance match was also confirmed by thermocouple data from the shroud support, and it verified that the anlytical modeling previously used was representative of the engine structure. Some relatively minor model parameter adjustments were necessary to force calculated results to more perfectly match measured clearance responses. The test verified the calculated importance of hot or warm rotor throttle rebursts on minimum blade tip clearance experienced in engine operation.

Calculations predicted that steady-state engine operating clearances are set at engine transient operating conditions. The worst case or minimum clearance condition, occurs during a hot rotor reburst which has idle dwell times less than 3 to 4 minutes.

The turbine shroud support member is considerably less massive than the turbine disk and, consequently, cools more quickly than the disk. The rapid reacceleration of the engine adds rotational stress growth and blade thermal growth to the still existent disk thermal growth. The net result is a hot blade tip radius greater than that of the supporting structure, which results in rubs.

A significant part of engine deterioration may be caused by "warm rotor rebursts" (idle times over 4 minutes) for which little data was available. Testing included several runs to provide data relative to this type engine operation. Rebursts with idle dwell times of 8, 6, 4, 2, 1 and 0.5 minutes were completed to quantitatively evaluate these significant round engine and out-of-roundness responses for which good analytical predictions were not available.

Out-of-roundness was evaluated for all the steady-state and transient operations tested. Pre-test predictions of out-of-roundness at steady-state conditions were compared to measured out-of-roundness. While the results compared well in magnitude, the shapes differed sufficiently to indicate that out-of-roundness driving phenomena existed which were not accounted for analytically.

However, test results did clearly indicate that the phenomena not accounted for in the analysis were thermal and not mechanical load induced. Testing and analysis accounted for mechanical distortions caused by engine thrust, redundant torque, and mount horizontal and vertical loads as well as manufacturing tolerances. These distortions agreed very well with analytical predictions. Test data also included temperature measurements for portions of the engine structure which were analytically determined to affect turbine stator roundness. They included combustor exit temperature profile, turbine mid-frame temperature distributions, shroud support temperature distributions and low pressure turbine case temperature distributions. These temperature measurements lead to improvements in out-of-roundness data correlation.

Comparison of test results, field-observed turbine deterioration rates, and rub locations agree well and verify that most turbine blade tip clearance deterioration is the result of combined stator out-of-roundness and warm rotor rebursts.

Concluding Remarks

As a result of this program accurate measurements of HP turbine clearances and out-of-roundness were obtained for steady-state operations as well as throttle bursts, chops and rebursts from various dwell times at ground idle. The measured average clearances agree well with the analytical predictions.

The learning gained from this testing and from subsequent analytical refinements is being applied to identify potential improvements in turbine efficiency for the CF6-50 and several other engine models. This is especially true in the development of enhanced ability to optimize HP turbine rotor-to-stator transient response rate and stator out-of-roundness calculations.

Instrumented Core Engine In Altitude Test Facility

FIGURE 1

Compressor Instrumentation/Flow Schematic

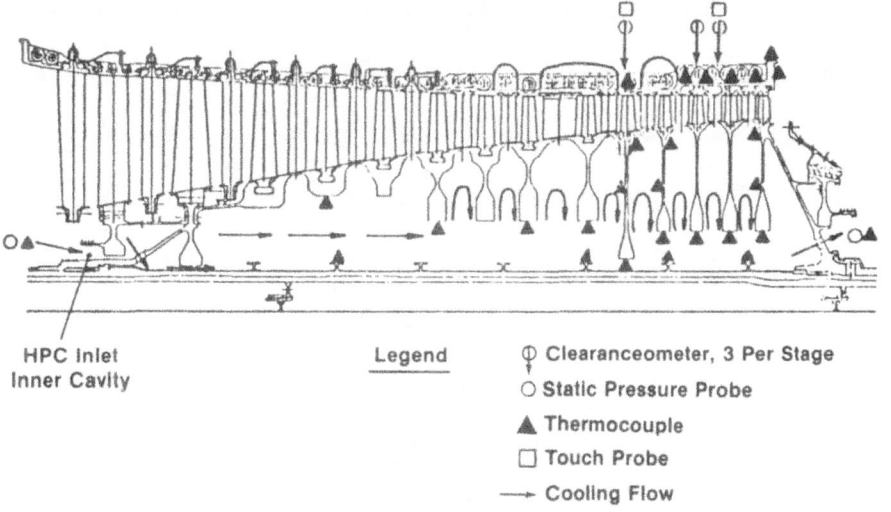

FIGURE 2

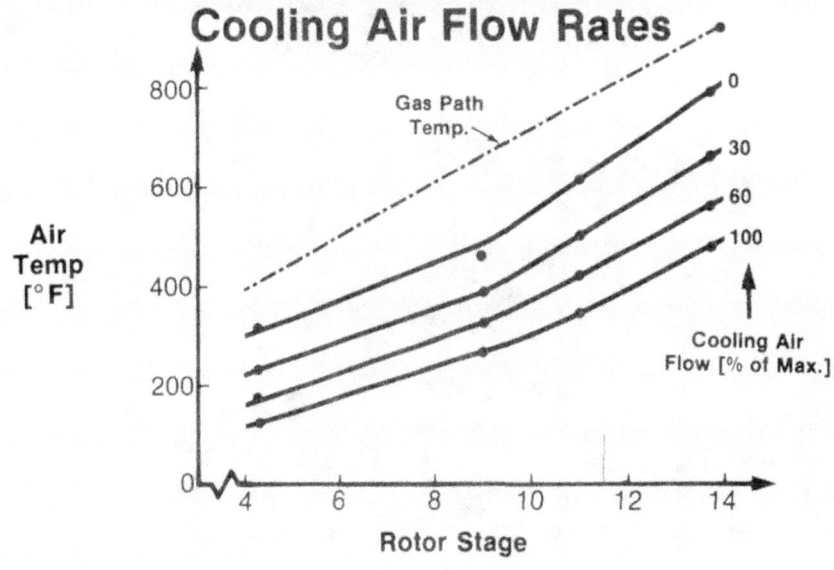

FIGURE 3

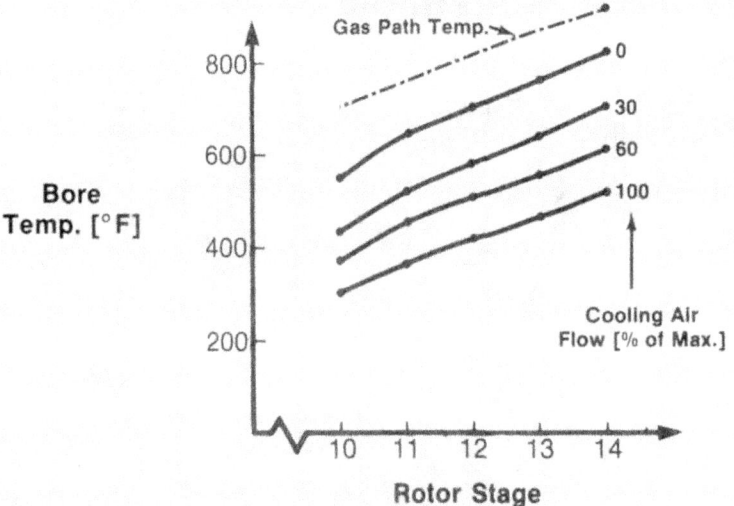

FIGURE 4

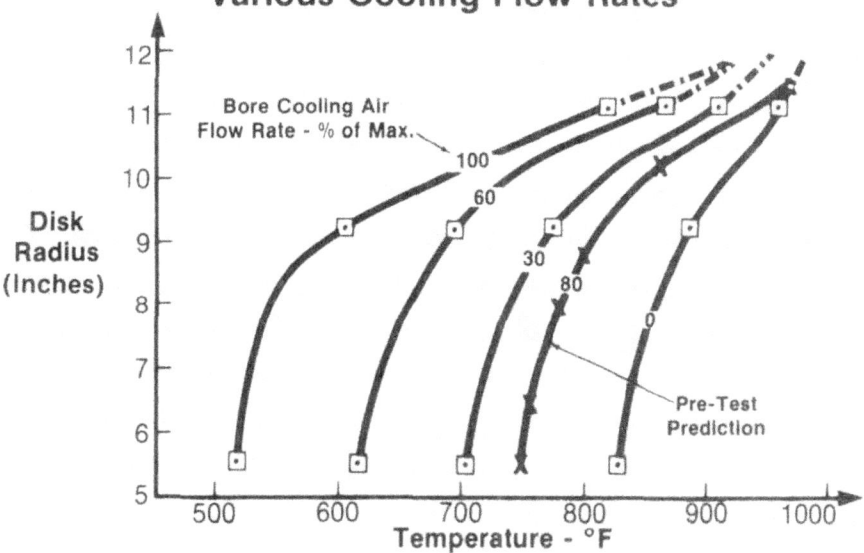

FIGURE 5

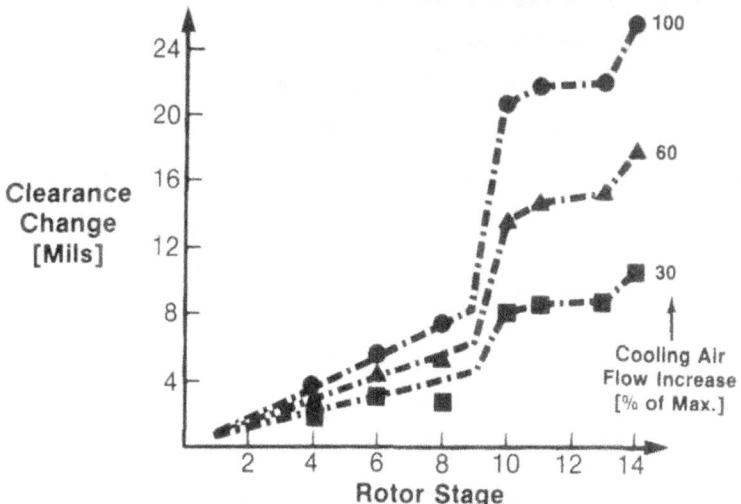

FIGURE 6

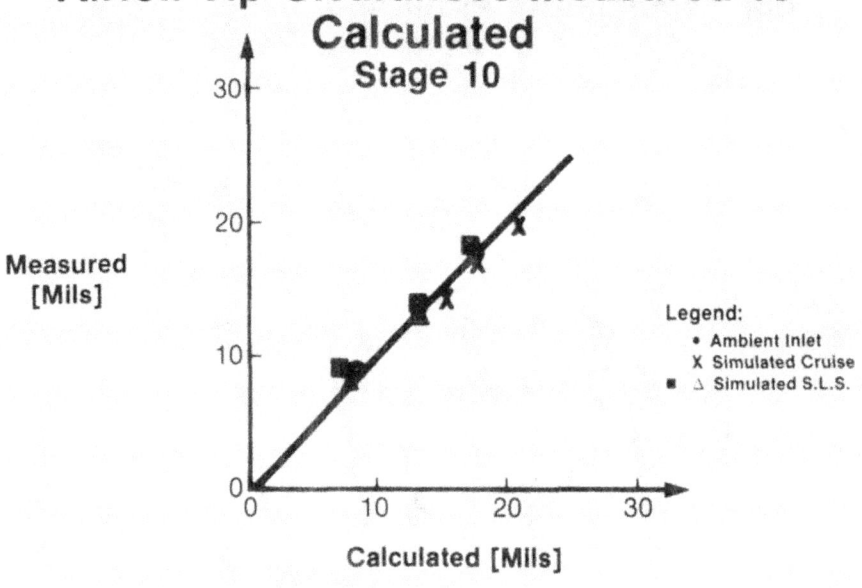

FIGURE 7

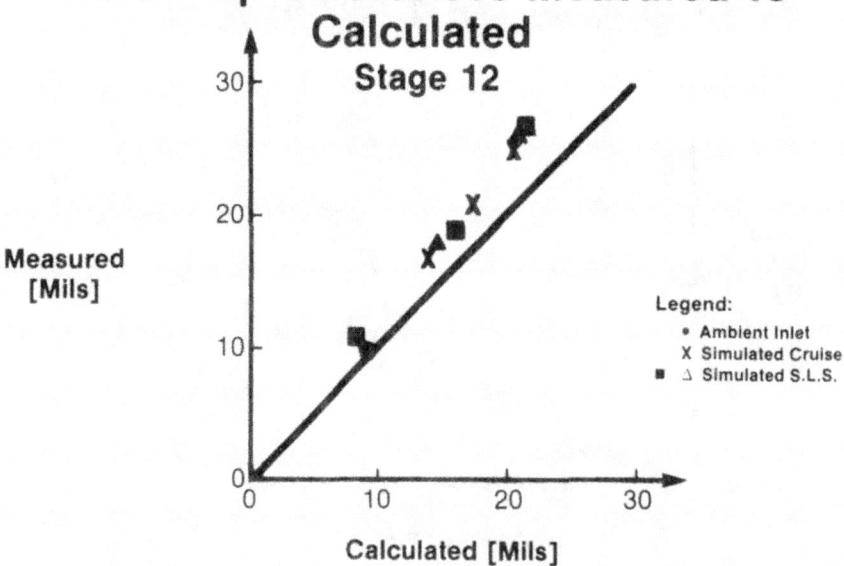

FIGURE 8

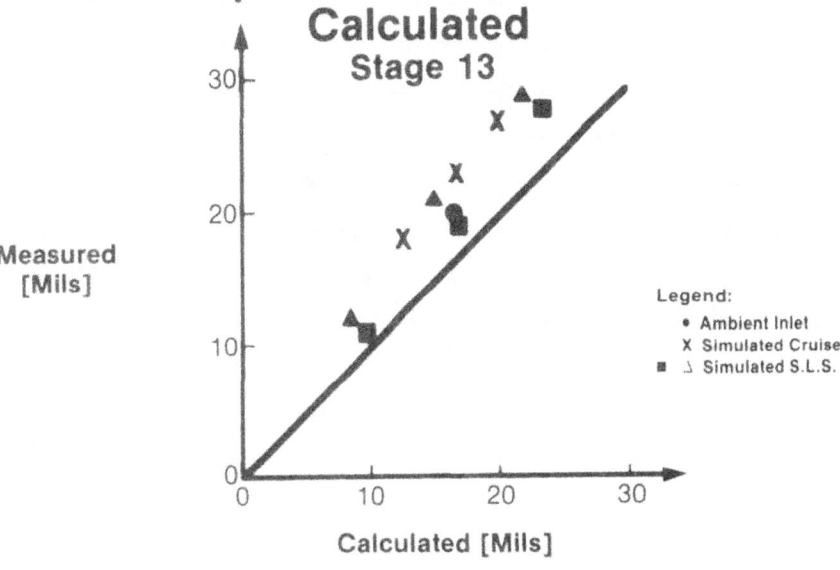

FIGURE 9

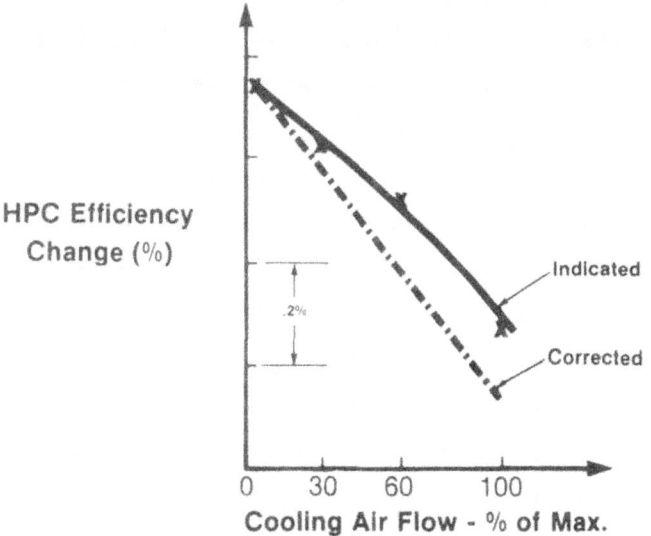

FIGURE 10

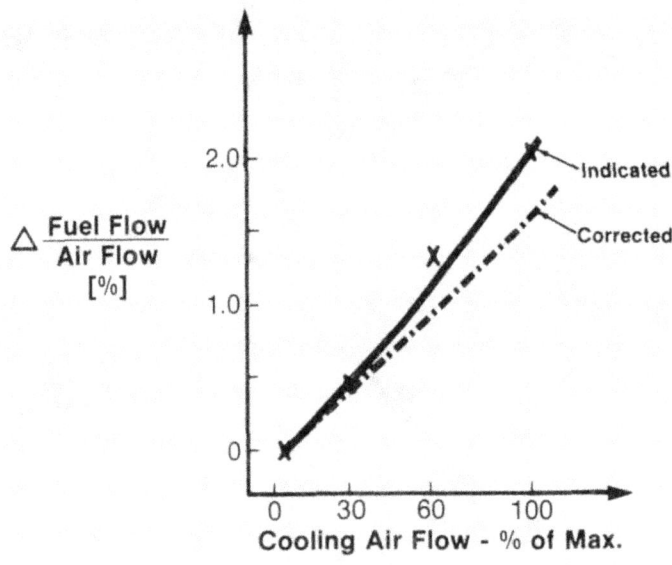

FIGURE 11

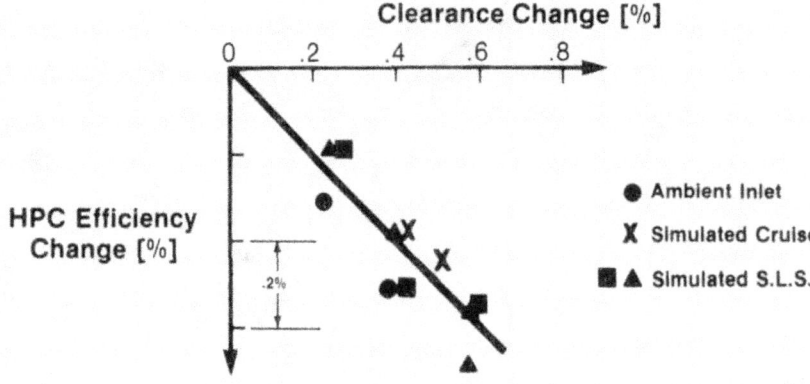

FIGURE 12

Fuel Flow Change vs Airfoil Tip Clearance Change

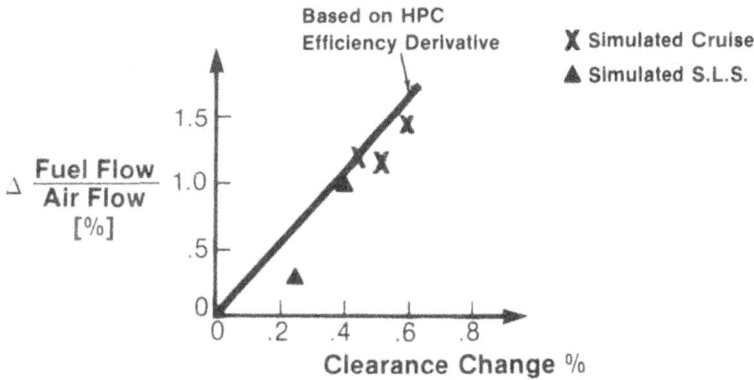

FIGURE 13

Effect of Casing Distortion

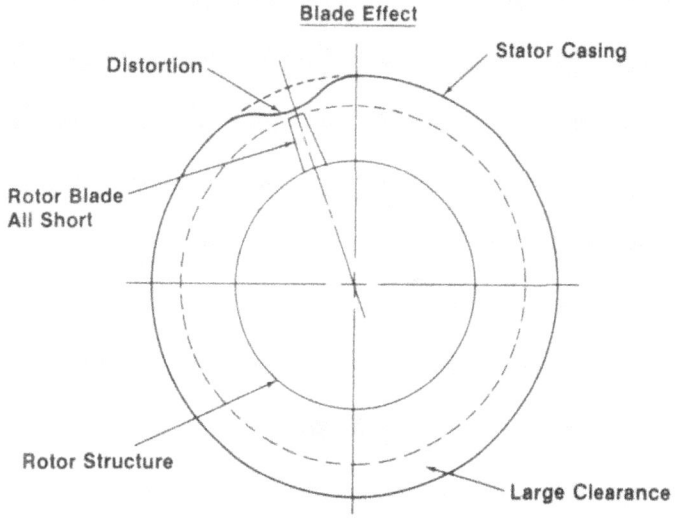

FIGURE 14

Effect of Casing Distortion

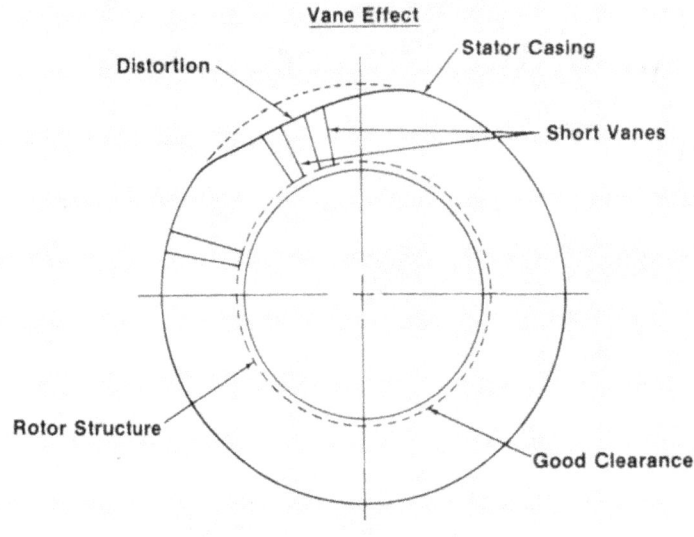

FIGURE 15

Casing Out of Roundness vs Stage

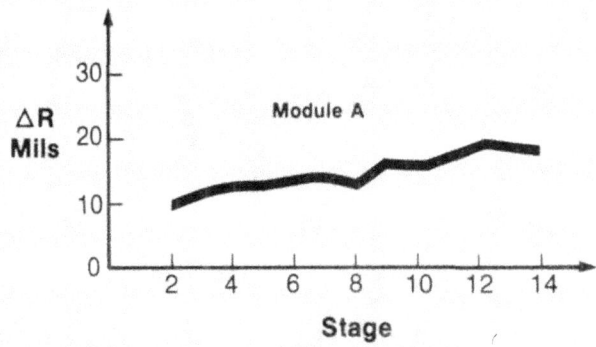

FIGURE 16

Instrumented CF6-50 Engine In Test Cell

FIGURE 19

Current CF6-50 HPT Cross Section

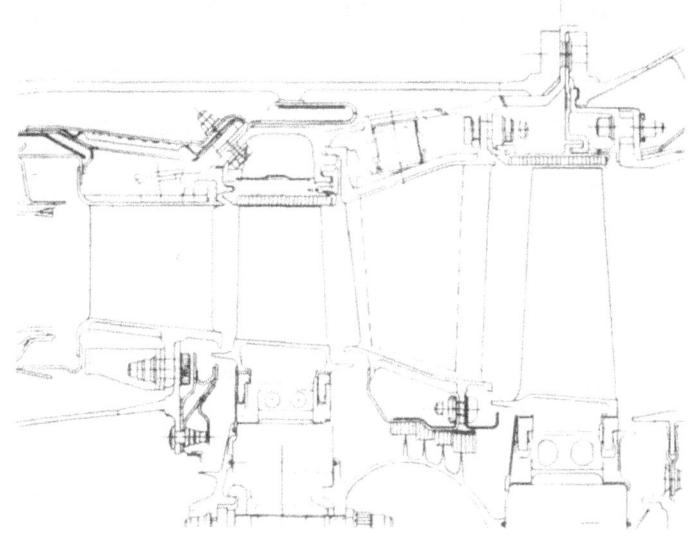

FIGURE 20

Hand Grinding of Blade Tips at Assy

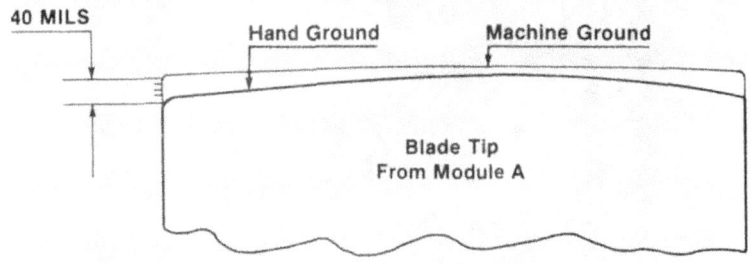

FIGURE 17

Casing Out of Roundness vs Stage

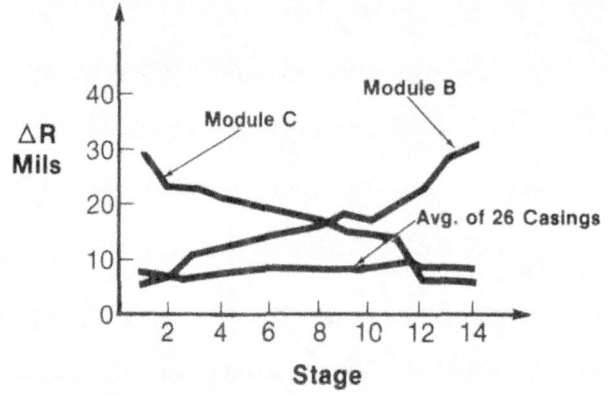

FIGURE 18

Turbine Modification to Receive Clearanceometer Probes

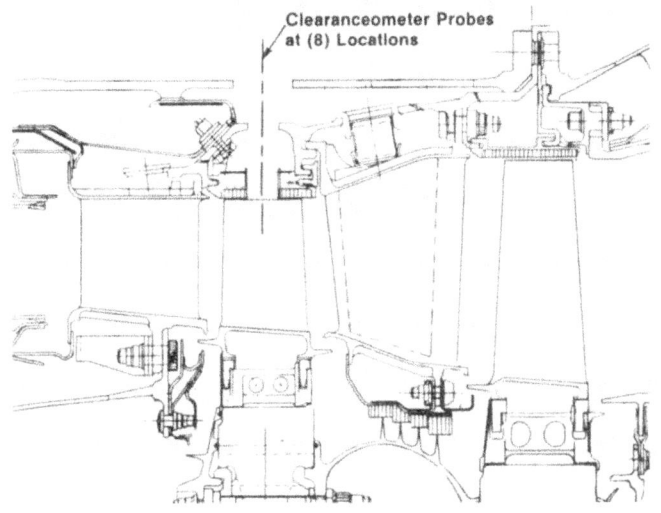

FIGURE 21

Probe Angular Position (ALF)

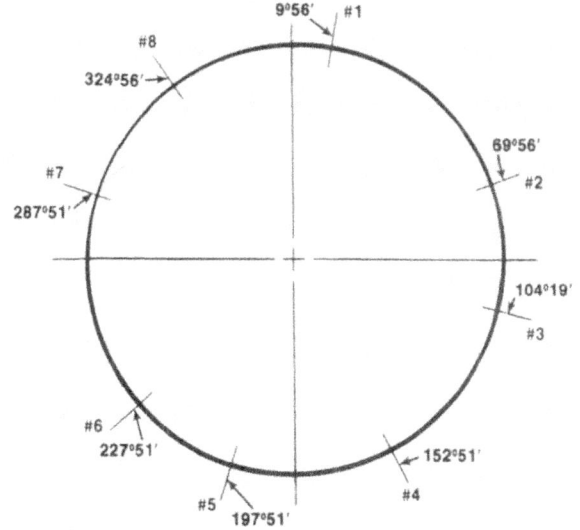

FIGURE 22

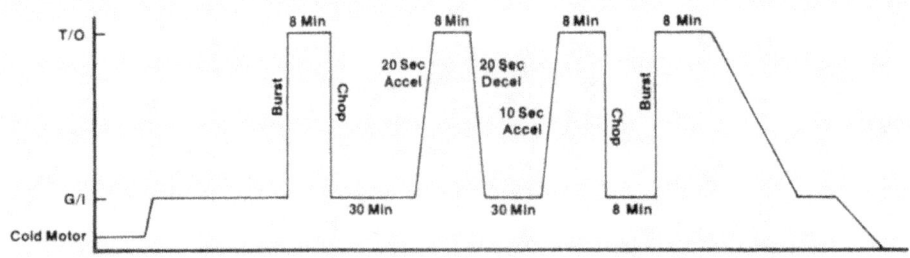

FIGURE 23

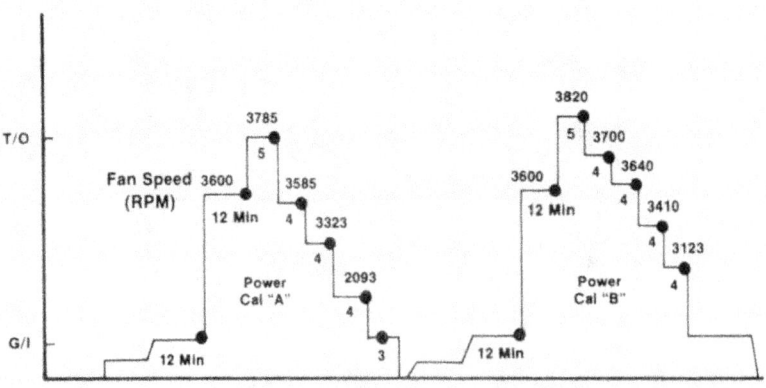

FIGURE 24

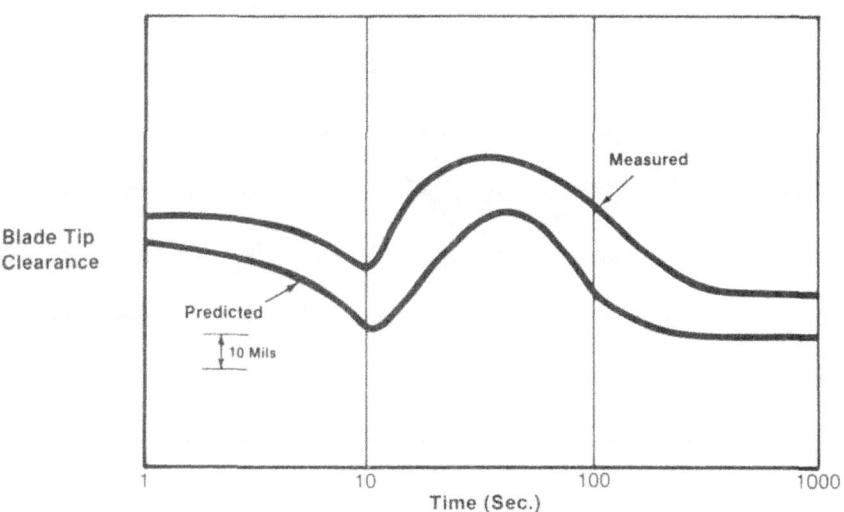

FIGURE 25

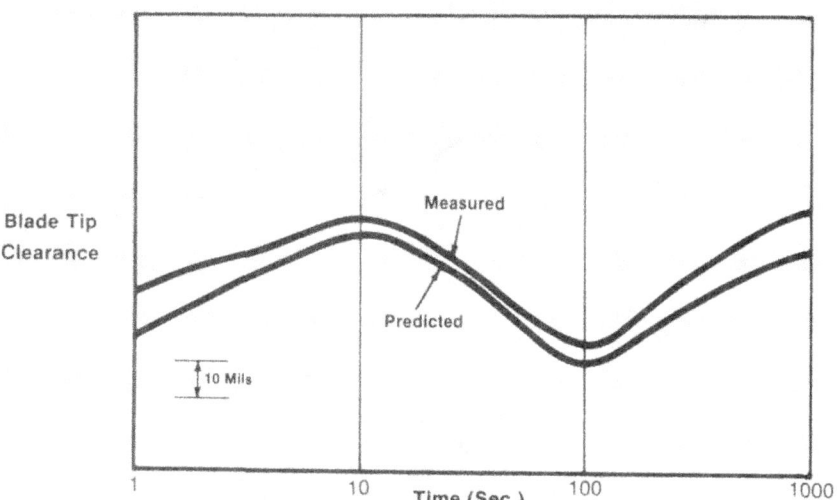

FIGURE 26

Stator Out-Of-Roundness — Throttle Burst

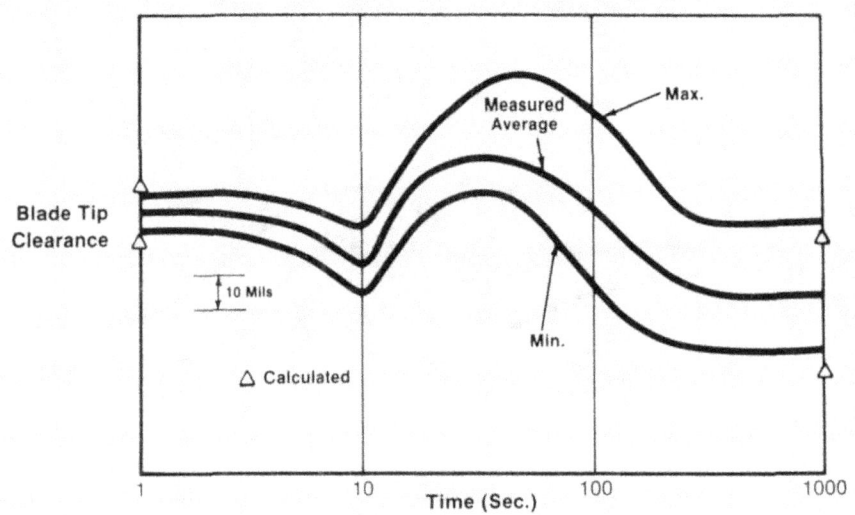

FIGURE 27

Stator Out-Of-Roundness — Throttle Chop

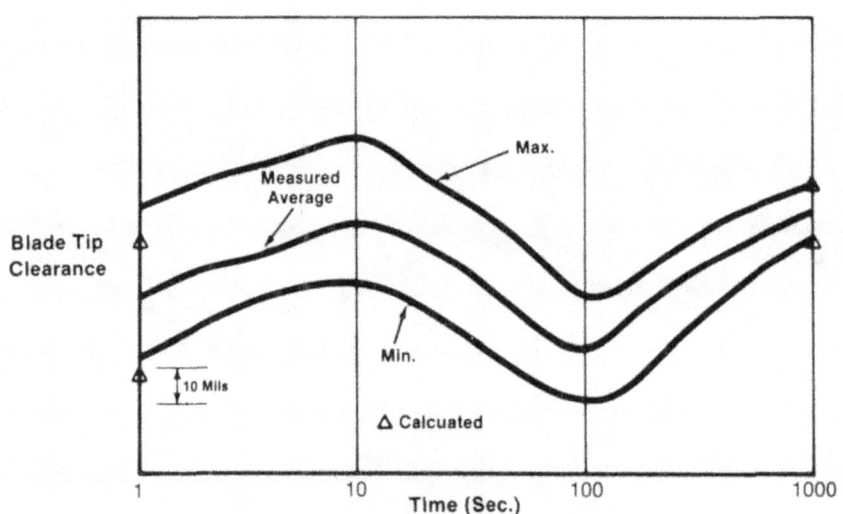

FIGURE 28

JT9D JET ENGINE DIAGNOSTICS PROGRAM*

W. J. Olsson and W. J. Stromberg
Pratt & Whitney Aircraft Group

SUMMARY

The NASA JT9D Engine Diagnostics Program has been a four-year effort to identify and quantify the various engine deterioration phenomena that affect JT9D performance retention and identify approaches to improve performance retention of current and future engines. The program has included surveys of historical data, monitoring of in-service engines, testing of instrumented engines, analysis, and analytical modeling. The Boeing Commercial Airplane Company, Douglas Aircraft Company, Trans World Airlines, Pan American World Airways, and Northwest Airlines participated as subcontractors in various phases of the program. Historical data was provided also by American Airlines.

The initial studies established that performance deterioration is made up of short- and long-term modes, both of which are flight cycle related phenomena. The later efforts provided additional data and refined and expanded on the initial conclusions.

The short-term deterioration occurs primarily during airplane acceptance testing prior to delivery to the airline. Therefore, it has small effect on revenue service performance retention. The long-term deterioration continues throughout engine life with a negative effect on performance retention.

The combined effect of the short- and long-term deterioration modes for the JT9D-7 is shown on figure 1. An increase of 2 percent in cruise thrust specific fuel consumption is typical after 2000 flight cycles of revenue service due to performance loss in unrepaired engines.

Short-term deterioration results from an increase in gas-path running clearances with resultant decreases in engine module efficiencies. This short-term effect is caused by flight-load induced engine deflections with resulting rubbing of airfoils and seals. Wearing of blades and seals occurs for the most part prior to revenue service during the various airplane maneuvers associated with the production acceptance testing of the airplane. This flight-load induced wear occurs in all modules. The results show a 0.8 percent increase in flight thrust specific fuel consumption during the predelivery airplane acceptance testing and an additional 0.3 percent increase during early revenue service.

Long-term performance deterioration is also a flight cycle related phenomenon. It is caused by erosion of airfoils and gas-path seals during

---

* This work was conducted by Pratt & Whitney Aircraft for the National Aeronautics and Space Administration under Contract NAS3-20632.

ground operation and take-off and by cyclic induced thermal distortion of the high-pressure turbine airfoils. Erosion primarily affects cold section efficiencies by blunting the blade leading edges, reducing airfoil chord, and further opening running clearances. Thermal distortion of airfoils results from high-temperature cycling of the airfoils with resultant gas-path leakage and loss of optimum airfoil shape.

The diagnostics program has shown that performance retention within 1 to 2 percent of initial revenue performance can be maintained with a proper program of hot section and cold section maintenance as shown on figure 2.

INTRODUCTION

The NASA JT9D Engine Diagnostics Program is a part of the NASA sponsored Engine Component Improvement (ECI) Project which is directed toward improving the fuel consumption of selected current high bypass ratio turbofan engines and their derivatives by 5 percent over the life of these engines. The ECI project is divided into two subprojects: Performance Improvement and Engine Diagnostics. Performance Improvement is directed toward developing fuel saving component technology which may be applied to current engines and their derivatives. Engine Diagnostics is directed toward identifying and quantifying engine performance losses that occur during the engine's service life and developing criteria for minimizing these losses, as shown on figure 3. The JT9D Jet Engine Diagnostics Program, which is now nearing completion, has successfully identified and quantified the various causes of JT9D performance deterioration and the possible approaches toward improved performance retention.

This paper will briefly describe the various approaches used in this project and the results and conclusions that have been reported to date.

APPROACH

The ideal approach for determining the cause and extent of engine deterioration would consist of tracking a large number of individual engines from production testing through extended revenue service, monitoring their flight performance with expanded instrumentation, closely tracking their maintenance histories, then correlating specific maintenance events, performance shifts, and operational history. For numerous reasons this procedure was not feasible. Therefore, we took the following approach.

The JT9D-7A engine was selected for the study since various models had been operating for a long time and some of these models were still in production; thus, both ample high-time and new engine data were available.

The first task was the collection of available historical data. These data included:

o Pratt & Whitney Aircraft production performance records to establish a base level.

o   Airframe manufacturers certification records to show early changes in performance.

o   Airline and Pratt & Whitney Aircraft prerepair and postrepair calibration test results and hardware inspection results to explain long-term changes.

o   In-flight engine monitoring data to establish the relation between ground performance and cruise performance changes.

Based on the analysis of these data, some preliminary conclusions were drawn:

o   There are four generic causes of engine performance deterioration, namely: 1) flight-load induced clearance changes; 2) erosion of fan and compressor airfoils and seals; 3) thermal distortion of hot section parts; and 4) variations in airline repair standards.

o   Performance deterioration trends may be divided into two distinct time periods: short-term and long-term deterioration. The prime cause of short-term deterioration is flight-load induced rubs which open gas-path clearances, thus reducing module efficiencies and influencing airflow. The analysis of the historical data as seen on figure 4 showed a 1 percent increase in thrust specific fuel consumption at sea level in the first few flights conducted by the airframe manufacturer prior to delivery of the airplane to the airlines.

o   Performance deterioration then occurs at a slower rate dominated by erosion of cold section airfoils and seals, with resulting blunting of airfoils and further opening of running clearances. This erosion and thermal cycle induced distortion of hot section airfoils results in loss of airfoil efficiency and increased secondary flow leakage. The historical data showed the sea level thrust specific fuel consumption to have increased to about 4 percent above production levels after acceptance testing and 3000 revenue flight cycles of an unrepaired engine operation as seen on figure 5.

o   The deterioration of the turbine airfoils and seals results from changes in their environment. These changing temperature and flow patterns are caused by deterioration in the compressor and combustor modules. Thus, more frequent cold section maintenance pays off in reduced deterioration in the higher priced hot sections.

o   A comparison of the fleet historical prerepair and postrepair calibration data showed an average performance recovery of 1 percent in sea level take-off thrust specific fuel consumption with a potential for 2 percent recovery with increased cold section and hot section refurbishment.

The first phase of the program provided an abundance of information but left numerous gaps in the data. The second phase, or in-service engine performance study, conducted jointly with Pan American World Airways, expanded the data base significantly by allowing us to monitor a controlled sample of 28 JT9D-7A engines in the Pan Am 747SP airplane fleet from preflight testing of the engines at Boeing through 2100 flight cycles of operation. The data collection included: installed engine ground calibrations before the first airplane flight and periodically during subsequent revenue service; in-flight engine calibrations during the flights immediately following the ground calibrations; a complete set of crew-collected engine flight condition monitoring data from the fleet; prerepair and postrepair calibrations and repair histories on each of these engines that came into the shop; and an expanded instrumentation calibration and complete analytical teardown of one of the engines after 141 flight cycles (see figure 6).

The results of this effort firmly established that the flight-loads induced short-term deterioration occurs in the first few flights prior to revenue service. It provided ample data for the refinement of the various engine module deterioration prediction models which were first developed on the basis of the historical data. Finally, it provided a correlation between performance retention at flight cruise conditions and performance change as measured by ground calibrations. The quality of the flight performance data was less than that of the ground tests due to the available instrumentation systems. However, the data sample was large enough that statistical trends could be drawn. One such set of data is 747SP engine conditioning monitoring (ECM) fuel flow data shown on figure 7. The data were recorded at cruise altitudes between 32,000 and 40,000 feet and corrected to 35,000 feet and constant engine pressure ratio (EPR). A trend line through the 1398 data points shows a 1.7 percent increase in fuel flow rate after 1500 revenue flight cycles from the start of airline service on engines with no repairs.

The short-term flight-load induced performance loss, though not significantly contributing to revenue service wear, does present a challenge. If it can be eliminated or significantly reduced, the new airplane could be delivered to the airline with up to 1 percent improved sea level thrust specific fuel consumption which is equivalent to 0.8 percent improved cruise thrust specific fuel consumption. Previous studies have estimated that more than 80 percent of the flight-load induced damage is caused by aerodynamic (pressure) loads applied to the fan cowl, and the remaining damage is caused by inertia loads from gusts and hard landings plus maneuver-induced gyroscopic loads.

The final two data gathering tasks of the JT9D diagnostics program were test programs directed toward a better understanding of this flight-load induced wear. The first of these was the Simulated Aerodynamic Loads Test conducted in a Pratt & Whitney Aircraft test stand. The objectives of this test program were to determine the changes in engine operating clearances and performance under (1) thrust and thermal loads; (2) static simulated aerodynamic flight loads, figure 8; and (3) the combination of thrust, thermal, and static aerodynamic loads during engine operation to permit validation of the levels, module distribution, and causes for short-term performance losses. In addition, the test program would validate or permit refinement of previous analytical study results on the impact of aerodynamic flight loads on performance losses. To accomplish these objectives, an engine was analytically built with average production clearances and new seals as well as extensive instrumentation to monitor performance, case temperatures, and clearance changes. A special loading device was designed and constructed to permit application of known moments and shear forces to the engine by the use of cables placed around the flight inlet. These loads simulated the estimated aerodynamic pressure distributions that occur on the inlet in various important segments of a typical airplane flight.

The test engine and loading device were installed in the Pratt & Whitney Aircraft X-Ray Test Facility, shown on figure 9, to permit the use of X-ray techniques in conjunction with laser probe clearance measuring instrumentation to monitor important engine clearance changes under both steady state and transient engine operating conditions. Upon completion of the simulated flight-load test program, the test engine was analytically disassembled and the condition of gas-path parts and final clearances was extensively documented.

The performance monitoring calibrations between tests indicated that the engine lost 1.1 percent in sea level take-off thrust specific fuel consumption due to permanent clearance changes caused by the application of these inlet loads. Another 0.2 percent change in thrust specific fuel consumption was produced by an increase in airfoil surface roughness in the low-pressure compressor and thermal distortion in the high-pressure turbine. This additional 0.2 percent was a result of the experimental nature of this test program and does not occur in early revenue service. Prior to the test program, the change in sea level performance due to clearance changes was predicted to be 0.9 percent. Therefore, the agreement between measured and predicted performance is considered to be satisfactory.

The overall engine performance loss was distributed among all modules; however, the low-pressure compressor and high-pressure turbine contributed the major portion of the loss. The major permanent clearance changes (seal rubs) occurred in the fan, high-pressure compressor, and high-pressure turbine and were found to be the direct result of the loads imposed. Table I compares these results with previous comparisons of module contribution to sea level performance changes with early usage.

Transient testing, conducted after completion of the simulated aerodynamic loading, indicated no additional performance losses associated with transient engine operation.

The flight loads test was the final phase of the JT9D Diagnostics Program. It was conducted as a joint effort with the Boeing Commercial Airplane Company. Boeing, under contract with NASA Langley, provided the test airplane and measured the flight loads on the instrumented engines. Pratt & Whitney Aircraft, under contract with NASA Lewis, provided the instrumented engines and measured the effects of the flight loads on the engines. The flight loads test was conducted to verify the simulated aerodynamic loads used in the X-ray load test program and to further expand on the flight conditions and flight load effects measured in that program. Specifically, the flight loads test objectives were as listed on figure 10.

The test approach was to install an analytically built and instrumented engine in position No. 3 on the Boeing test 747 (RA001) airplane and an analytically built and instrumented fan case on the position No. 4 engine, figure 11. The analytically built engine was calibrated at Pratt & Whitney Aircraft before delivery and then again after installation in the aircraft, prior to flight testing.

A series of flight tests was conducted with progressively increasing flight loads. Continuous, simultaneous measurements with all data systems were recorded to accurately document the cause and effect relationship of flight loads to engine deterioriation. Performance and fan clearances were documented after each flight by calibrations and rub measurements to determine the effect of increasing loads.

A final postflight calibration at Pratt & Whitney Aircraft and an analytical teardown of the analytically built engine was conducted following the flight program to quantify the effects of the flight loads, figure 12.

Instrumentation included pressure taps on the positions No. 3 and 4 nacelles to measure the aerodynamic loads, accelerometers on both engines to measure inertia loads, and rate gyros on both engines to measure gyroscopic loads. Clearance closures were monitored by laser probes on the high-pressure turbine of the position No. 3 engine and on both fans. Thermocouples on the high-pressure turbine of the position No. 3 engine measured the transient and steady state thermal effects on the running clearances. Finally, expanded performance instrumentation on position No. 3 engine permitted closer performance monitoring since it was the prime data source. The position No. 4 engine was instrumented sufficiently to identify clearance and load differences due to its position on the wing.

The flight loads testing was successfully completed, and the test data analysis is now in process. The clearance closures at actual flight conditions generally repeated the measured closures in the X-ray load test program under simulated flight loads, see figure 13. This program also confirmed that aerodynamic loads occurring during high power operation, that is, take-off rotation, airplane power-on stalls, and high G maneuvers, are the prime cause of fan performance deterioration. In addition, the preliminary analysis indicates that the combination of mechanical loads and transient thermal expansion during the extended high power climb is the prime cause of short-term deterioration in the high-pressure turbine.

The final report of the flight test program will be issued this fall and will include a final refinement of the performance deterioration models, including analytical results of the X-ray load test and the flight load test.

ENGINE PERFORMANCE RETENTION PREDICTION MODELS

One of the major objectives of this program has been the development and refinement of analytic models for predicting the deterioration with engine usage of both the complete JT9D engine and the individual modules. These models consist of families of curves which define the changes in the performance parameters (efficiency, flow capacity) with usage for each of the engine modules. These various parameter changes are applied to the JT9D performance analysis program to determine the predicted performance change with usage of an average engine. The preliminary models were prepared based on analysis of the performance, engine usage, and replaced parts condition data collected in the first phase of the program. All the in-service data collected on the Pan American 747SP fleet was used for the first refinement of the models. This effort was followed by a just-completed second refinement of the short-term deterioration predictions based on the X-ray load test results. Table I compares representative results from the different data sources. Figures 14 and 15 show the thrust specific fuel consumption changes at sea level for an average engine versus usage as predicted by the latest model. For that model, it was assumed that high-pressure turbine performance had been stabilized at a constant level after 1000 flight cycles and the low-pressure turbine after 2000 cycles by a hot section maintenance program. Figure 14 subdivides the predicted deterioration by module. As seen, the low-pressure compressor and high-pressure turbine are most sensitive to early flight-load induced deterioration. Erosion of airfoils and seals is the prime contributor to long-term deterioration in the cold section as shown on figure 15, while thermal distortion is the prime contributor in the hot section. One more refinement of these models will be made after completion of the flight loads test data analysis.

To validate the models at cruise conditions, it was first necessary to establish actual in-flight average performance. The engine condition monitoring and in-flight calibration data collected on unrepaired Pan American 747SP/JT9D-7A engines from start of revenue service to 1500 flight cycles provided this performance data. Performance at cruise conditions was determined to be less sensitive to component deterioration than at sea level. This reduced sensitivity results from the fact that the ram pressure ratio increases the nozzle pressure ratio at cruise and, thus, makes performance less sensitive to gas generator losses. This effect has been demonstrated in the Pratt & Whitney Aircraft (Willgoos) altitude test facility. The result is that the increase in cruise thrust specific fuel consumption due to component deterioration is about 75 percent of the increase at sea level. The JT9D performance retention model supports the results and was used to develop the curves on figures 1 and 2. Evaluation of cruise performance data from the flight loads test will permit a further refinement of the cruise performance retention model.

CONCLUSIONS

Performance deterioration in the JT9D-7 is a flight sensitive phenomenon caused by a short-term and two long-term wear modes. The short-term deterioration occurs primarily during airplane acceptance testing and, therefore, does not affect airline operation. The long-term wear takes place continuously over the engine life so that the performance loss can be minimized by a sound maintenance program. Short-term deterioration is primarily due to flight-load induced blade and gas-path seal wear which result in increased gas-path running clearances. The wear occurs in all engine modules but has the most deleterious effect on the low-pressure compressor and high-pressure turbine performance. The wear occurs during conditions that combine minimum axisymmetric running clearances and maximum engine distortion or asymmetric closure.

Minimum axisymmetric clearance occurs during high power operation due to the combined effect of centrifugal forces and high metal temperatures. Maximum asymmetric closure is caused by airplane maneuver induced aerodynamic loads and thrust induced engine bending loads. Thus, short-term deterioration is a cyclic effect in that it occurs during take-off and climb and other maneuvers which combine high aerodynamic loads and high engine power. Cruise, approach, and landing do not contribute to short-term deterioration.

Long-term deterioration is also flight cycle dependent. It is caused by erosion of the airfoils and seals, which cause airfoil roughness, bluntness, chord loss, and increased gas-path clearances, and by thermal distortion of turbine airfoils which reduces their efficiency and increases leakage.

Ingestion of foreign matter during taxi, take-off, and landing operations is the primary cause of erosion. Changing gas-flow patterns caused by erosion plus thermal cycling of the engine are the prime causes of thermal distortion of the turbine airfoils. The split of deterioration by module on figure 14 shows the increasing importance of high-pressure compressor and low-pressure turbine in long-term deterioration.

The refined performance deterioration prediction model shows a 2.1 percent increase in cruise thrust specific fuel consumption in the first 1500 revenue flight cycles. The program results also show that a good program of both hot section and cold section maintenance can maintain cruise performance between 1 and 2 percent of that at start of revenue service.

RECOMMENDATIONS

Based on the results of the completed phases of the JT9D Diagnostics Program and the preliminary results of the Flight Loads Test Program, the following recommendations, summarized on figure 16, are made toward improved performance retention in current and future propulsion systems.

Performance retention in current engines can best be maintained by following improved maintenance practices which have been developed jointly with the airlines, based on the early findings of this and industry-sponsored programs. These improved practices provide both cold-section and hot-section refurbishment, and the potential results are summarized on figure 17.

Performance retention in future propulsion systems will benefit from the following:

o Performance deterioration caused by flight loads and thrust-induced loads will be minimized by integrated engine and nacelle designs that consider the effects of both flight loads and thrust loads.

o Further development of gas-path clearance control systems and abradable rub strips will provide closer running clearance control in the high-pressure turbine.

o Erosion effects on cold section airfoils and seals will be minimized by improved coatings and materials and the consideration of refined gas-path designs for reducing the ingestion of erosive material.

o Erosion effects on hot section airfoils and seals will be minimized by improved high temperature materials.

o Thermal distortion effects will be minimized by refined gas-path designs and improved maintenance programs that reduce temperature profile shifts.

TABLE I

COMPARISON OF MODULE CONTRIBUTION TO
SEA LEVEL TSFC DETERIORATION

	Historical Data Analysis (149 Cycles)	In-Service Engine Analysis (150 Cycles)	P&WA Testing of P-695743 (141 Cycles)	Simulated Aerodynamic Loads Test of P-662211
	Change in TSFC (%) at Sea Level Static Take-Off Thrust			
Fan	+0.1	+0.2	+0.1	+0.2
Low-Pressure Compressor	+0.2	+0.4	+0.4	+0.3
High-Pressure Compressor	+0.3	+0.2	+0.3	+0.2
High-Pressure Turbine	+0.4	+0.4	+0.6	+0.5
Low-Pressure Turbine	+0.5	+0.1	+0.1	+0.1
TOTAL	+1.5	+1.3	+1.5	+1.3

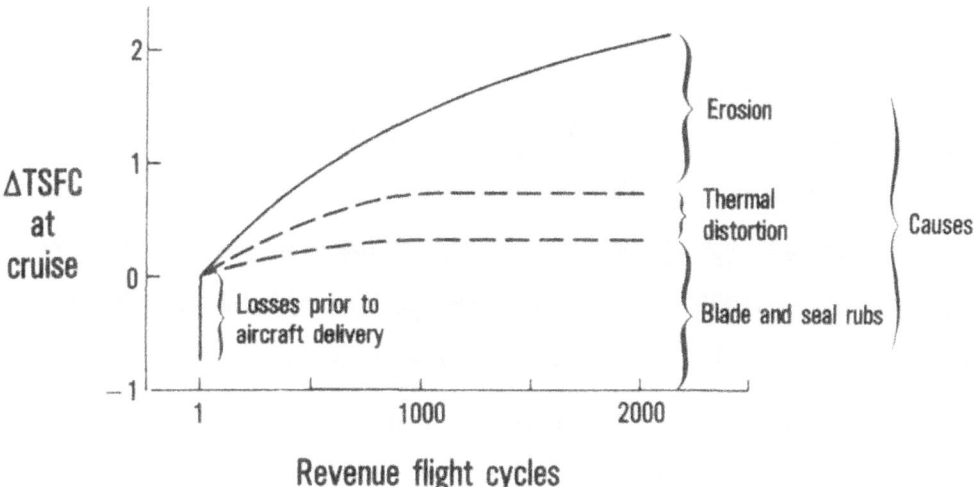

Figure 1   JT9D-7A In-Service Engine Performance Deterioration at Altitude Cruise Conditions.   (J24873-2)

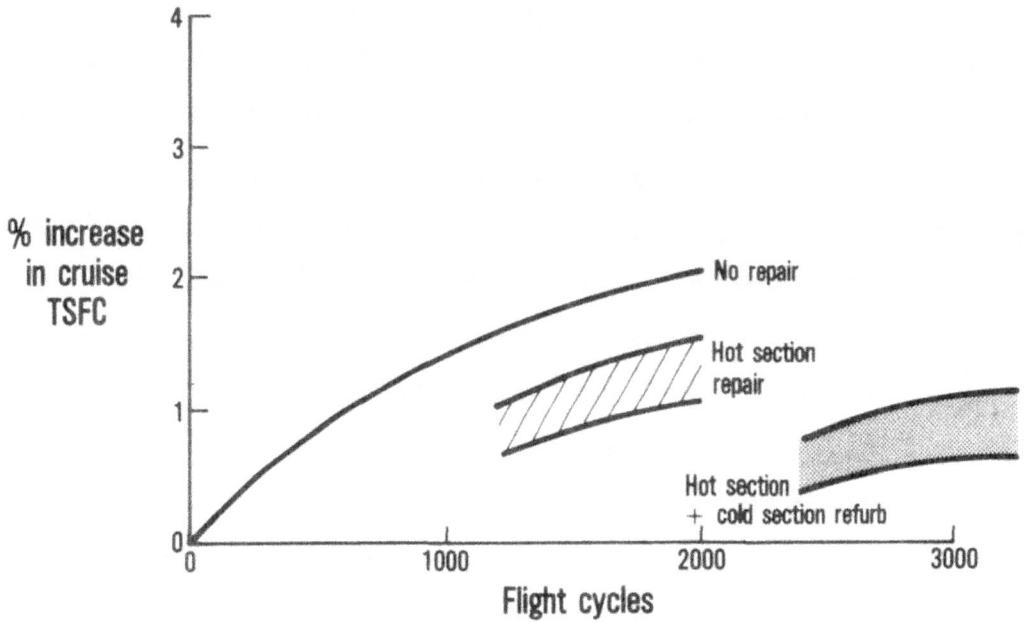

Figure 2   Effect of Repair on JT9D-7A Engine Cruise Thrust Specific Fuel Consumption.   (J24603-24)

1. Define scope of JT9D performance deterioration

2. Identify and quantify sources of JT9D performance deterioration

3. Determine sensitivity of component performance to engine parts deterioration

4. Develop analytical model of JT9D performance deterioration

5. Recommend performance retention techniques for current and future engines

Figure 3 JT9D Diagnostics Program Objectives. (J23878-4)

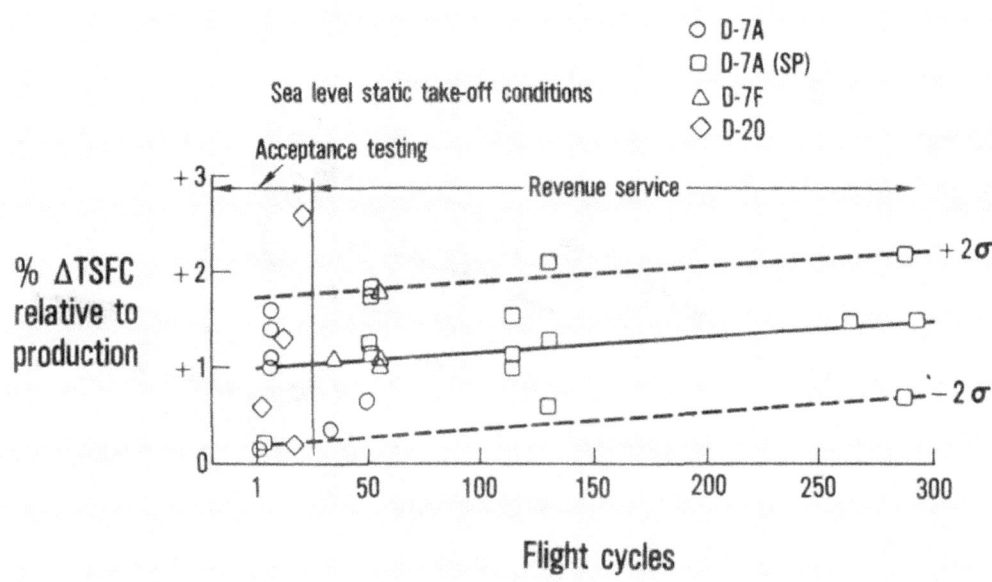

Figure 4 Historical Short-Term Deterioration Data. (J24873-4)

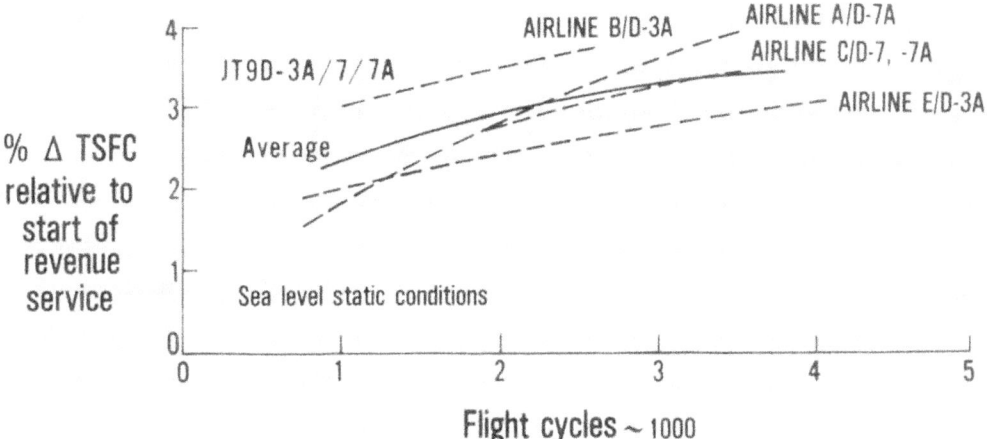

Figure 5 Historical Long-Term Deterioration Data for Unrepaired Engines.
(J24603-8)

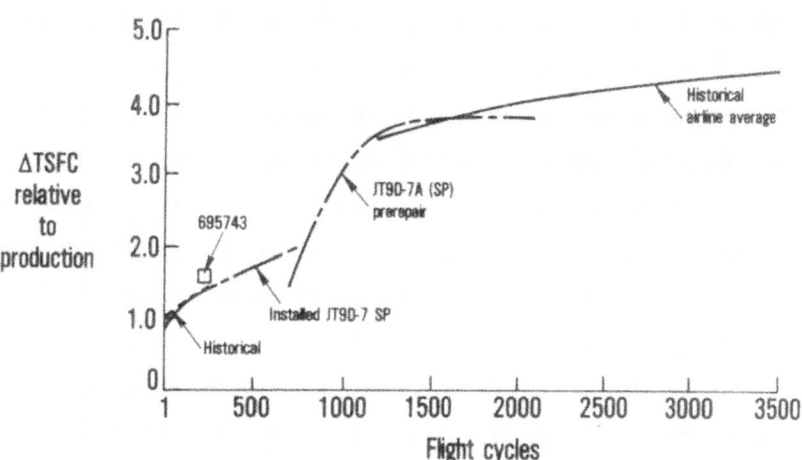

- Installed ground test from 0 − 1100 flight cycles
- Expanded testing and analytic teardown at 141 cycles
- Pre and post repair calibrations

Figure 6 Pan American 747SP/JT9D-7A In-Service Engine Performance Data.
(J24873-6)

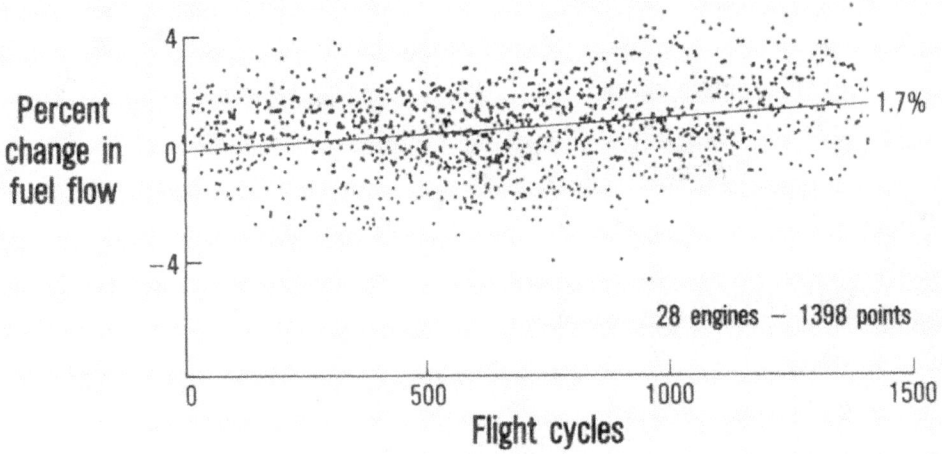

Figure 7 Cruise Fuel Flow Trend with Usage for Pan American 747SP/JT9D-7A Unrepaired Engines. (J24873-7)

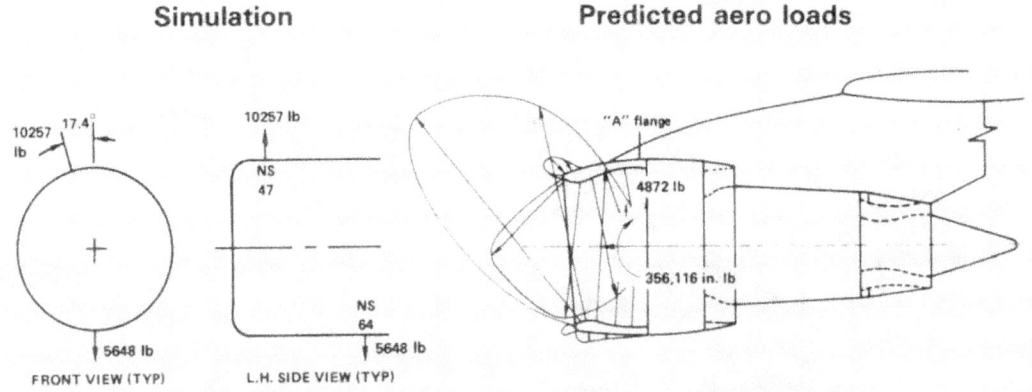

Figure 8 Inlet Air Loads at Take-Off Rotation. (J21704-193)

Figure 9 X-Ray Facility with Test Engine Installed. (J24603-15)

- Measure typical aerodynamic and inertia loads during acceptance test and revenue service
- Explore effects of gross weight, sink rate, pitch angle, and maneuvers on nacelle loads
- Measure engine clearance closures and engine performance resulting from the airplane maneuvers
- Provide data for improved propulsion system designs

Figure 10 Flight Loads Test Program Objectives. (J24603-3)

Figure 11 Instrumented JT9D-7A Engine Installed on RA001 Airplane for Flight Loads Test. (J24603-22)

- Analytical build of instrumented fan and high pressure turbine
- Initial engine calibration in test cell
- Installed engine ground calibration
- Production acceptance test flight at 550,000 lb takeoff gross weight
- Installed engine ground calibration
- Wind-up turns to 2Gs
- Installed engine ground calibration
- Heavy gross weight takeoff, maximum dynamic pressure and maximum Mach number
- Installed engine ground calibration
- Final engine calibration in test cell
- Analytical teardown of fan and high pressure turbine

Figure 12 Test and Inspection Sequence for Flight Loads Test Program. (J24018-6)

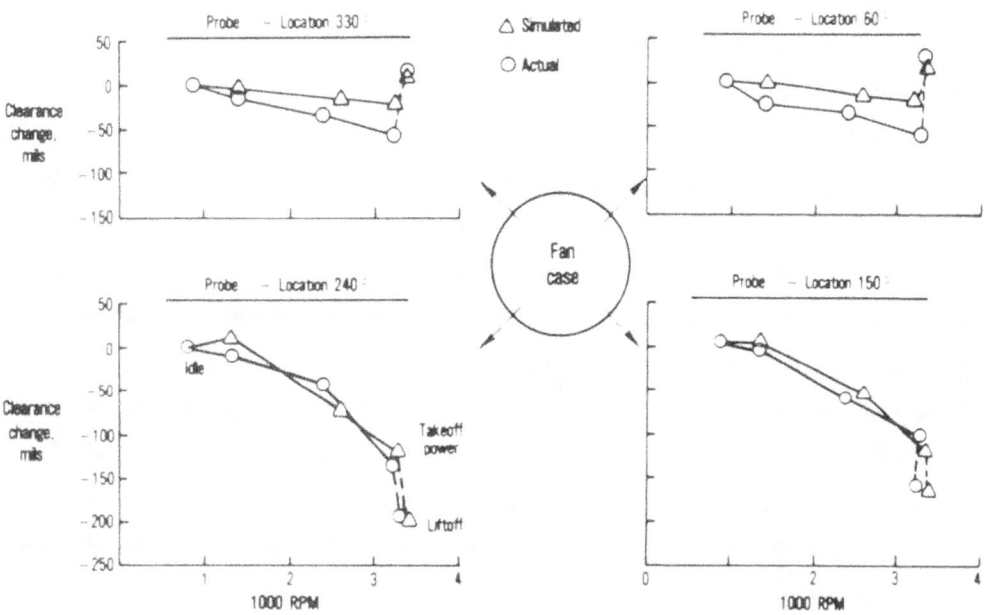

Figure 13 Comparisons of Simulated and Actual Flight Test Fan Closures; Acceleration to Take-Off. (J24873-10)

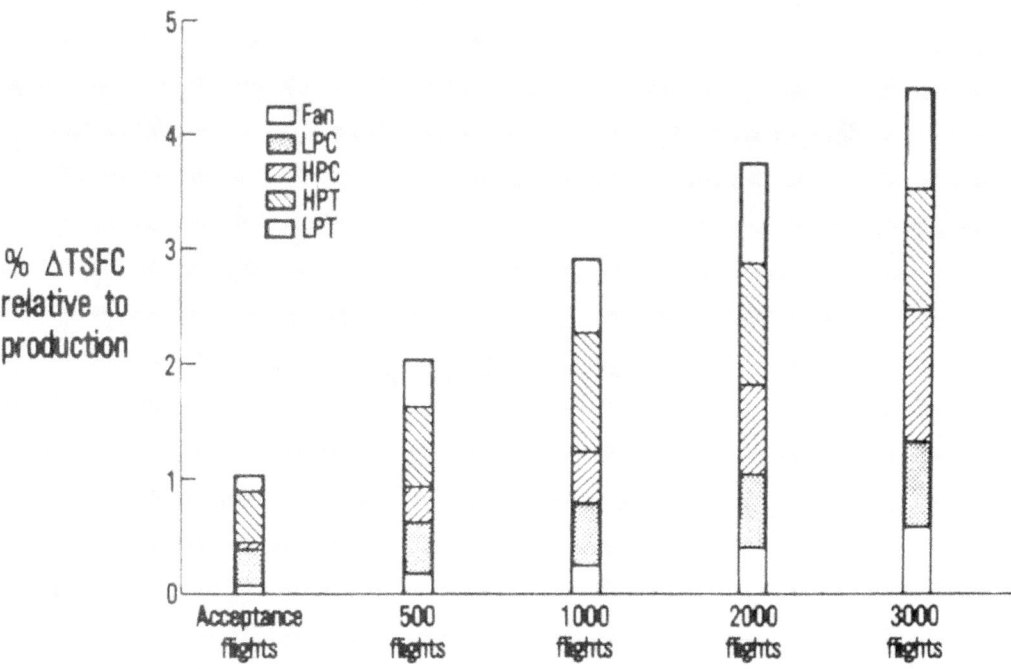

Figure 14 JT9D-7 Sea Level Perfomance Deterioration Distribution by Engine Module. (J24603-21)

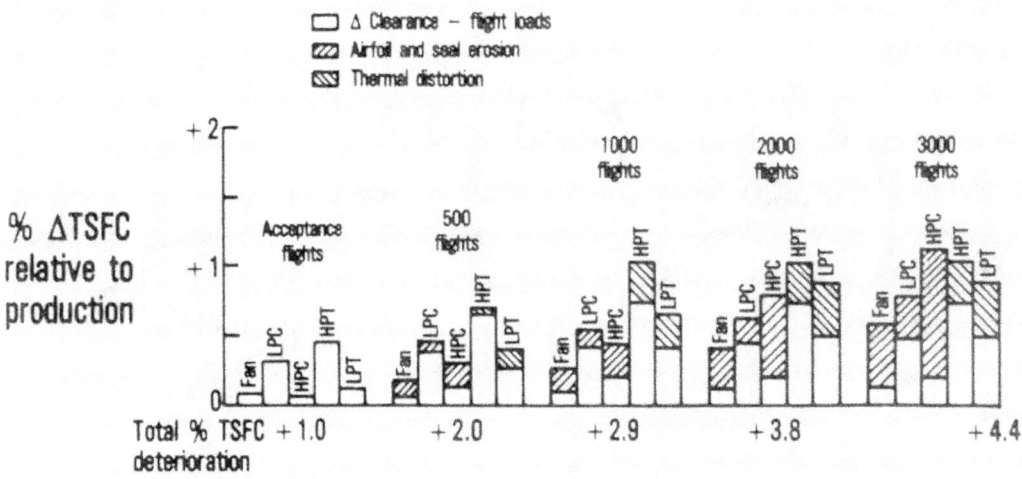

Figure 15 JT9D-7 Sea Level Performance Deterioration Distribution by Cause.
(J24603-23)

Current engines:

- Cold section refurbishment

Future engines:

- Flight-load resistant propulsion systems
- Erosion resistant design and materials
- Thermal distortion resistant design and materials

Figure 16 Flight Loads Test Program Recommendations.   (J24765-17)

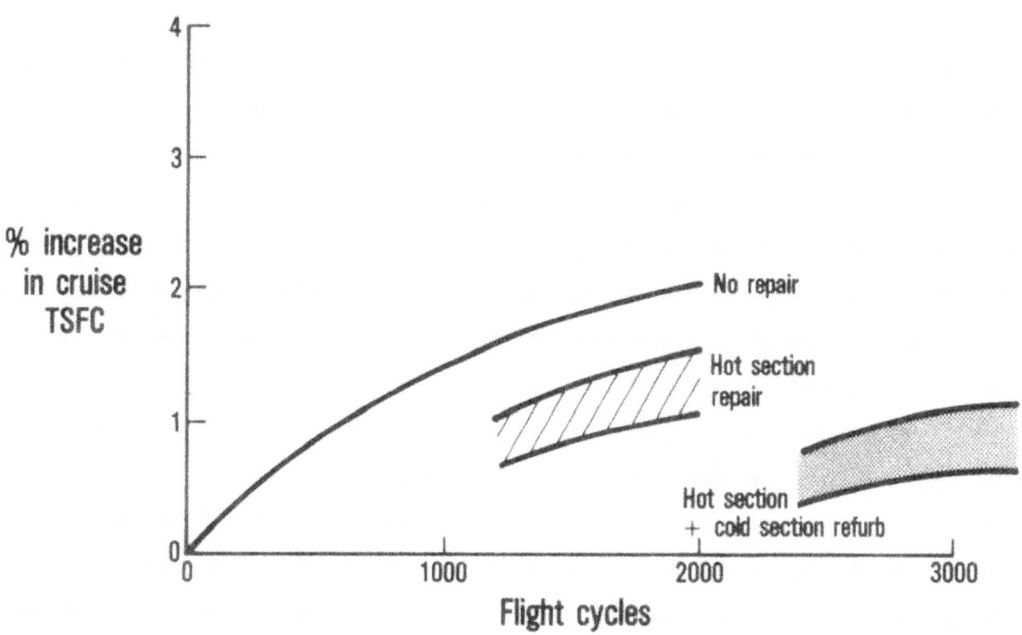

Figure 17 Effect of Repair on JT9D-7A Engine Cruise Thrust Specific Fuel Consumption. (J24603-24)

Page intentionally left blank

JT8D ENGINE PERFORMANCE RETENTION

Albert D. James and David R. Weisel
Pratt & Whitney Aircraft Group

SUMMARY

The attractive performance retention characteristics of the JT8D engine are described. Because of its moderate bypass ratio and turbine temperature, and stiff structural design, the performance retention versus flight cycles of the JT8D engine sets a standard that is difficult for other engines to equal.

In addition, the significant benefits of refurbishment of the JT8D engine are presented. Cold section refurbishment offers thrust specific fuel consumption improvements of up to 2 percent and payback in less than a year, making a very attractive investment option for the airlines.

INTRODUCTION

The ability of an aircraft powerplant to retain its marketed performance is one of the primary considerations in its development or selection. Escalating price of fuel during the last decade has placed greater emphasis on defining the performance retention characteristics of current, as well as new powerplants. Any performance loss can be directly related to increased operating costs, and fuel is an ever increasing portion of those operating costs.

The Pratt & Whitney Aircraft JT8D engine is a first generation turbofan engine, i.e., a low bypass ratio, dual spool, axial flow turbofan that enjoys the distinction of being the most widely used engine in commercial service. To date, over 10,000 units have been delivered and power more than 3,000 aircraft for 175 airline operators worldwide. To assist these operators in minimizing fuel costs and maximizing time on-wing, Pratt & Whitney Aircraft has conducted an extensive analysis of the JT8D engine in order to define the industry potential for performance recovery, the modes of performance loss within the component modules and cost effective means of recovering that lost performance.

PERFORMANCE CHARACTERISTICS

Quantifying performance losses as a function of engine service time has been an industry problem since the initiation of commercial service. Manufacturer development processes include many tests with cyclic power adjustments that provide advanced notice of problems arising from thermal distress and fatigue. They do not have the means of accelerating the normal erosion process that the gas path hardware is subjected to daily in commercial service.

However, operators have all these conditions occurring continuously, but usually do not have the facilities, manpower, financial commitment or time to do the extra testing that would provide this basic knowledge.

Reams of inflight cruise monitoring data could be reviewed. Almost all airlines require the flight crew to record one data point during cruise on each engine during each flight leg. Five performance parameters are recorded: engine pressure ratio, high and low spool rotor speeds, fuel flow and exhaust gas temperature. The data quality suffers due to the use of aircraft instrumentation. The principal use of the data is to monitor the engine parameters for abrupt changes that would signal a potential incidence of part failure. Data scatter is significant. Considerable massaging is required to smooth out the data enough to generate trending information that will identify a severe deviation thereby making the identification of small performance losses very difficult. It was, therefore, decided to base the on-wing (pre-repair) deterioration characteristic on ground engine cell data acquired from engines run prior to repair.

Figure 1 provides the average performance retention characteristics of the JT8D engine and the individual sea level static data points used as the basis for this curve. The average curve is described by a second order curve fit through all the data, and is considered representative for all JT8D-1 through JT8D-17 models.

Figure 2 shows this characteristic in relation to the industry average post repair performance levels. Three levels of performance are shown that exist in commercial service today. The uppermost curve represents the average on-wing (pre-repair) deterioration. Individual operators may experience deterioration rates significantly different from this curve, either better or worse, since there are a multitude of factors that can influence the deterioration rate. These factors will be discussed in the later paragraphs.

Average industry performance levels after repair are represented by the middle curve. Industry repair practices have reflected the philosophy of on-condition maintenance, which were directed at minimizing engine maintenance cost. Cheap fuel prices and known operating limits have allowed operation under this maintenance philosophy. Engines are operating today that have accumulated operating times in the 30,000 hour or cycle region with periodic hot section repair and only occasional minor cold section work to repair airfoil damage from Bill of Material or foreign objects. Refurbishment of the compressors to recover performance loss (particularly specific fuel consumption) was dismissed by operators as not being cost effective. These components were generally refurbished only when incidences of compressor surge were encountered.

This curve then reflects the results of on-condition maintenance philosophy with an implied penalty because of limited cold section repair. The curve flattens to a relatively constant loss resulting from the imposition of the operating temperature limit. Gas path deterioration and increasing exhaust gas temperature coexist. Generally, the JT8D engine has not had a problem with the exhaust gas temperature measuring system indicating temperatures that are falsely high or low. Therefore, as the gas path deteriorates, the exhaust gas temperature limit restricts takeoff power setting, forcing the engine back to the repair shop. The practice of minimum repair then manifests itself in ever shorter periods of operation between repair cycles and exhaust gas temperature limiting operation.

Fuel shortages and the continuing escalation of fuel prices have refocused the industry emphasis on fuel economy, both in operation and maintenance. Practices that were not cost effective a few years ago, now have reasonable payback periods.

In the interest of identifying effective refurbishment, Pratt & Whitney Aircraft funded a JT8D Engine Maintenance Technology study to define the primary modes of deterioration in the engine, identify the modules where maximum performance restoration could be attained and develop cost effective methods of recovering that lost performance. Cooperation of four major airlines (three domestic and one foreign), two years of studying hardware, scrap records and available data, and an in-house testing program using a loaned high time service engine, have identified and demonstrated that significant performance could be recovered through fan and high pressure compressor refurbishment.

The lower curve (Figure 2) represents the average level of performance attainable through a revised maintenance philosophy that recognizes periodic cold section refurbishment as a major part of performance recovery.

It would appear that a significant improvement is still available, since the average remains approximately 2.0 percent above the new engine baseline. The curve should not be interpreted to conclude that refurbishment to the "as new" performance level cannot be achieved. A few of the engines came within 0.5 percent of their production acceptance levels. Current production performance levels were used as a consistent base for all the post repair tests and the average engine performance has improved over the years. This would tend to make the difference appear larger than it actually was. Realistically, achievement of "as new" performance can be accomplished through refurbishment. Although material lost to erosion cannot be restored, airfoil shape and surface finish, which are the prime performance factors, can be restored with only a residual reduction in performance life.

FACTORS INFLUENCING PERFORMANCE RETENTION

Assuming large amounts of usable data were available to define the performance retention characteristics for each operator, a wide variation in operator average retention and engine to engine variations about each average could be expected. Engine deterioration rates are dependent on many factors other than engine to engine differences of gas path geometry and cycle matching. Of major influence are operational environment, operational philosophy and maintenance philosophy, which are particular to each individual airline.

Gaspath Geometry

The JT8D engine is a first generation turbofan engine, i. e., a dual spool, axial flow, low bypass turbofan. It is constructed with a full length annular fan duct that directs fan air rearward to mix with primary air in a common convergent exhaust nozzle. This results in rigid case construction so the installed engine is not adversely affected by the axial bending forces exerted by inlet air loads during aircraft rotation and maneuvering.

In comparison to more current state of the art engines, the gas path is built with relatively loose clearances between rotating and stationary parts, therefore, compressor and turbine airfoil tip rub is not a factor. The primary forms of performance loss are hot section thermal distress, compressor airfoil roughness and compressor airfoil erosion.

Airfoil roughness is the result of environmental contamination and generally occurs in the first 1,000 hours of operation after installation or cleaning. Contamination is the minor mode of compressor performance loss that appears to reach its maximum and then remains relatively constant throughout the operational phase.

High gas velocities and thinner airfoils in the high pressure compressor account for a large measure of the performance impact. In addition to the leading edge erosion exhibited in the fan and low pressure compressor, blade pressure side erosion also occurs. Both leading edge and pressure side erosion occur concurrently until approximately 4.0 percent chord is lost near the blade tip. At this point, the blade trailing edge has become excessively thin and begins to tear away. Figure 3 graphically shows the relative performance loss of each deterioration mode and the accelerated rate for performance loss that takes places when the trailing edge shreds.

Operational Environment

Operational route structure and geographical area have a definite impact on the deterioration rate. The JT8D engine is used in the short to medium range segment of the industry. Aircraft cyclic times, the time between takeoff and landing, vary between 20 minutes to two hours per cycle. The industry average is about one hour per cycle, which is considerably shorter than the typical long range aircraft which run four to seven hours per cycle.

Hot section distress, the primary mode of engine performance loss in all engines, is cyclic dependent. Time at high temperature is the key parameter in determining turbine hardware lives. Since maximum temperature is achieved during takeoff, deterioration as a function of engine cycles becomes most significant. On this basis, the JT8D engine performance retention characteristic is very attractive.

Cold section airfoil erosion, the principal mode of performance loss in the compressor, is cyclic dependent, and also dependent on geographical area. A Middle East operator typically exhibits more airfoil erosion due to sand ingestion, in 2,000 cycles than a large domestic operator may see in 30,000 cycles. Figure 4 illustrates the variation in erosion rates by assuming that the rates are linear.

Other geographical areas have their own particular problems, although not as severe as the desert operations. Operators in Alaska also have an erosion problem with volcanic ash and operators near salt water or highly industrialized areas have problems with corrosion and sulphidation.

Operational Philosophy

Power setting procedures that reduce the impact of hot section thermal shock during takeoff have been defined by engine and airframe manufacturers and are available to the operators. The procedures simply require the flight crew to compute the required power necessary to get airborne based on the ambient temperature, aircraft gross weight and runway length. Dependent on these variables, reduced power may be used instead of the full rating. This results in the engine operating at a lower turbine inlet temperature. A significant reduction in turbine distress and, consequently, a reduced deterioration rate can be realized if reduced power takeoffs are used whenever possible.

Figure 5 shows an analytical assessment of maintenance material cost resulting from reduced power takeoffs. The relative cost is shown as a function of engine time per cycle with lines of constant percentage of power reduction. Maintenance material savings are directly related to a reduction in turbine distress. Similar savings can be realized by using reduced climb power whenever aircraft loading and routing permit.

Maintenance Philosophy

Maintenance policy has been an evolving process as illustrated in Figure 6. During the early years of operation, maintenance was on a hard time basis. Engines were removed, inspected and repaired at specific intervals. As confidence in the hardware integrity became proven, the time intervals were extended based on operator experience.

Logical progression led to on-condition maintenance. This was considered at one time to be the ultimate policy. Engines stay on-wing until removal is forced by a fault or the inability to set power because of reaching an operational limit. Some parts still retain hard time limits based on maximum cycles for useful life, but generally these are long term limits. All interim maintenance is based on a problem developing or the engine reaching the exhaust gas temperature limit prior to meeting the power setting requirement. This was an acceptable maintenance policy during an era where fuel cost was a small portion of total airline operating cost.

Minimal maintenance results in increased fuel consumption and turbine temperature levels. Turbine distress occurs with increasing regularity and is readily evident during disassembly. Compressor deterioration is not as apparent, therefore, the extent of cold section maintenance was to blend repair obvious Bill of Material and foreign object damage.

The tightened world supply of fuels and strategic materials highlighted the need to change philosophies. Fuel became a major part of the airline operating cost and the maintenance emphasis was redirected toward reducing fleet fuel consumption. The impact of the cold section, fan and high pressure compressor particularly, became more obvious, forcing maintenance philosophy to swing back towards scheduled cold section refurbishment.

PERFORMANCE OPTIMIZATION

With the maintenance philosophy shift towards scheduled compressor refurbishment, the next logical progression is towards complete engine maintenance management. The deterioration rates of each module, and the potential to recover lost performance are recognized in this concept. This utilizes a cost effective mix of repair techniques and new parts to attain a service goal.

This also must address on-wing maintenance to retain the performance recovered through refurbishment for as long as possible. Periodic engine water wash and fuel nozzle cleaning should be included in any engine management program.

Compressor airfoil roughness has been found to occur in the first 1,000 hours of operation. Experience has shown that long term water washing is not very effective in removing compressor contamination. However, short term periodic washing (250 hour intervals) has demonstrated that performance loss due to contamination can be held off for 4,000 to 5,000 cycles. Intervals recommended for compressor water wash have been in 1,000 hour increments or less, but as more data is accumulated, the most significant results are associated with the shorter time interval wash procedures.

Considerable data from hot section inspections have correlated burned turbine nozzle vanes and streaked combustion chambers with coked fuel nozzles. Non-uniform fuel flow from plugged nozzles results in excessive temperature and burned vanes. On-wing fuel nozzle cleaning procedures have been proven effective in eliminating moderate coke deposits. Therefore, periodic cleaning in the recommended 1,000 hour intervals will maintain coke free operation and optimize turbine hardware performance.

PERFORMANCE RETENTION MODELING

In order to determine more cost and fuel efficient maintenance practices, an accurate model of engine performance loss must be constructed. This model can be used to quantify performance loss for a particular engine model as a function of usage. Additionally, a complete model will identify losses both by module and cause, such as erosion of airfoils, thermal distortion of hot section parts, and clearance increases between rotating and stationary parts due to erosion or rubs.

To begin the process of model construction, appropriate data sources must be selected. Inflight monitoring data is available, but its usefulness as a primary source is limited. The quality limitations of flight data have been discussed previously; large amounts of data must be trended in order to be meaningful. In addition, flight data furnishes far fewer parameters than the number of module losses to be defined. In particular, the lack of thrust measurement makes determination of low spool losses very difficult. Finally, flight data is affected by aircraft systems, particularly the pneumatic system, which makes analysis of the data even more difficult.

Two data sources were used to construct the JT8D engine performance retention model. The first of these was pre-repair data obtained from airline test cell runs. Because of the nature of airline operations, particularly with a well-proven engine like the JT8D and prevailing on-condition maintenance practices, pre-repair data is not normally obtained. However, there is a limited amount of first run data available. This data is relatively accurate and furnishes pressure ratios and thrust, in addition to more accurate definition of those parameters measured inflight. By comparing the pre-repair data for a particular engine to its production run data, it is possible to use a computer simulation of the engine to calculate efficiency and flow capacity change for every module. This analysis depends on use of known relationships between cold sections efficiency and flow capacity changes, referred to as "coupling" relationships.

The other data source used for the model was teardown inspection data. Used parts with known service times were collected and analyzed to determine performance changes from the new part configuration. Such data furnishes knowledge of the causes, as well as the magnitude of module performance loss. From the estimates of individual module performance changes from new, total engine performance change from new can be synthesized with the engine computer simulation. The results of this analysis can then be compared to the pre-repair analysis and then both analyses can be iterated until reasonable closure is obtained, as shown in Figure 7. Since each analytical approach has its limitations, such iteration enhances the validity of the solution.

This process was performed for the JT8D-9 engine. Figure 8 shows a comparison of increase in sea level takeoff exhaust gas temperature from production as predicted by the model, compared to the pre-repair data. Figure 9 shows the same comparison for thrust specific fuel consumption. Reasonably good correlation between the average of the data and the model is shown. The used part analysis showed that the component deterioration for a typical operator was most strongly related to flight cycles for reasons previously discussed. However, there can be considerable operator-to-operator variation associated with operating environment differences.

Modules losses at 4,000 and 8,000 cycles that result from the analysis are shown in Figure 10. Cold section losses are dominated by erosion and roughness damage, while hot section losses occur primarily in the high pressure turbine and are largely the result of thermal distortion (vane bow). There are no module losses due to clearance increases. The low bypass ratio of the JT8D engine minimizes thrust bending loads, and the long, stiff one piece fan duct effectively isolates the internal engine cases from nacelle aerodynamic loads. These features, plus the moderate hot section temperatures, result in a standard of performance retention that is difficult for other engines to equal.

Figures 11 and 12 show the overall impact of cold section versus hot section losses on exhaust gas temperature and thrust specific fuel consumption, measured at sea level takeoff conditions. The hot section dominates exhaust gas temperature increase, while cold section losses have greater thrust specific fuel consumption impact. Historically, engine overhaul has been directed primarily toward restoring exhaust gas temperature margin, and airline efforts were accordingly concentrated on hot-section repair. However, in the current era of constantly escalating fuel prices, Figure 12 shows the importance of periodic cold section refurbishment to minimize fuel burned.

In order to analyze the potential benefits of performance restoration on fuel burned, fuel consumption at altitude conditions must be evaluated. This can be done with the engine computer simulation. Analysis shows that there is an effect of both flight condition and power setting on thrust specific fuel consumption change due to component performance changes. A weighted thrust specific fuel consumption change has been defined which combines both climb and cruise thrust specific fuel consumption at typical altitude conditions, in proportion to the fuel consumed during a typical mission. For the JT8D engine in a typical 727 application, the weighted thrust specific fuel consumption change is approximately the average of climb and cruise changes.

Takeoff exhaust gas temperature and weighted thrust specific fuel consumption increases are shown by module and cause in Figures 13 and 14. Exhaust gas temperature increase is primarily controlled by the high pressure turbine; however, high pressure compressor losses also contribute significantly. Thrust specific fuel consumption increases are dominated by the fan and high pressure turbine; the high pressure compressor again contributing significantly.

The performance retention model can also be used to predict the impact of hot section repair, and project performance losses for multiple run engines. Figures 15 and 16 show the impact of hot section repair only. Typical first removal is shown at about 6,000 cycles, which would probably be for foreign object damage or hot section inspection. At this point, a change in high pressure turbine outer airseal from bill-of-material knife edge to honeycomb (typical airline practice) is shown. Erosion of the replacement outer airseal plus worse bow for repaired vanes results in more rapid exhaust gas temperature and thrust specific fuel consumption increase with flight cycles, so that an exhaust gas temperature-limited condition may be encountered at about 10,000 cycles. The effect of hot section only repair is that the interval between the exhaust gas temperature-limited shop visits decreases with successive shop visits. This is because of the underlying cold section damage which has not been corrected.

IMPACT OF COLD SECTION REFURBISHMENT

The assumed workscope for cold section restoration is shown in Figure 17. If cold section refurbishment is accomplished at "soft" time intervals of 12,000 to 17,000 cycles (nearest convenient time when the engine is in the shop), the model predicts the results of Figure 18. Fan and low pressure compressor restoration are accomplished at 13,500 cycles, along with high pressure turbine repair. At 17,000 cycles, the high pressure compressor and turbine are repaired. The benefits in improved thrust specific fuel consumption are readily apparent.

A number of studies have illustrated the cost-effectiveness of cold section refurbishment for the JT8D engine. Figure 19 illustrates summary results from one study. The study showed fan and low pressure compressor refurbishment to be most cost effective (earliest payback). Refurbishment of fan, low pressure compressor and high pressure compressor combined, resulted in 1.6 percent weighted thrust specific fuel consumption recovery at refurbishment, 21°F exhaust gas temperature recovery and payback in less than a year for a typical operator, based on 50 cents per gallon of fuel. Current fuel costs would further enhance cost effectiveness of cold section refurbishment, notwithstanding increased labor and parts costs since this study was conducted.

REFERENCES

1. Sallee, G. P., JT8D Maintenance Technology Study, United Technologies Corporation, Pratt & Whitney Aircraft Group, Commercial Products Division, June 1980.

2. 1980 JT9D Engine Regional Engineering and Maintenance Conference, Pratt & Whitney Aircraft Group, Commercial Products Division.
    October 1980, Zurich, Switzerland and Washington, D.C.
    November 1980, San Diego, California and Singapore

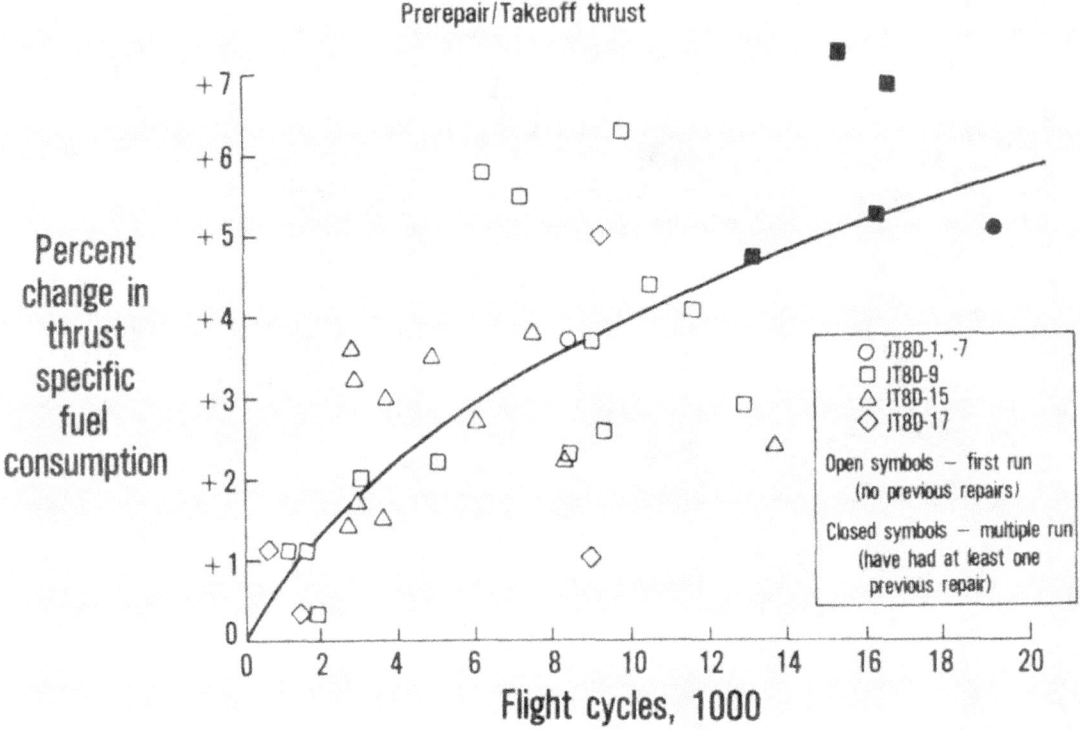

Figure 1   Sea Level Static Test Cell Data JT8D Engine Performance Retention

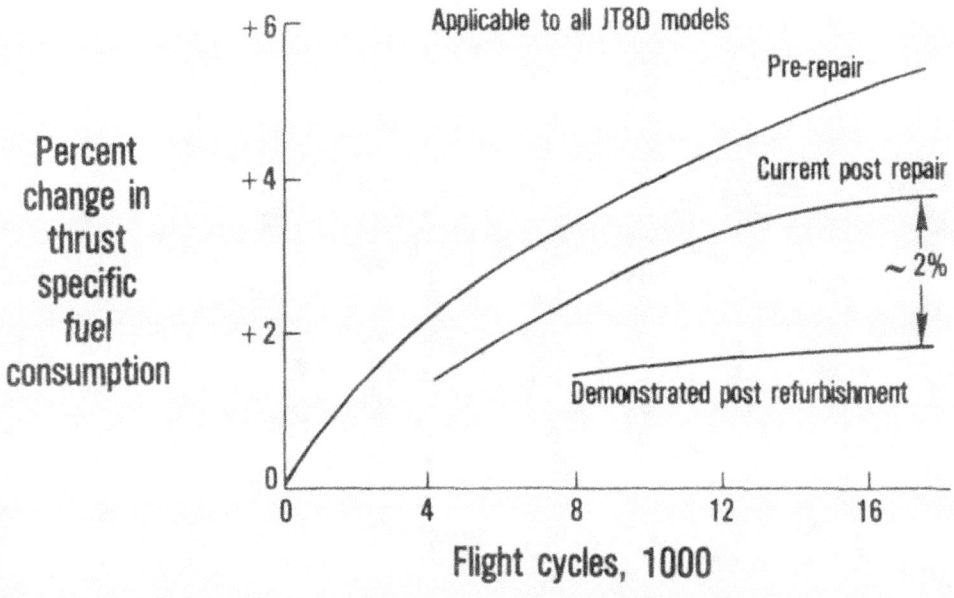

Figure 2   Industry Average Post Repair Level Shows Significant Potential for Performance Recovery

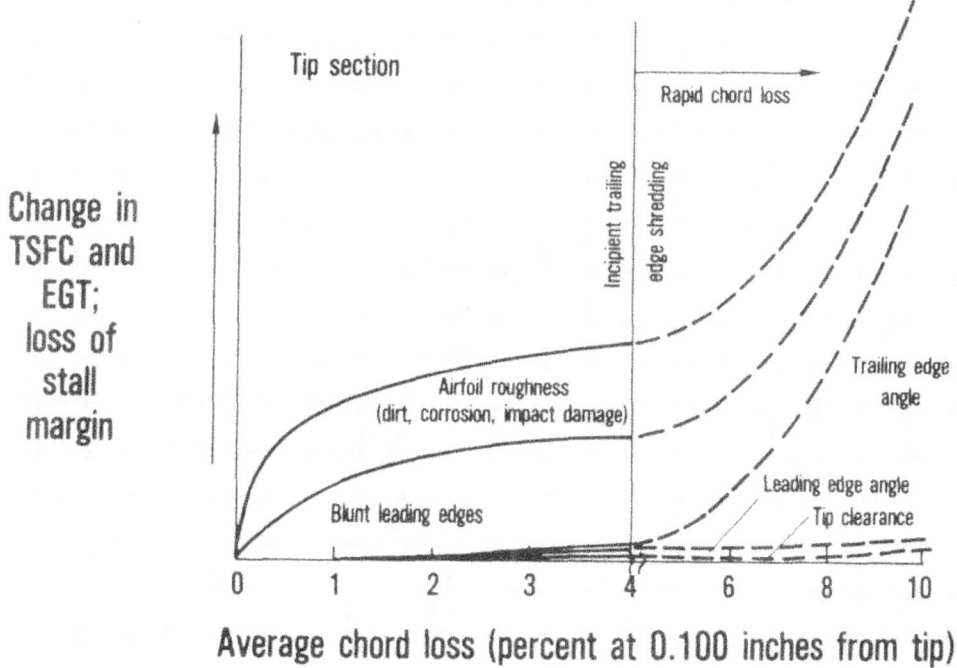

Figure 3   High Pressure Compressor Performance Retention Directly Related to Chord Loss

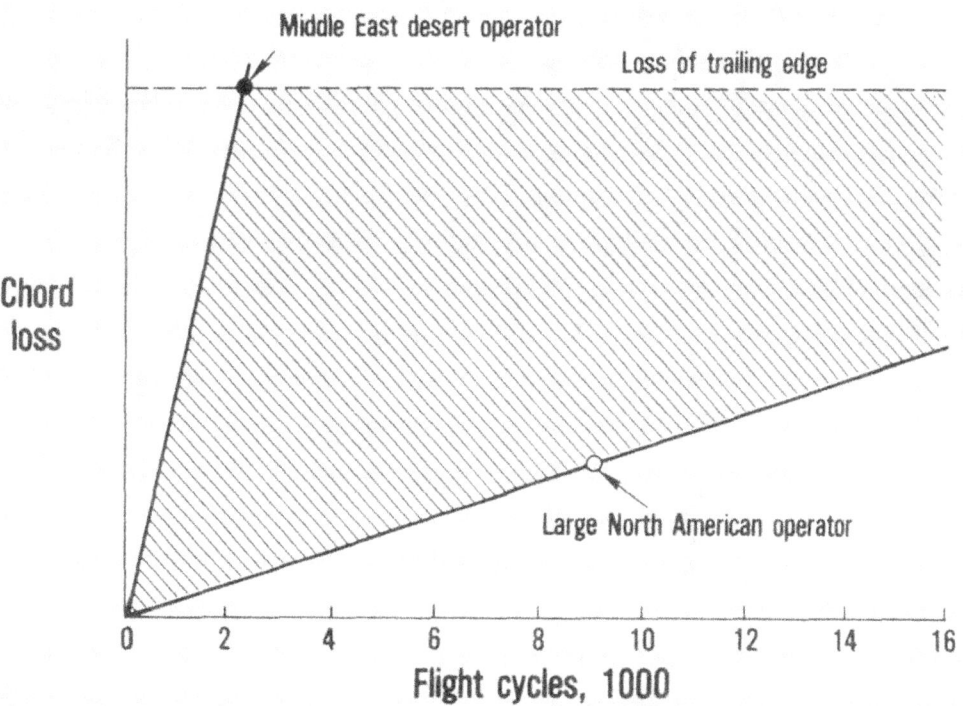

Figure 4   High Pressure Compressor Blade Chord Loss Rate Varies Greatly Between Operators

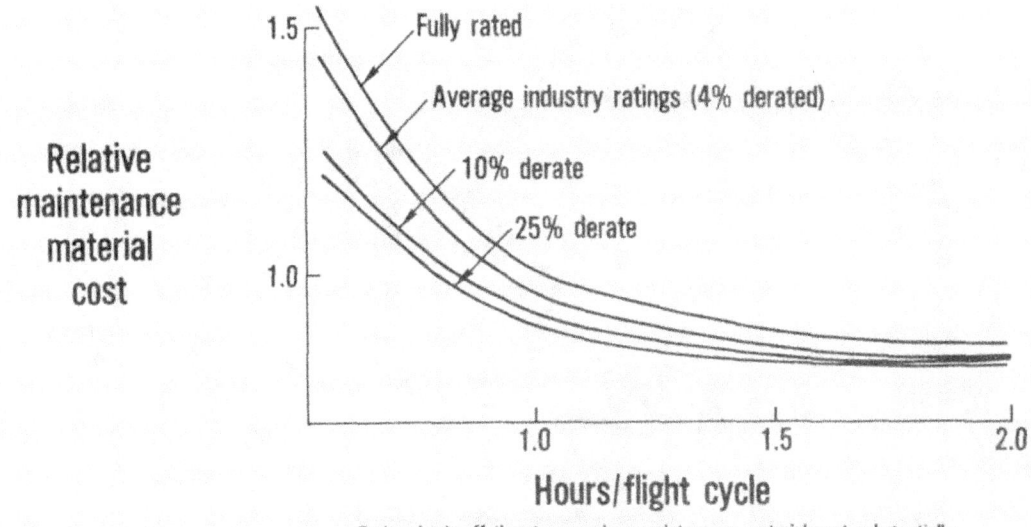

Figure 5   Maintenance Cost Reduced by Thrust Derate Procedures

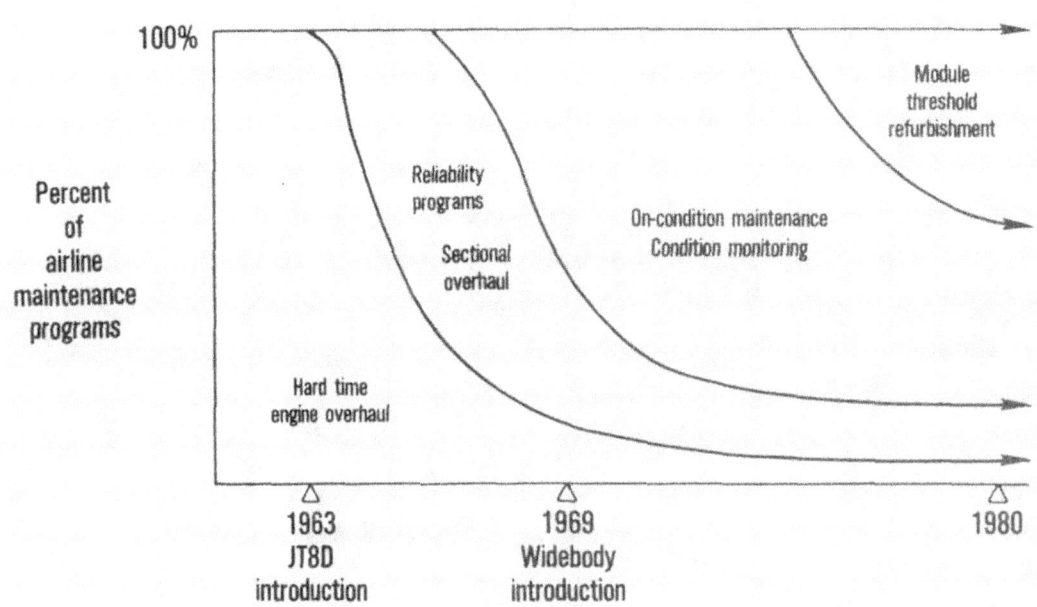

Figure 6   Engine Maintenance Concepts are Changing

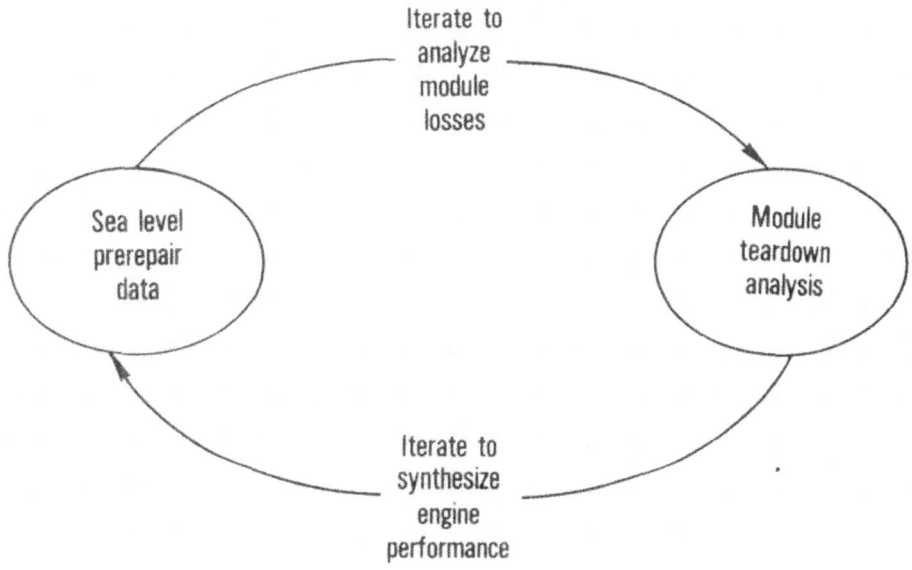

Figure 7   JT8D Engine Performance Retention Model-Approach

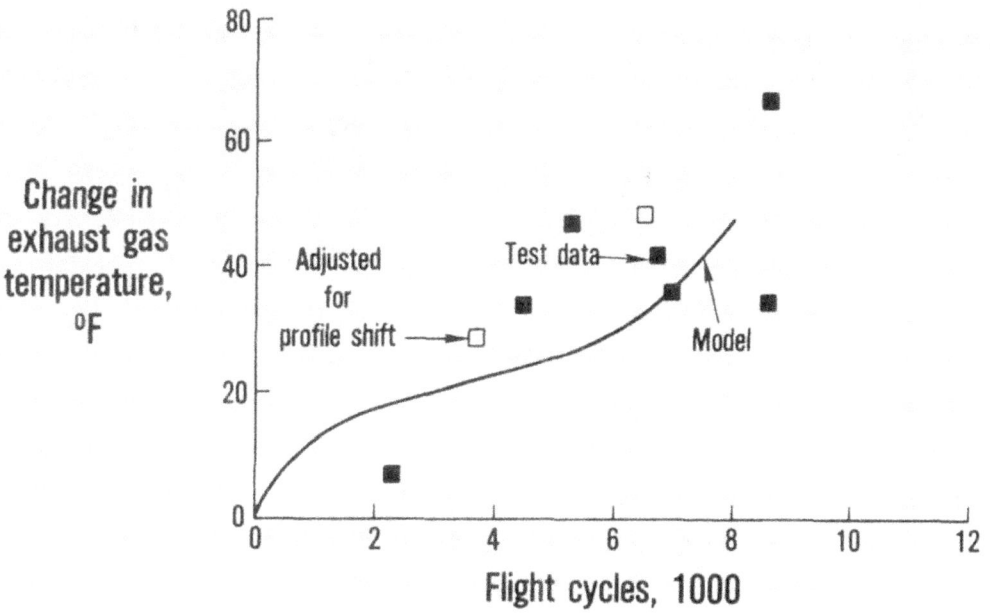

Figure 8   Model Agrees with Test Data

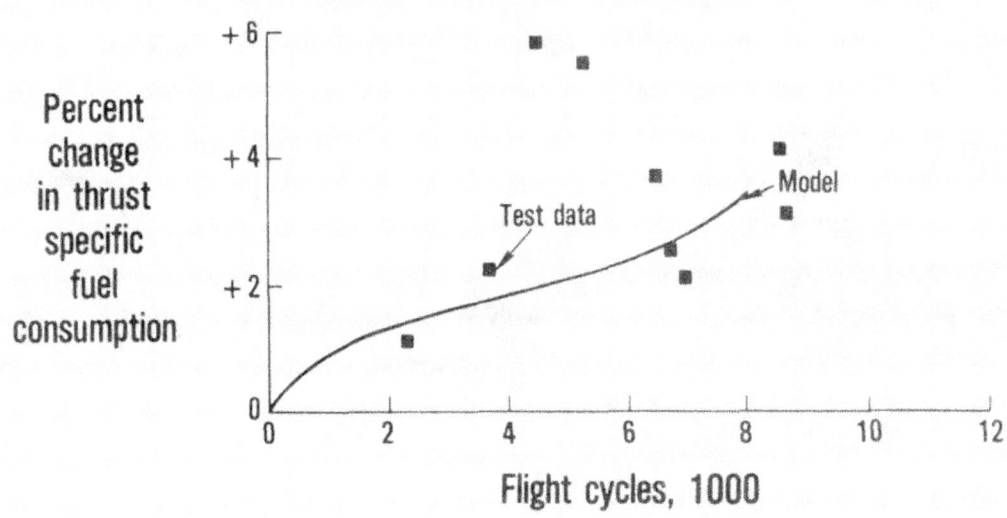

Figure 9   Model Agrees with Test Data

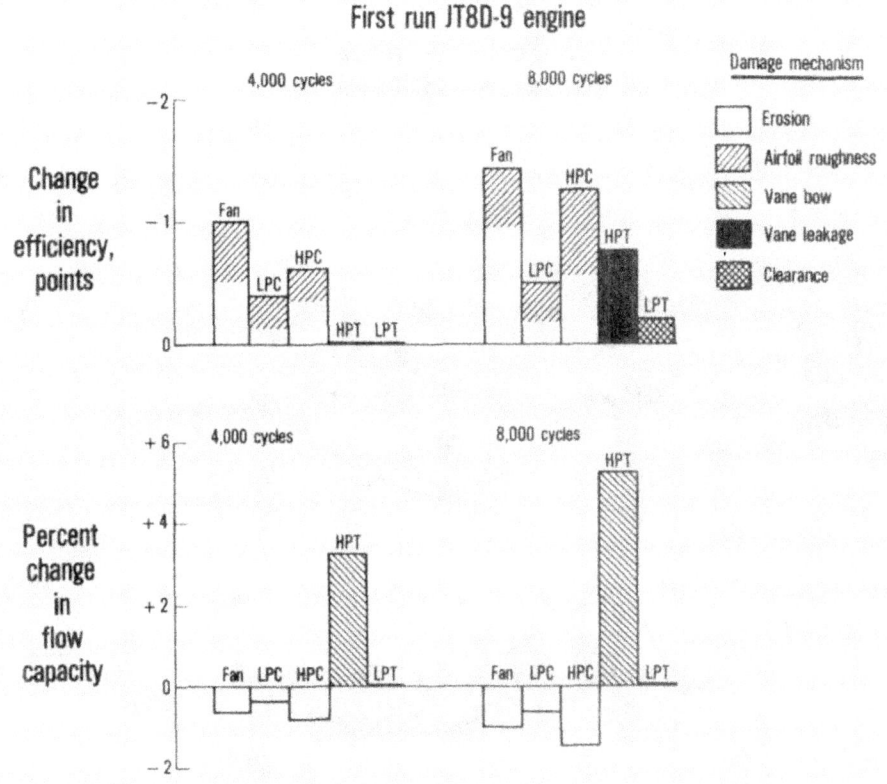

Figure 10   Model Losses Derived from Teardown Data

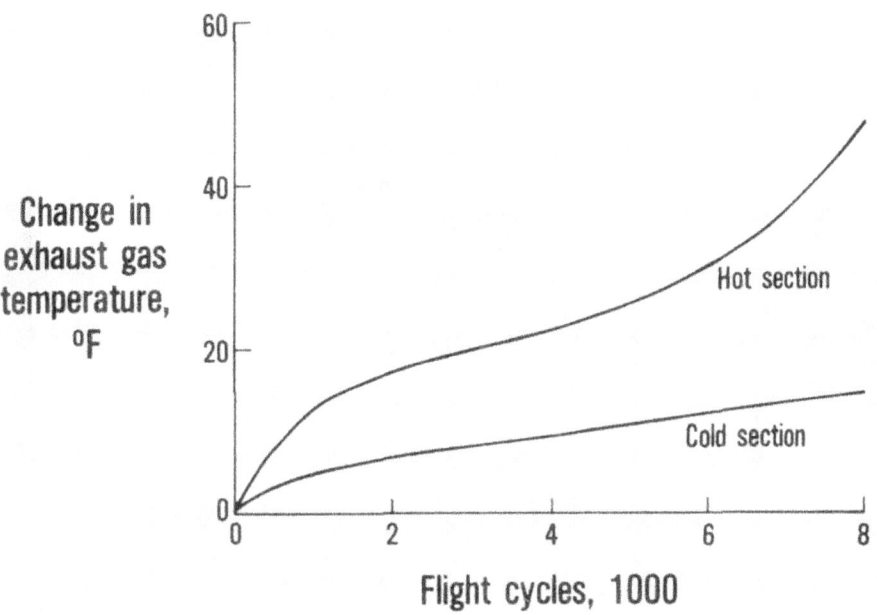

Figure 11   Exhaust Gas Temperature Increase Dominated by Hot Section

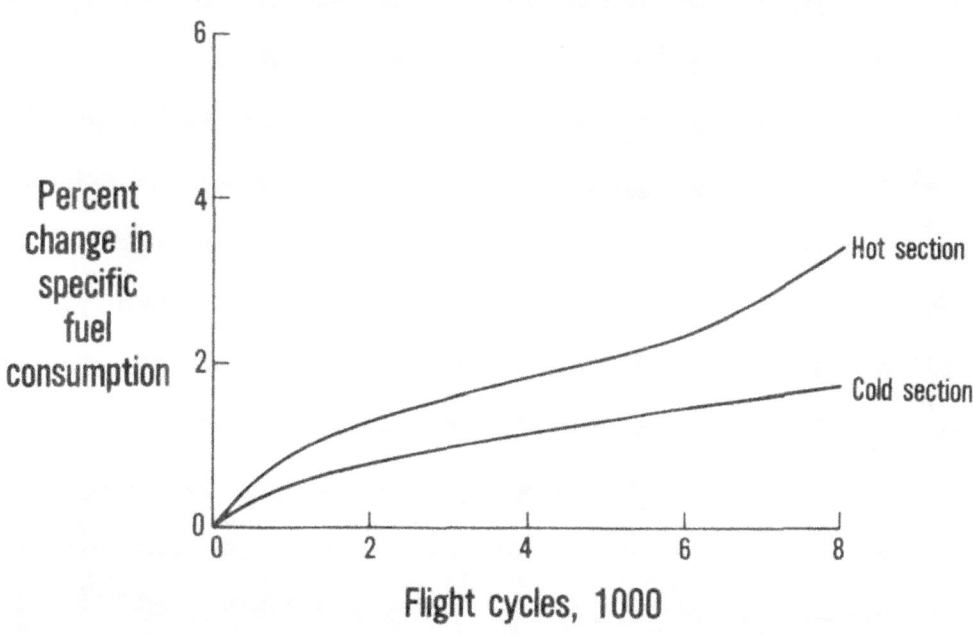

Figure 12   Thrust Specific Fuel Consumption Increase Controlled by Cold Section

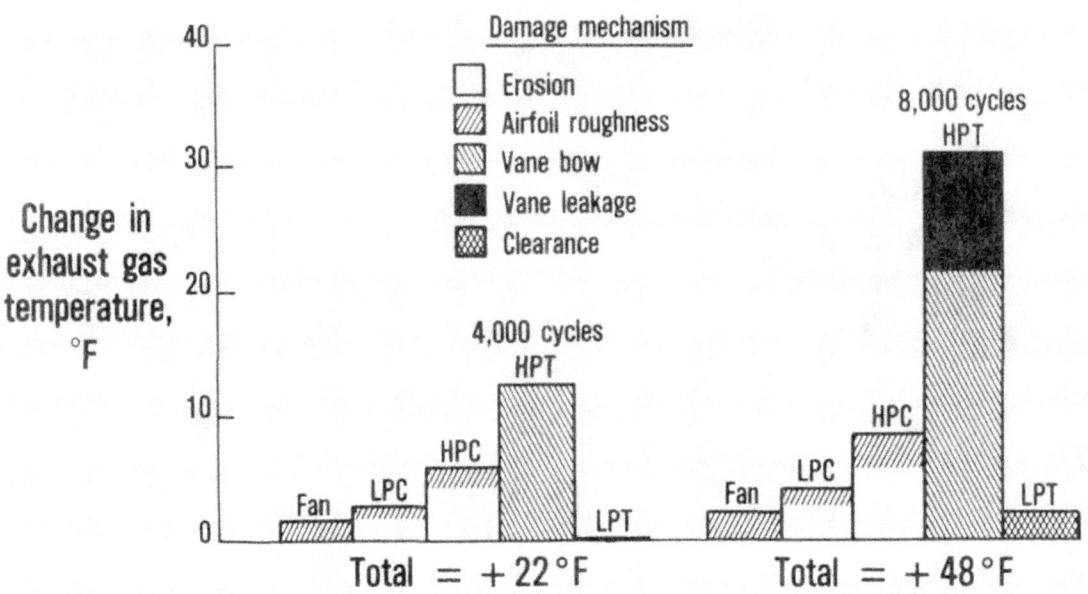

Figure 13　High Pressure Turbine Losses Major Cause of Exhaust Gas Temperature Increase

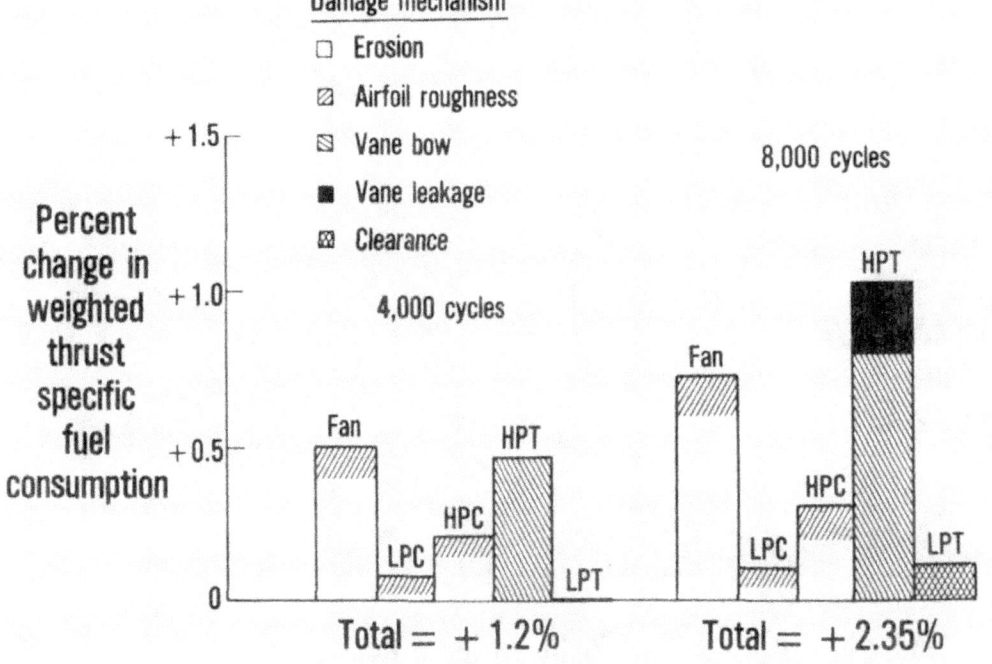

Figure 14　Cold Section Erosion and High Pressure Turbine Vane Bow Major Causes of Thrust Specific Fuel Consumption Loss

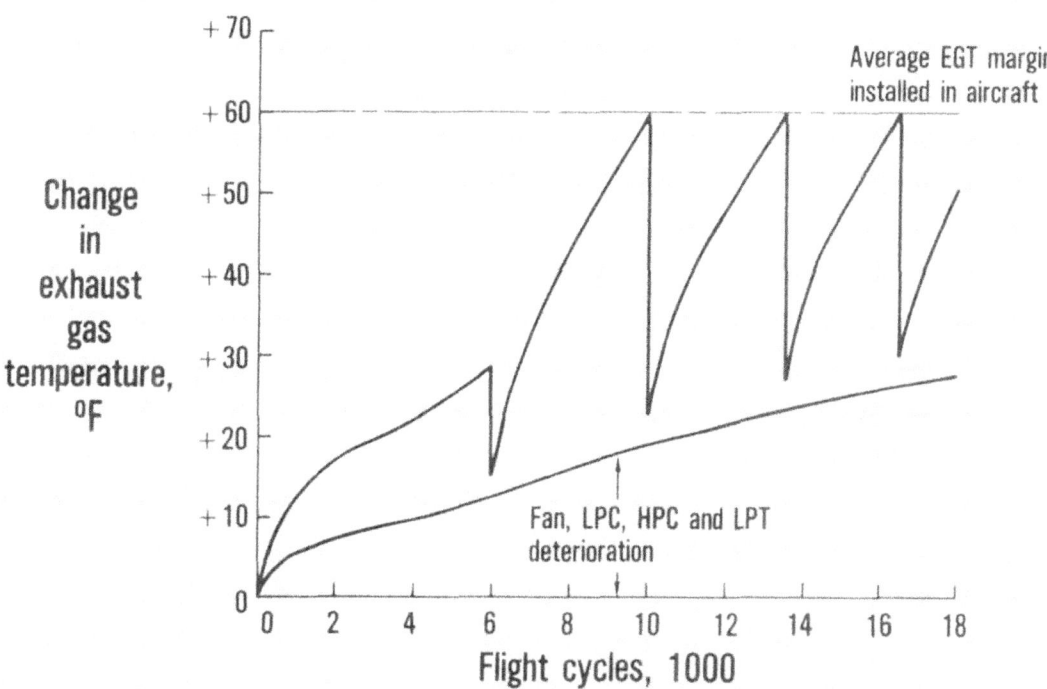

Figure 15 JT8D-9 Engine Multi-Run Model Projects Realistic Exhaust Gas Temperature-Limited Removals

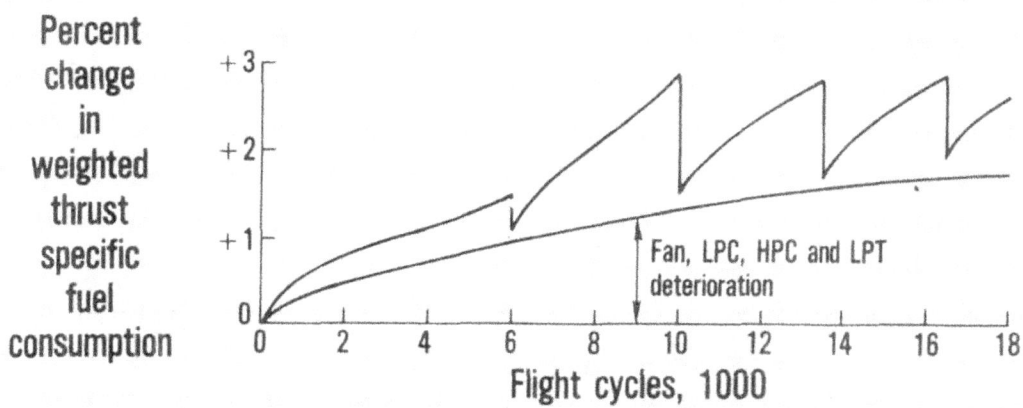

Figure 16 JT8D-9 Engine Multi-Run Model Shows Importance of Cold Section Losses

Module	Workscope
Fan	— All blades — SWECO clean, chamfer cut
Low pressure compressor	— All blades — SWECO clean, chord check, leading edge radius restoration, replace as necessary
	— All stators — vapor clean, check vane angle, re-angle as necessary
High pressure compressor	— All blades — SWECO clean, chord check, leading edge radius restoration, replace as necessary
	— All stators — vapor clean, check vane angle, re-angle as necessary

Figure 17   Assumed Workscope for Compressor Restoration

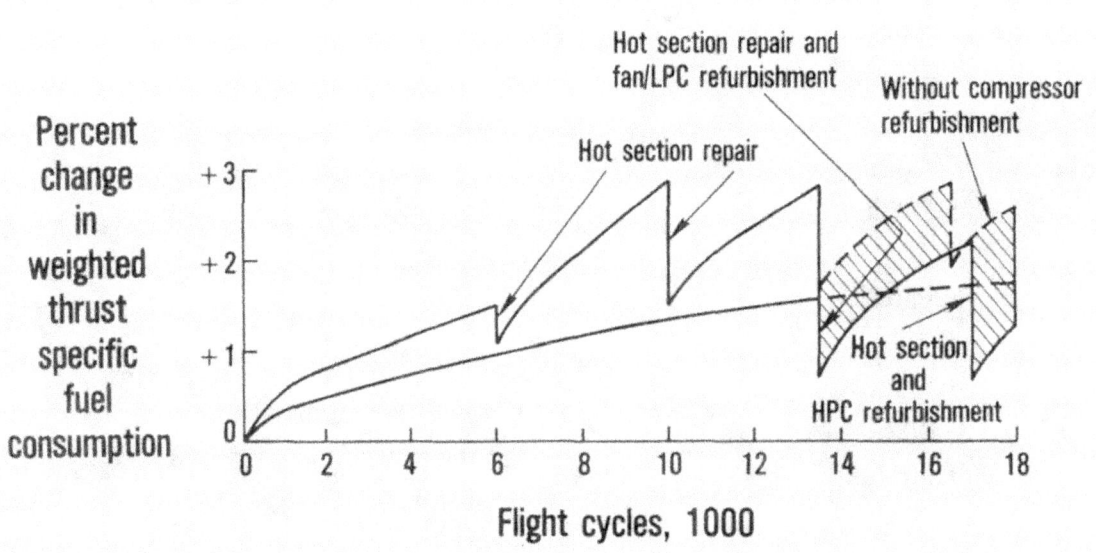

Figure 18   JT8D-9 Engine Multi-Run Model Shows Benefits of Cold Section Refurbishment

- Weighted TSFC recovery at refurbishment = 1.6% ⎫
  ⎬ Fan, LPC, HPC only
- EGT recovery at refurbishment = 21°F ⎭

- Payback period less than one year (typical operator)

    Figure 19   Cold Section Refurbishment Cost Benefit Analysis

Page intentionally left blank

PERFORMANCE RETENTION OF THE RB211 POWERPLANT IN SERVICE

B. L. Astridge
Rolls-Royce Limited

J. T. Pinder
Rolls-Royce Incorporated

INTRODUCTION

It is perhaps a statement of the obvious, but an understanding of the mechanisms of deterioration is essential in order that features to counteract performance degradation can be built into the basic design of an engine and nacelle. Furthermore, the interpretation must be continued in service for effective feedback to provide modifications which may be necessary in maintaining a satisfactory performance retention program.

The in-service assessment must, therefore, be accurate as to magnitude and causes and this requires consideration of:

1. The powerplant as a complete entity, i.e. the engine components and nacelle including the thrust reverser.

2. Measurement of performance in flight rather than by sole reliance on the scaling of test cell data to flight conditions (although some correlation should be possible).

3. The relationship of engine parts condition to overhaul performance and in-flight deterioration level of that engine.

Hence a performance retention program covers design, feedback to design, measurement and analysis of in-service experience and continuous review of the condition of engine components. These aspects are addressed by consideration of the RB211 engine in service in both the Lockheed L1011 Tristar and Boeing 747 aircraft.

PERFORMANCE RETENTION DESIGN FEATURES

Basic Design Features

Performance deficiency will arise when main gas path airflow, either past blade tip seals or through internal cooling air passages or leakage overboard, is excessive. Further major sources of performance loss will arise in the event that significant blade damage or erosion or aerofoil contamination arising because of dirt deposition on oil contaminated components takes place.

The RB211 was designed with features which address these problems, as indicated in Figure 1. Blade tip clearance control and internal cooling air passage seal clearance are considered in the structural design, thermal matching and the use of shrouded blades and stators. Overboard leakage is considered in both main engine casing design and powerplant sealing features such as the reverser seal. Limitation of core engine contamination and erosion is featured in the core engine intake configuration where the generous spacing of the core splitter relative to the fan, together with the wide chord fan, allows the majority of the ingested particles to be centrifuged through the bypass duct.

Details of the particular aspects relating to these features in the RB211 engine are shown in the following illustrations. Blade tip clearance control at all conditions is enhanced by the features illustrated in Figure 2. The three shaft configuration allows a short engine with stiff shafts and casings. Furthermore, the features which determine the gas path are separated from the structural casings and mounted with radial flexibility so that inertia and thrust forces in flight have minimum effect on tip clearances. The effectiveness of these unique features is reflected in the lack of initial performance lost in flight relative to the test bed passoff performance, the RB211 having demonstrated a deterioration including any losses associated with the first flight, of less than 0.3% SFC (specific fuel consumption) in 100 flights from new.

Thermal matching of static and rotating engine components is illustrated in Figures 3 and 4. Engines are, even in normal operation, subject to rapid changes in operating temperatures. Serious mismatch in expansion of the static and rotating components will result in large compressor and turbine blade tip and cooling air seal component excursions and, hence, wear such that, at stabilized conditions, large clearances will occur with a resulting serious performance deficiency. Engine axial matching is illustrated in Figure 3 and shows the HP (high pressure) turbine static and rotating seal members are located to a common datum, the HP location bearing and with load path structures exposed to substantially the same temperatures thus ensuring common expansion. The equally important radial matching feature in the turbines is shown in Figure 4. The HP turbine features a thermal control ring to restrict thermal growth of the static seal member to match the growth of the rotating member, the expansion of which is restricted to a low rate due to the mass of the turbine disc. The IP (intermediate pressure) and LP (low pressure) turbine static seal members are located in the turbine casings and their growth is matched to that of the discs and blades by a combination of casing insulation and external cooling with undercowl ventilation air. X-ray data obtained at various conditions has been extensively used to optimize the design of the features which fix the relative position of the seals and the blade tip fins and this has been achieved without requiring introduction of a scheduled turbine casing cooling air control system.

The air used to cool the turbine casing is also used to ventilate undercowl zones and is no more than the amount required for that purpose. An inherent performance loss does not, therefore, require debiting to the system. However, this powerplant air is taken from the fan stream and excessive leakage of this air, either through the ventilation system or directly overboard via reverser seals, etc., has a powerful effect on cruise SFC and a much smaller, and hence, more difficult to detect effect at static conditions. Consideration of the RB211 nacelle as a complete powerplant has included development of a powerplant leakage check for use in development programs and as a check of service experience. The efficiency of powerplant sealing with time has, therefore, been checked against new engine behavior as shown in Figure 5.

Feedback From Service

No resume of design features would be complete without consideration of the essential feedback of experience from service. There are many examples arising from RB211 experience and a few have been chosen as typical of the aspects discussed hitherto. The first is in the area of tip clearance control where experience with the RB211-22B dictated extensive changes to the HP turbine sealing arrangements for the RB211-524. The second relates to the control of compressor contamination resulting from oil leakage and the third to elimination of excessive variation in powerplant ventilation air, both of which have resulted in more minor but nevertheless, significant changes.

Figure 6 shows the changes in HP turbine shroud segment design for the RB211-524 version as a result of experience gained with the RB211-22B. In the earlier design the leakage paths shown arise in time from component distortion and wear, the NGV (nozzle guide vane) support ring distorting under axial load and the shroud segment leading edge fretting and allowing leakage and loss of axial location. In the RB211-524 design the offending joint in the NGV support ring is eliminated and the shroud segment location is changed from axial to radial under which conditions the sealing is reinforced by aerodynamic loading. It should be further noted that the RB211-524 shroud segment is of such a form that the segment base protects the leading edge of the honeycomb from erosion.

A modification which was designed to prevent oil contamination is sketched in Figure 7. Experience has demonstrated that excessive front bearing housing oil leakage was arising from an over generous hydraulic oil seal chamber volume.

On shutdown the excess oil occupying this chamber leaks into the IP compressor and acts as a base for contamination. A simple modification which has alleviated the situation was the introduction of a liner to reduce the oil chamber volume.

The third category of modification which has been introduced because of service experience is associated with the sealing of the powerplant undercowl ventilation air. Excess airflow was arising from variation in the birdmouth of the seal between turbine section undercowl ventilation zones. The addition of a more positive seal, as illustrated in Figure 8, has cured this problem.

Integrated Exhaust Pressure Ratio (EPR) System

In order to monitor a performance retention program it is necessary to be in a position to measure, accurately, the in-flight performance of the engine. Fundamental to this problem is the ability to measure the thrust setting of the engine. This is achieved in the RB211 by means of the integrated EPR system.

Thrust is fundamentally related to nozzle pressure ratio and it, therefore follows that, provided it can be measured reliably, exhaust pressure ratio must be the most accurate means of power setting. However, this accuracy can only be achieved in a two nozzle engine by sampling the pressure in both streams and then deriving a mean of the two pressures which is weighted in proportion to the areas of the two streams. The RB211 integrated EPR system fits this requirement and its basic nature relating thrust and pressure ratio means that this relationship is virtually unaffected by deterioration. Hence, measurement of in flight performance is possible. The total system is shown diagrammatically in Figure 9 and shows the fan air to be sampled with three five point rakes and the hot nozzle by five four point rakes. Integration of the two nozzle pressures is aerodynamic by the simple block shown in Figure 10, the integrated pressure being sampled between two appropriately sized orifices. It should be noted that the fan nozzle pressure is always higher than the turbine exhaust pressure and the system flow is, therefore, always from fan to turbine exhaust. Carbon or other contamination risk is, therefore, minimal.

As a further refinement, the RB211 integrated EPR system is trimmed, by electrical means, in order to produce a relationship between integrated EPR and thrust which is common to all engines. This is required to take into account the change in the integrated EPR/thrust relationship brought about by variation between units within build tolerances, of the factors which influence this characteristic. Figure 11 illustrates the principle of trimming. The test integrated EPR/thrust relationship is checked to be within quality limits determined from experience, and the trimmer is selected to adjust the relationship, within trimmer steps, to a common characteristic for all engines. The pilot uses the control lever to select an integrated EPR and the trimming procedure previously described ensures that each engine produces the required thrust at the selected power lever setting. The assurance of minimum thrust with alternative setting procedures leads to some engines producing more thrust than required with consequent deleterious effect on component lives.

MEASUREMENT OF DETERIORATION

In Flight

The integrated EPR system described above, with the inherent insensitivity of its relationship to thrust with engine deterioration, allows consistent measurement of in flight performance. However, the reliability of in flight data and its method of interpretation can only be understood with some knowledge of the potential inaccuracies which can occur. At the risk of some over simplification Table 1 lists inherent inaccuracies arising from measuring instrument inaccuracies, both aircraft and engine, and, another prime source of variability, the normal aircraft Environmental Control System (ECS). The latter source of variability comes about because engine air is supplied to a common manifold such that engine to engine and duct loss variations can result in differing amounts of bleed being extracted from each engine in the installation. If the integrated EPR/thrust relationship were not substantially insensitive to deterioration a further variable for power setting at the basic parameter, EPR, would be necessary in attempting to interpret fuel flow changes as changes in fuel flow at a thrust, i.e., SFC.

Nevertheless it can be seen that up to ±2.2% variation in fuel flow at an integrated EPR could arise between engines due to the measurement inaccuracies. In a fleet of aircraft it would be expected that the incidence of inaccuracy would, in most cases, be random and that measurement of fleet performance, as an average, be possible, but that individual engine performance would be somewhat less reliable. Mean fleet data for the RB211-524 engine is shown on Figure 12. The flight monitor line is that established as representative of RB211-524 engines using both averages of larger quantities of flight crew recorded data and some specific aircraft audits. Variation in deterioration rates will arise from differences in operation procedures and rates as low as that shown in Figure 12 for specific audits have been recorded with confidence.

Having made these points, it should be noted that many of the instrument inaccuracies of Table 1 may remain consistent during a particular installed life and, hence, become systematic for the engine in question. This will become further reinforced if data can be recorded with the subject engines either operating without ECS bleed or at least isolated to a potentially reliable source, i.e., engine isolated to one ECS pack. If, indeed, the errors are systematic, it is possible to measure individual performance changes during an installed life, i.e., individual engine deterioration, and reasonable success has been achieved in that individually assessed engines have demonstrated characteristics similar to that established by fleet mean evaluation of flight data and comparison of pre and post-installation test bed data.

Test Cell Confirmation

It has been noted that the relationship between test cell and flight performance can be distorted in the event of large variations in powerplant leakage overboard, the effect at cruise being 2.5 times that at sea level static. In addition, the effect of component efficiency changes at cruise are smaller than at sea level and, therefore, test cell measurements of deterioration must be scaled down by the order of 25% for comparison with flight. This reduction has been derived analytically from engine models and proven in sea level/altitude test cell comparisons.

However, control of powerplant leakage has been a consistent feature of the RB211 design, development and operation and the relationship of measured flight deterioration to that indicated by pre and post service test cell checks could be expected to show consistency only with application of the latter scale.

This is confirmed by Figure 13 showing the current summary of RB211-524 experience. The curve is the current mean of considerable flight data and represents the difference between first flight performance and the performance at any point during the installation. The test points shown represent the difference in SFC, on the test bed, between the pre-installation pass-off test and post-installation performance checks scaled by 25% to represent component efficiency changes only.

Overhaul Deterioration

The total deterioration picture for a fleet is a composite of the rate of deterioration of installed engines and the amount of performance recovered by overhaul. This is usually illustrated by the familiar "saw tooth" plot. The relationship of overhaul performance to new engine performance is a function of the degree of rework applied. The modular design of the RB211 engine permits overhaul of individual modules to be carried out without stripping the whole engine and, therefore, it is important to understand the sources of deterioration such that the appropriate emphasis can be applied as the opportunity arises during shop visits. Examples of the type of overhaul practices pursued for the RB211 follow but as introduction Figure 14 can be shown as illustration of the increased scatter, relative to new, which is inevitable in modular overhaul. The mean level associated with this is, however, a not unreasonable 0.5 to 1.0% above that of new engines.

OVERHAUL PRACTICES

No discussion on performance retention would be complete without some reference to overhaul practices. The amount of overhaul work carried out at each maintenance shop visit for purely performance reasons needs to be judged against the economic return which can be expected from the work carried out. This is a continually changing picture. As the price of fuel continues to rise it now becomes feasible to carry out work which would not previously have been considered to be worthwhile.

This final section of the paper illustrates the effect which various levels of overhaul can have on the performance of the RB211.

RB211 Fan Leading Edge Restoration

The RB211 fan, in common with all other fans, suffers from leading edge erosion which, if allowed to carry on unchecked, eventually leads to the fan having a 'square' leading edge with resulting loss in performance. Tests carried out with ex-service fans which have this amount of leading edge erosion indicate that if the leading edge is restored correctly, by means of careful rounding of the leading edge, as shown in the upper illustration of Figure 15, and improvement in SFC of 0.4% can be obtained. If, however, the restoration is applied to a fan which has an eroded leading edge which is greater than 1.40 mm (.055 inches) thick it is essential that the blade is thinned and blended over the first inch of chord in order that the correct radius of leading edge can be applied. If this procedure is not carried out and the restoration is done as shown in the lower half of Figure 15, the benefits obtained will be minimal.

RB211 IP Turbine Shroud Segments

The maintenance of the minimum tip clearances at vital performance conditions, such as climb and cruise ratings, is essential if the best economy of operation is to be attained. A particular area where tip clearance control has a very powerful effect on performance is the turbine. As previously explained, the RB211 utilizes a particular structure and turbine casing cooling air system which ensures that the axial and radial growth of the rotors and cases are closely matched during the important parts, from the performance viewpoint, of the flight.

All RB211 turbines utilize a honeycomb static tip seal segment and many of these honeycomb seals are pre-profiled. The shape of this pre-profile is designed to be closely similar to, but slightly smaller than, the shape obtained from service wear patterns and a careful running in procedure is then carried out prior to the engine carrying out its performance acceptance test in order that the turbine can machine out the excess honeycomb in a controlled way, thus ensuring a precise matching of the turbine to the honeycomb seals. The benefits of this careful pre-profiling and running in procedure are that the turbine knife edge wear is minimized, thus allowing the turbine to be used again without the necessity for regular knife edge rework and that the performance obtained after this is that which will be attainable by the customer.

In overhauling these seal segments, however, it is vitally important that the honeycomb is correctly positioned on the carrier plate since, if this is not done, the correct relationship between the turbine rotor and its tip seals cannot be maintained. While this may appear to be a statement of the obvious on several occasions the quality of some of these reworked seals has not always been perfect. Figure 16 shows on outline of an IP turbine shroud segment and many examples have been observed where the honeycomb has been too short or out of position by significant amounts and tests carried out using new IP turbine shroud segments made with honeycomb mis-positioned by 1 cell (0.068 inches) width have indicated that a loss of SFC of approximately 1.5% can be expected. This loss in performance is directly attributable to the turbine becoming disengaged from its seal segment.

RB211 HP System Tolerances

In the days when fuel was relatively inexpensive many tolerances were written into overhaul manuals which, although satisfactory from the mechanical standpoint, were not always optimized for fuel economy. Examples which are shown here concern the amount and type of damage to compressor blades, rotating air seal and rotor tip clearances. The following examples indicate the amount of improvement which can be achieved when the acceptable limits are modified to take into account the need for economy rather than considering mechanical integrity alone.

<u>RB211 Rear HP Turbine Stepped Seal</u>. On introduction of the HP feed standard of HP turbine, it was necessary to raise the chamber pressure to the rear of the HP turbine to maintain the correct bearing load. This was achieved by introduction of a balance piston seal and removal of HP3 flow restriction upstream. The flow of the cooling airflow is then controlled by the balance piston seal and the stepped seal and is critically dependent upon the quality of these seals. The cooling air to the rear of the HP turbine is illustrated in Figure 17.

In investigating the poor standard of overhaul of one operator it was discovered that the operator was consistently working to the maximum allowable radial clearance allowed by the manual for the stepped seal. The maximum limit had been set at a time when fuel was much cheaper and it presented no mechanical hazards. The limit was reduced to an amount which is more appropriate to fuel economy and, as shown on the CUSUM trend plot, Figure 18, the average level of TGT (Turbine Gas Temperature) has been reduced by an average of 3.6°C, equivalent to a reduction in SFC of approximately 0.35%.

<u>Blade Dressing</u>. In investigating the effect of overhaul manual limits on performance, with the intention of making economical modifications to the overhaul manual limits, the HP system illustrated on Figure 19 was removed from an ex-service engine which had just undergone a performance evaluation. It was decided to arbitrarily reduce the amount of blade dressing associated with the HP compressor to one-half that allowed in the overhaul manual and to tighten the HP compressor tip clearance to within book minimum + 0.005 inches. In addition to this all HP turbine seals were checked for compliance with the book. The inspection revealed that, with the exception of excessive dressing on one stage, always a subjective judgment, the machine was within book limits but not the stated goal. In order to achieve these goals two stages were rebladed to reduce tip clearance and dressing to the new limits and a few blades were changed for similar reasons on 3 of the other 4 stages. In addition to this, although within limits, it was decided to fit new static air seals to the HP turbine. In the event the assessment was that the amount of dressing was reduced to a little over one-third of that deemed acceptable by the overhaul manual. The net effect of these changes was to reduce the level of TGT by 12°C and the SFC by 1.1%, an amount which is consistent with an improvement in HP system efficiency of 1.7% together with a reduction in cooling airflow of 0.8%.

The economics of these changes are now being assessed.

The design of successive members of the RB211 family of engines has been continually modified to take into account experience gained from service engines, but each has retained the original unique features of separation of structural and gas path casings and modularity together with the ability, by use of integrated EPR, to have its performance monitored accurately in flight. With the current price of fuel, a 1% sustained reduction in SFC is worth approximately $30,000/year/engine and it is easy to see that fairly major changes in overhaul practice, which were once uneconomical, are now becoming eminently desirable and will become more so as the price of fuel continues to advance. The structural features of the RB211 ensure that the rate of deterioration in performance is low and that the cost of overhauling the engine is minimized.

TABLE I

RB211 ENGINES

ACCURACY OF IN FLIGHT MEASUREMENT OF FUEL FLOW AT AN EPR

SOURCE OF VARIABILITY	POTENTIAL INACCURACY OF MEASUREMENT	EFFECT OF FUEL FLOW AT AN EPR
FLIGHT CONDITION		
MACH NUMBER	±0.01	±1.06%
TAT	±2.0°C	±0.46%
ALTITUDE	±30 METRES	±0.46%
ENGINE INSTRUMENTATION		
EPR TRANSMITTER AND GAUGE	±0.45%	±1.7%
FUEL FLOW	±0.5%	±0.5%
ENVIRONMENTAL CONTROL SYSTEM FLOW VARIATION BETWEEN ENGINES	±0.25%	±0.5%
RMS ACCURACY		±2.2%

RB211 Performance retention features

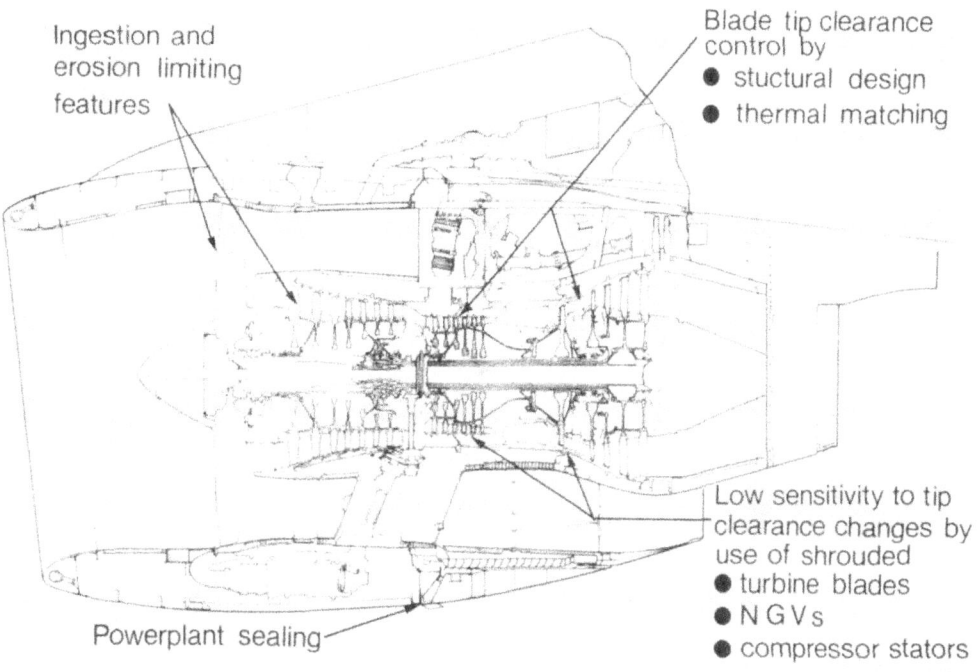

FIGURE 1

RB211 Load carrying structure

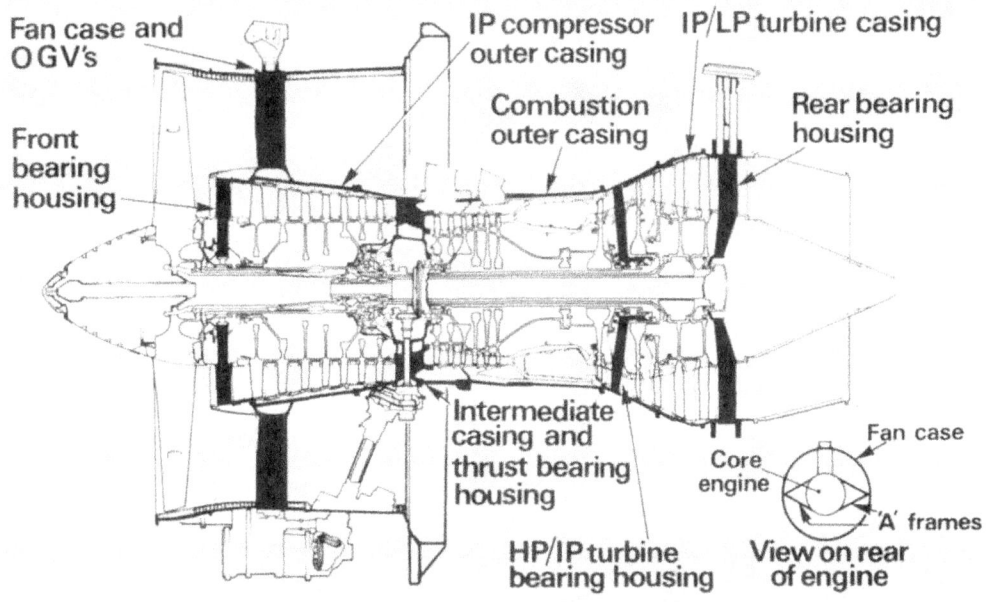

FIGURE 2

RB 211 HP turbine tip seal matching

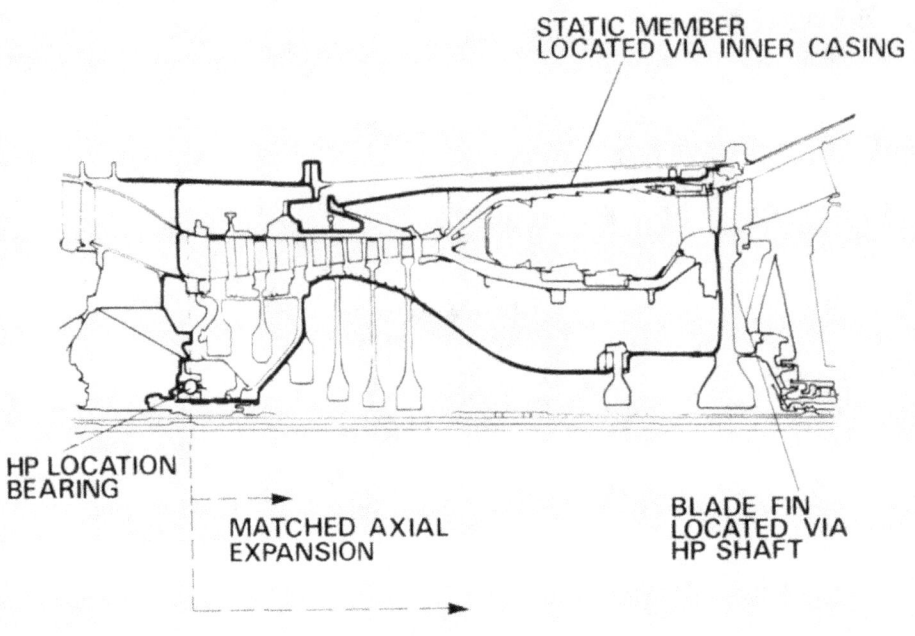

FIGURE 3

RB 211 524 Turbine tip seals

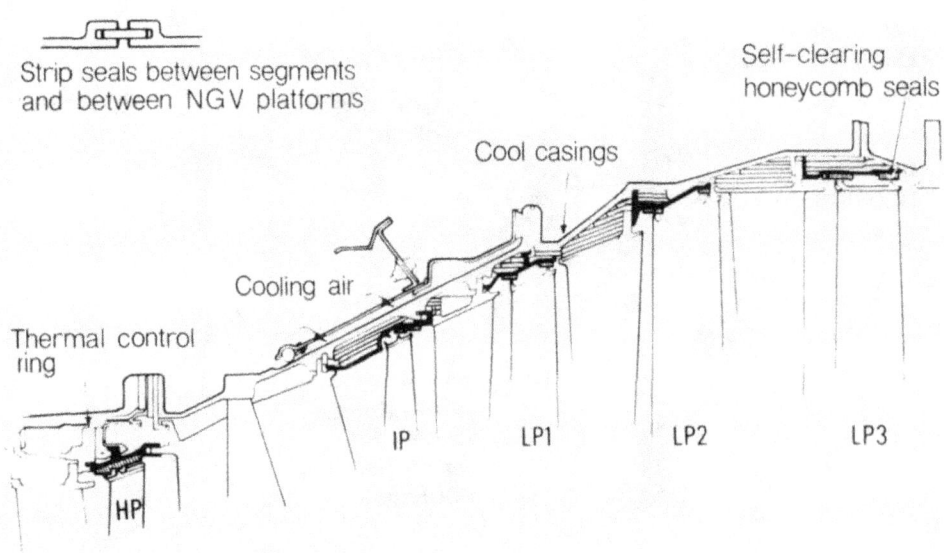

FIGURE 4

RB 211 powerplant leakage

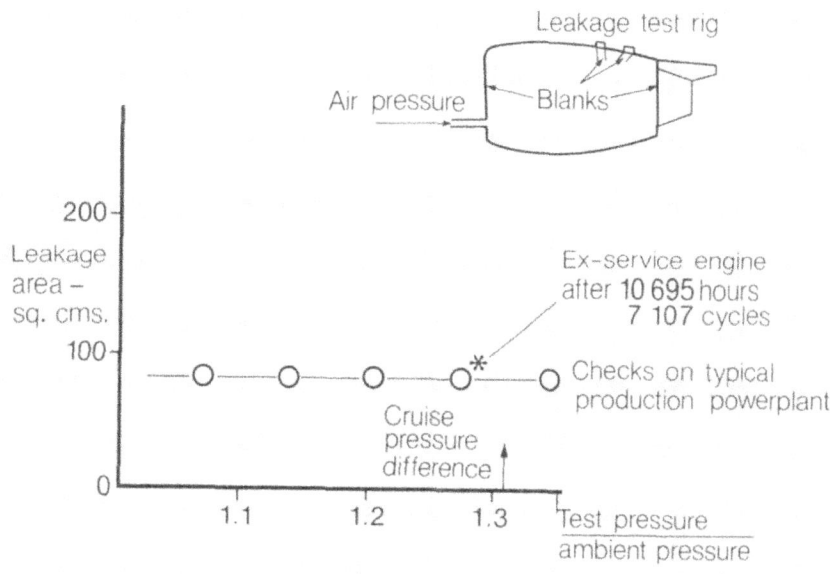

FIGURE 5

RB 211 Revision to HP Turbine shroud segment design

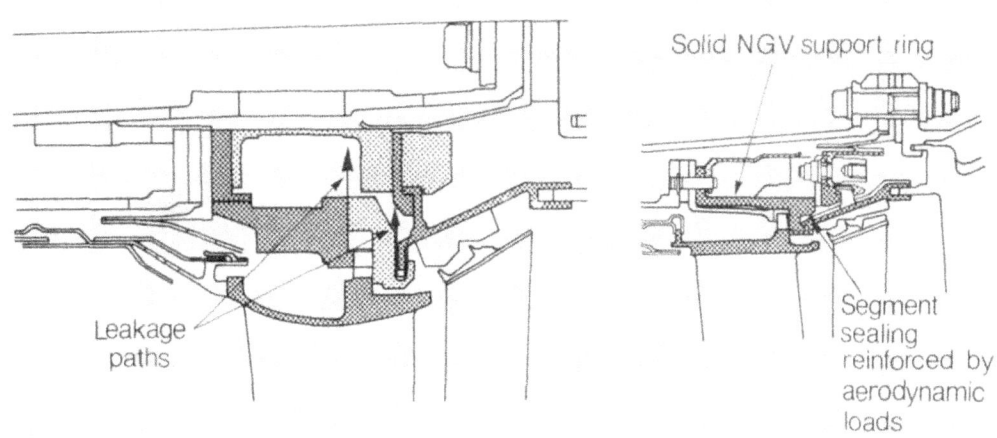

RB 211-22

RB 211-524

FIGURE 6

RB211 IP compressor rotor front stub shaft hydraulic oil seal

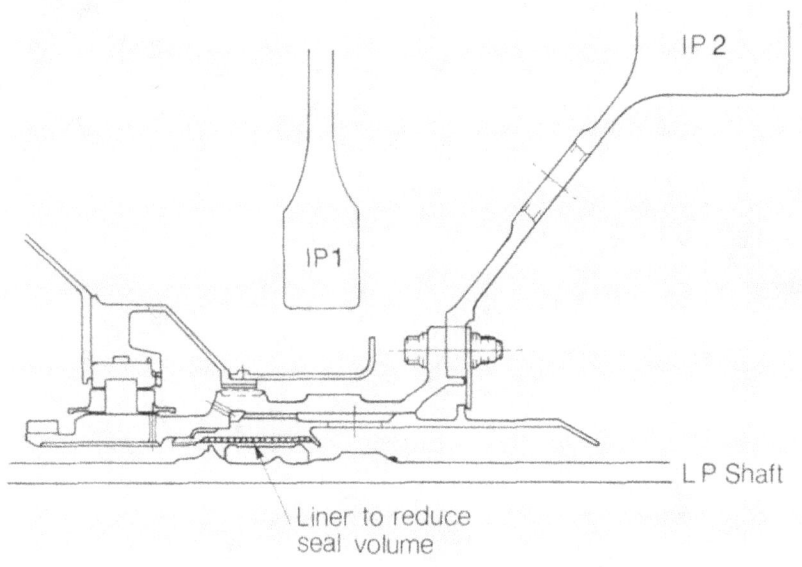

FIGURE 7

RB 211 Control of pod ventilation Zone 4A/4B bulkhead

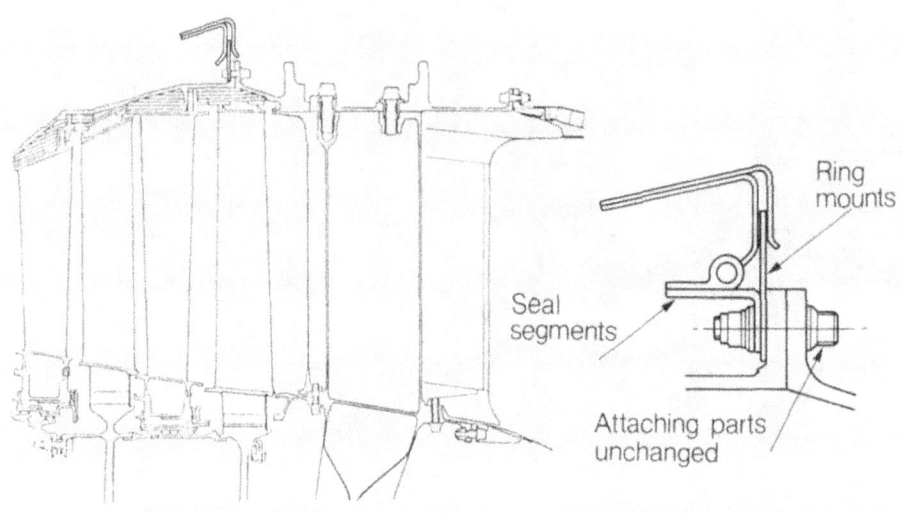

FIGURE 8

RB 211-22 IEPR system general arrangement

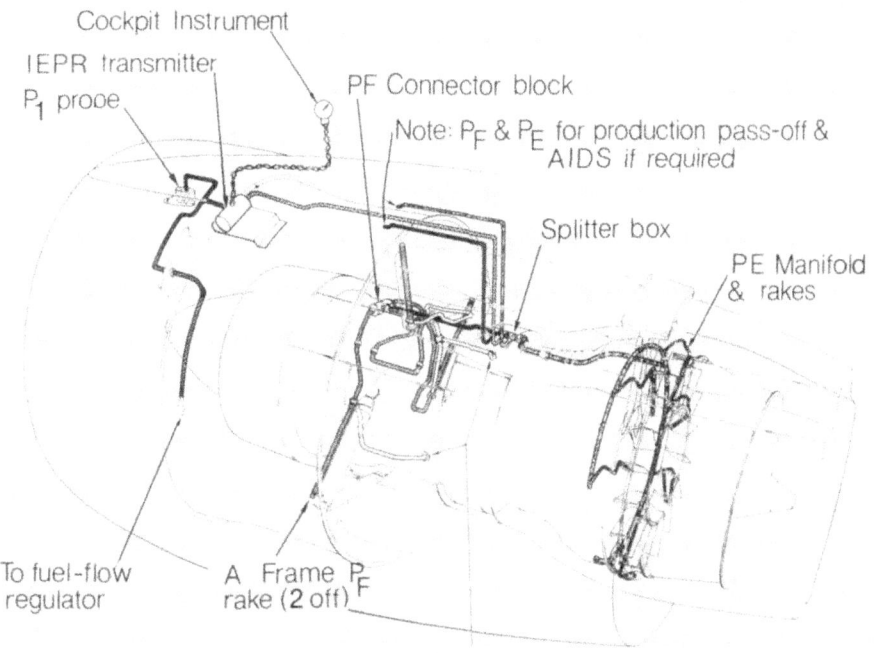

FIGURE 9

RB211 EPR System integrator block

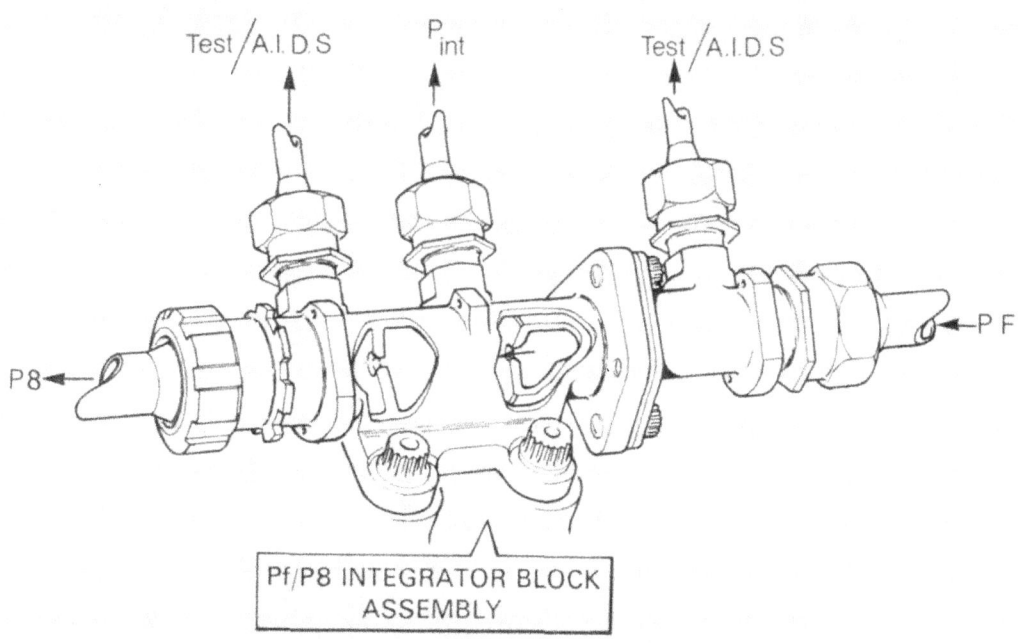

FIGURE 10

Selection of EPR trimmer on pass off

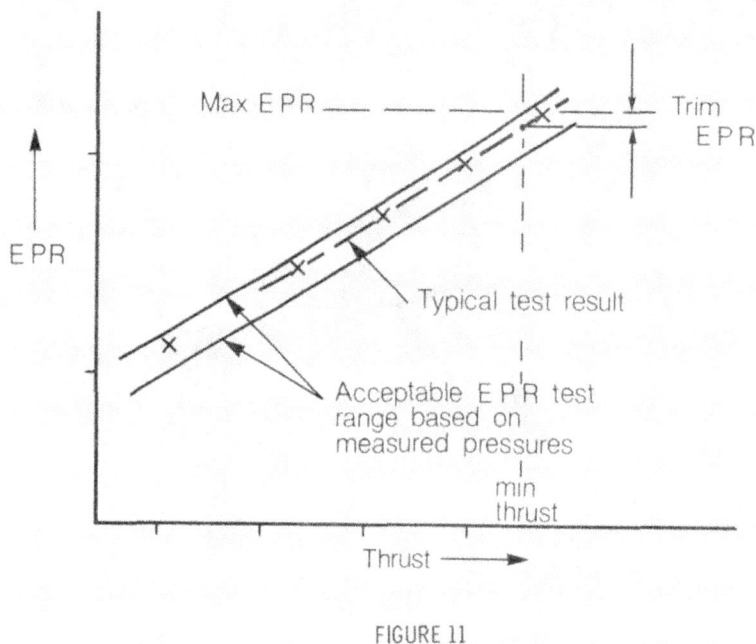

FIGURE 11

RB211-524 In flight deterioration

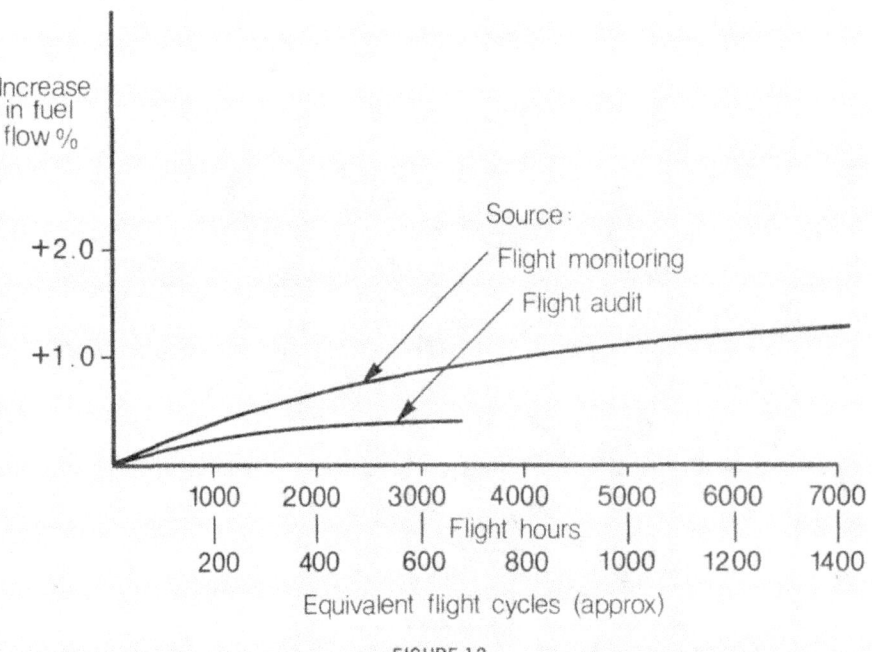

FIGURE 12

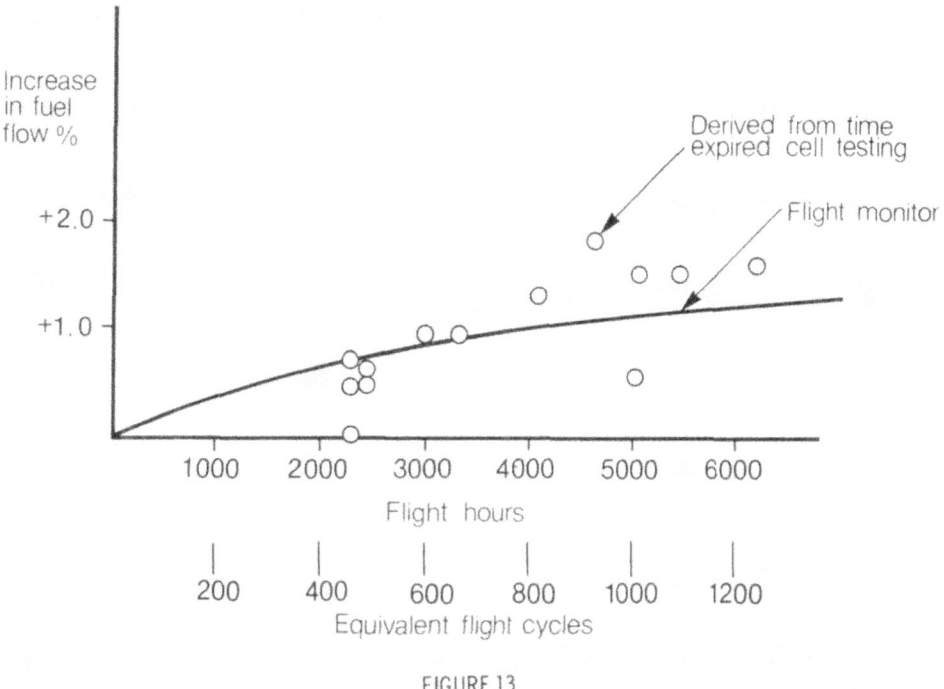

FIGURE 13

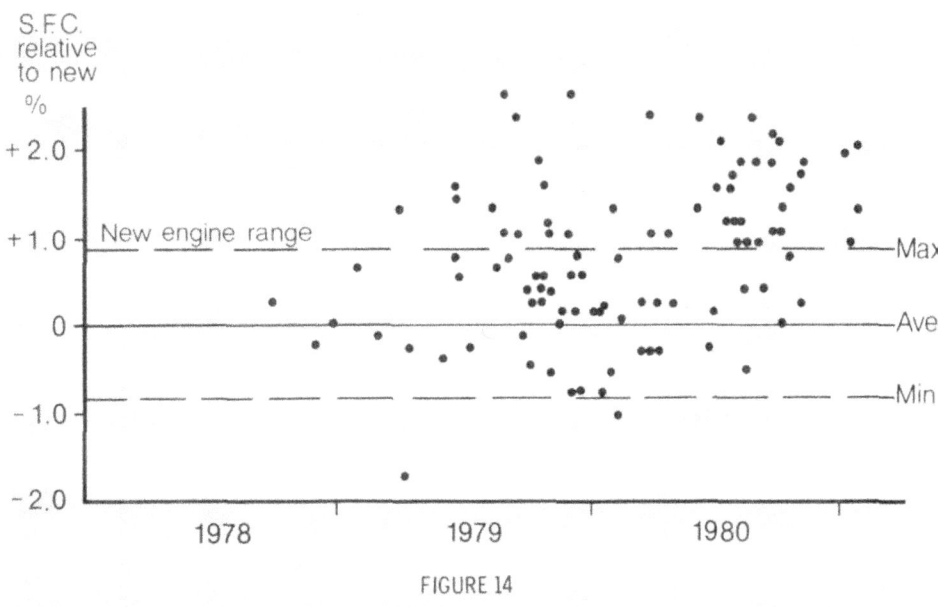

FIGURE 14

RB 211 Fan leading edge restoration

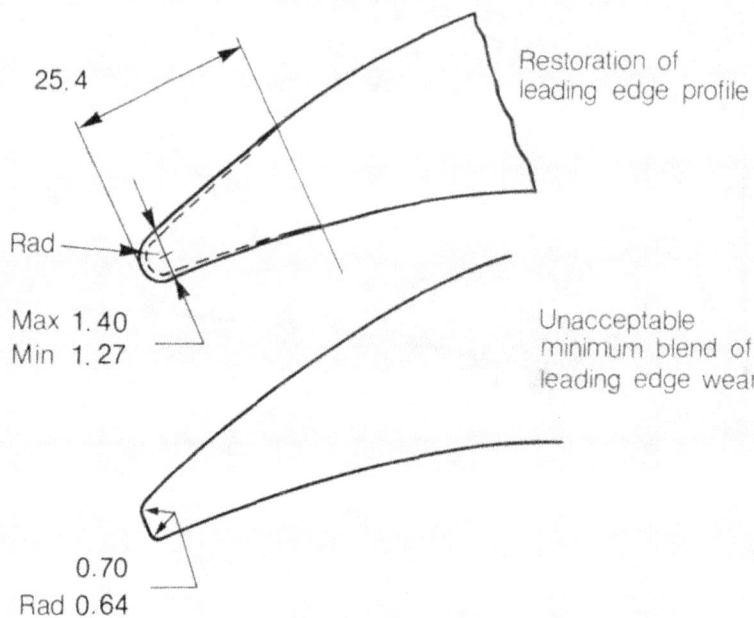

FIGURE 15

RB211 I.P. Turbine shroud segment

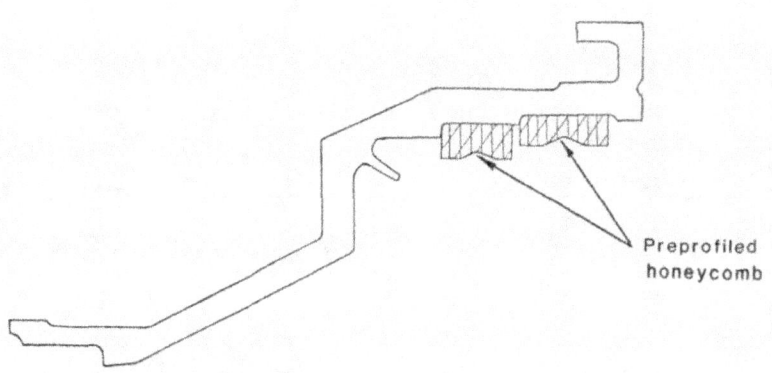

FIGURE 16

RB211 Turbine internal cooling airflow

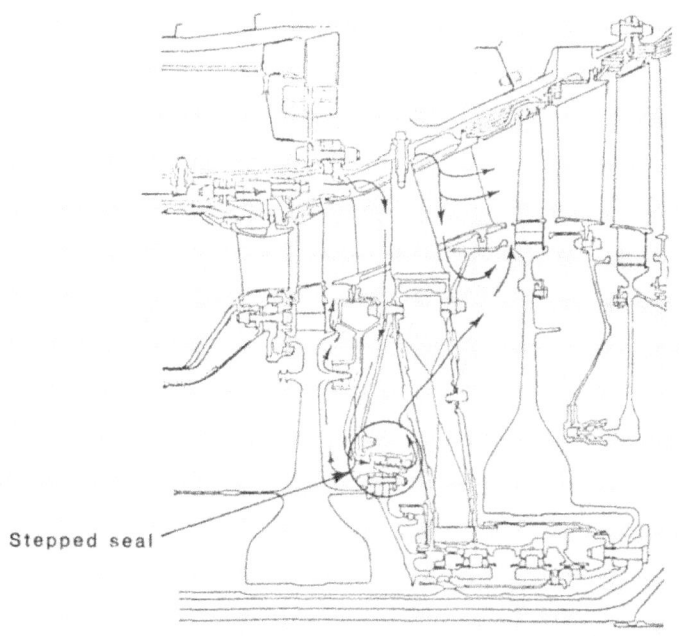

FIGURE 17

RB211-22B Test bed takeoff T.G.T. cusum plot

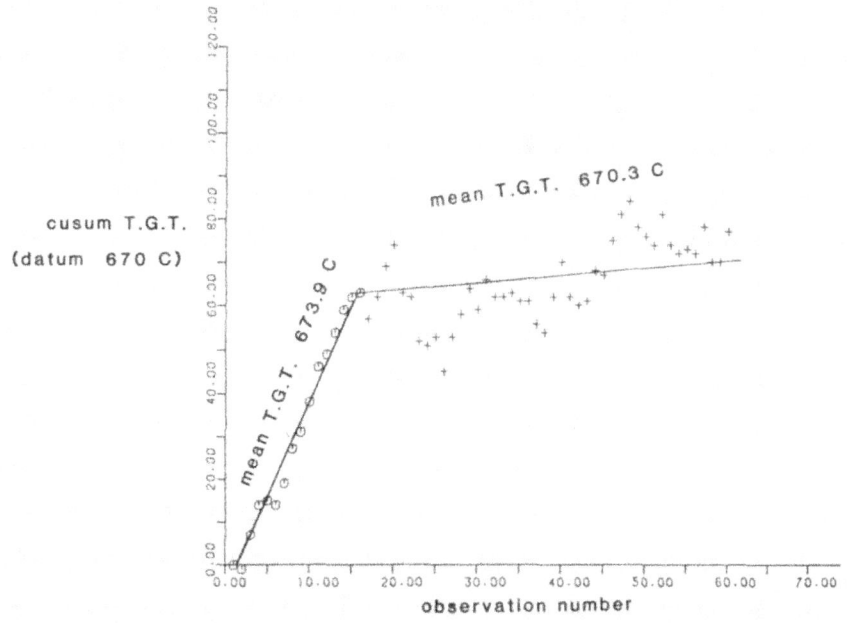

FIGURE 18

RB211 H.P. System

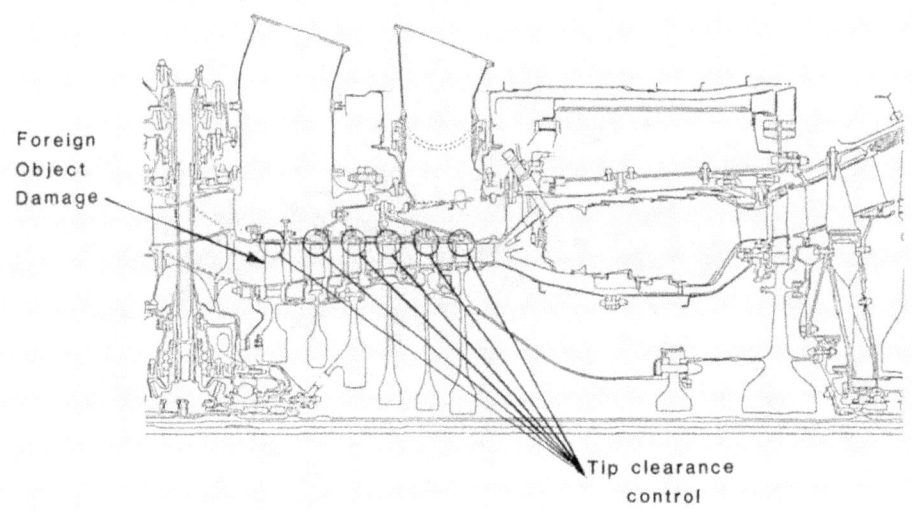

FIGURE 19

PERFORMANCE DETERIORATION - AN AIRLINE PERSPECTIVE

Niels B. Andersen
Pan American World Airways, Inc.

SUMMARY

Due to the drastic energy cost increases which have been with us since the 1973 oil embargo, Pan Am, along with most of the world's airlines, has become painfully aware of the impact on costs and operations caused by the steadily deteriorating fuel efficiencies that characterize the aircraft we operate.

We estimate that the fuel efficiency of our original 747 fleet is down 6 to 6 1/2% from when it was new, of which about 1 1/2 to 2% is airframe deterioration and the balance of 4 to 5% is in the engines.

Although the NASA engine diagnostics program recommends periodic refurbishment as a technique for reducing deterioration, our own experience with this approach (which we adopted for improved reliability rather than performance restoration) has been disappointing.

Pan Am has consistently held that efforts for improvement of existing engines to achieve reduced fuel consumption should be in the direction of retaining the performance already in the engines rather than developing sophisticated design changes to reduce fuel burned.

Furthermore, we have always stressed the necessity for retrofitability in a practical and cost effective sense of any fuel savings feature.

Additional on-board engine instrumentation to allow component performance analysis using data from actual flight conditions is a most desirable feature for new transport aircraft designs. This will allow us to define for our shops which parts of the engine need attention to restore excessive performance losses. Effective measurement of one very important engine parameter continues to elude us, namely thrust. We need a thrust meter.

Main engine bearing configuration (number and location), cowl load sharing, inlet reactive loading along with practically any other design and/or installation feature that will stiffen the engine will, in our view, have significant payoffs in retaining engine performance efficiencies and reducing fuel consumption.

INTRODUCTION

Like most of the world's airlines today, Pan American World Airways has become painfully aware of the impact on costs and operations caused by the steadily deteriorating fuel efficiencies that characterize the kinds of transport aircraft we all operate.

This awareness was spawned by the 1973 oil embargo, which precipitated the sharp, relentless energy cost escalations that have been with us since that time and seems destined to continue with no relief in sight.

Fuel prices have increased nine-fold since pre-embargo days. Pan Am's fuel costs for 1973 for a fleet of 142 aircraft (30 747's and 112 narrow-body aircraft) was $170 million (1.2 billion gallons) for an average price of approximately 14¢/gallon. For 1981 Pan Am's fuel budget is $1.36 billion (1.1 billion gallons) for a fleet of 112 aircraft (64 wide-body and 57 narrow-body aircraft) for an average price of $1.24/gallon. Fuel costs have risen from 25% of direct operating cost in 1973 to nearly 50% currently.

After just over 11 years of operation, we find that our 747-100 aircraft have deteriorated to the point where fuel efficiency is down 6 to 6 1/2% from when they were new. Our fleet of 747SP aircraft which entered service in the period 1976 to 1979 has deteriorated to a point where on the average fuel efficiency is down about 5 to 5 1/2% from when they were new.

Based on our own performance monitoring effort along with what has been learned from the NASA deterioration studies, we feel pretty confident that we can isolate about 1 1/2 to 2% of that deterioration to the airframe. The remaining 4 to 5% performance loss is attributable to the engines - this loss appears in spite of a number performance improvement modifications which we have incorporated in the JT9D over the years.

At current fuel prices, recovering or retaining just 1% fuel efficiency amounts to $7.6 million (6.9 million gallons of fuel) saved for the year just for our 747 fleet alone. Across the entire fleet, savings for a 1% improvement would exceed $10 million.

With this basic background information, it is not difficult to understand why we are so concerned about performance deterioration.

Pan Am's commitment to finding the causes of and cures for these punishing performance losses is reflected at least in part by its enthusiastic support for and extensive participation in the NASA Engine Component Improvement Program which commenced in 1977. We served as reviewers for both the Performance Improvement Program and for the Engine Diagnostics Program. In addition, Pan Am was under subcontract to Pratt & Whitney Aircraft to provide extensive historical engine performance data as well as making available certain JT9D engines in a program of special test cell, on-wing and in-flight tests to determine the mechanisms of deterioration in the JT9D-7A engine.

While our experience and effort to date have focused on the JT9D engine, we are closely monitoring the performance of our newly-acquired L1011-500 aircraft powered by the Rolls Royce RB211-524B engines. Furthermore, our efforts will now broaden to incorporate the CF6-6 and CF6-50 along with a sizable JT8D contingent.

As a result of our efforts and concerns, we have formed certain views and ideas about various aspects of engine performance deterioration and retention. These thoughts and ideas are set out hereunder.

NASA ENGINE DIAGNOSTICS AND PERFORMANCE IMPROVEMENT PROGRAM

At the outset of the NASA Engine Diagnostics and Performance Improvement Programs we strongly urged, in our capacity as program reviewers, that emphasis be placed on finding ways to retain performance with particular stress on retrofitability. We have consistently maintained that from the standpoint of reducing overall fuel consumption, the potential payoff is greater if we are can retain performance that is already in the engines we operate rather than to develop sophisticated design changes to reduce fuel burned.

We have always stressed the importance of retrofitability of any modification, whether for performance improvement or for performance retention. Unless the now more than 3,000 CF6 and JT9D engines in service can play a part, it is doubtful in our minds at least that significant fuel savings will be realized for these model engines.

As it has turned out so far, very few of the concepts developed in the Performance Improvement Program are retrofitable in any practical sense.

ENGINE REFURBISHMENT

One of the principal recommendations to come from the Engine Diagnostics Program - and one that was somewhat disappointing to us - was that operators should periodically refurbish the compressor section as well as the turbine section as an effective means of partially overcoming deterioration.

We have always known that new parts will improve engine performance. However, this is a very costly way to gain performance, and as long as the basic design of the parts is unchanged the deterioration characteristics are fundamentally unchanged. At best this technique restores some performance for a limited period of time, but performance retention is really not improved.

Quite coincidentally, in 1978 as the Engine Diagnostics Program was well under way, Pan Am initiated a major change in its engine maintenance philosophy, changing from the long-popular on-condition maintenance concept to a periodic refurbishment program of the kind recommended by the engine diagnostics study for its JT9D engines.

This refurbishment program was adopted at Pan Am specifically to achieve improved engine reliability, with reduced fuel consumption as an anticipated secondary benefit.

The new maintenance program has been quite successful from the standpoint of improving reliability of the JT9D. In addition, TSFC of completely refurbished engines are on the average 1 to 1 1/2% lower than all other engines when measured in the test cell after repair.

However, over the 3 year period the program has been in effect, we have been unable to see any real improvement in fuel consumption attributable to refurbishment based on our routine aircraft and engine performance monitoring procedures. The most we might be able to say is that further deterioration may have been somewhat arrested. This has been an unexpected and disappointing result, for which we have no good explanation at this time.

ENGINE INSTRUMENTATION

The day is approaching, at least at Pan Am, where engines with high fuel consumption may occasionally be removed for that reason. Hitherto unscheduled removals have largely been associated with high EGT, mechanical failure or boroscope inspection revealing incipient failure.

As we approach an economic environment where high fuel consumption becomes a cause for engine removal, it is becoming increasingly apparent to us that the current variety of on-board engine instrumentation, which has changed little during the some twenty year that jet transport aircraft have been operating is inadequate.

We believe that new generations of transport aircraft should incorporate expanded on-board engine instrumentation to allow comprehensive engine component analysis using data from actual flight conditions rather than having to rely on sea level, static test cell data. When an engine is removed we must be able to specify to the shops with confidence which parts of the engine require attention to recover valuable fuel efficiency. We believe this is feasible.

Specifically, additional instrumentation should probably include at least pressure and temperature between engine stages. Where variable vanes are featured, vane angle should also be displayed in the cockpit.

Such additional engine parameters probably need not be displayed continuously. One approach would be to have one set of gages installed on the engineer's panel with a selector switch to display one engine at a time. With the advent of sophisticated performance management and flight management systems, there should be all sorts of possibilities for automatic recording of data on command from the flight engineer.

A discussion of engine instrumentation would be incomplete without mentioning thrust meters. Such an instrument has been the dream of people like us for many years. Thrust remains one of the two or three most important performance parameters for jet aircraft, yet its accurate and reliable measurement in flight continues to elude us.

Until a good thrust meter is developed, we feel that improvements can and must be made in the two most popular thrust-setting parameters: engine pressure ratio, as on the JT9D, and low-spool RPM, as on the CF6. As a reliable, accurate measure of actual thrust, especially under cruise conditions, we believe both systems have some serious flaws. In both systems there are what appear to us to be unexplained shifts in their relationships to net thrust so that we are not necessarily getting the thrust we think we are getting when we set EPR or N1. At this point common sense tells us that the integrated engine pressure ratio system used on the RB211 engine is probably superior to either of the other two systems. However, since the RB211 is quite new to us, we will have to withhold judgment for a while.

ENGINE DESIGN AND INSTALLATION FEATURES

Certain features of engine design and installation are clearly demonstrating important advantages in engine performance retention.

Bearing arrangement no doubt has an important role in performance retention. Four bearings seem to be insufficient while six are probably more than are required. A well-designed 5-bearing system would seem to be an optimum configuration.

Bearings with over-hung components such as fans, should be designed to minimize such over-hang to limit associated wobble, which in turn leads to shroud rub, or to allow closer running clearances.

We are convinced that almost any effort to improve stiffness and generally reduce flexing of the engine structure will pay off significantly in performance retention - even at a weight penalty. For this reason we favor cowl load sharing to provide additional rigidity at a relatively small cost in additional weight and complexity. Studies by the manufacturers are presently under way in this area for the JT9D-7 installation on the 747s. The approach under development is particularly attractive in that it looks very promising and cost effective for retrofit. Pan Am has indicated a strong interest in this program and we have offered to participate in any meaningful way, such as perhaps a service test program.

Along these same lines, Pan Am is planning to participate with the Boeing company this year in a service test program of a device designed to react against flight loads on the engine inlet of the 747, thereby reducing fan and low compressor shroud rub. This too is very attractive to us because of the retrofit potential, which is indeed what will be done for the service test program.

These three areas, bearing location and number, cowl load sharing, and inlet reactive loading are, in our view, key areas in the battle to retain engine performance efficiencies - particularly since the performance which is ordinarily lost when an engine flexes is lost during the first flight or two and has been largely considered unrecoverable.

CONCLUSIONS

The foregoing points up Pan Am's great concern about maintaining the fuel efficiencies of its fleet of aircraft and engines. We have actively supported past programs to determine causes of and cures for engine performance deterioration and will continue to pursue efforts to apply this valuable knowledge effectively to current as well as future engine designs.

There are over 3,000 JT9D and CF6 engines in service at this time with the number growing slowly but steadily. These engines can be expected to remain in service for a good many years to come. The challenge therefore remains to develop practical, retrofitable performance retention features that can save significant quantities of fuel on this very large body of engines in the 1980's and no doubt the 1990's.

IMPROVING TURBINE ENGINE COMPRESSOR PERFORMANCE RETENTION THROUGH AIRFOIL COATINGS*

L.A. Friedrich
Pratt & Whitney Aircraft Group

Introduction:

Alteration of compressor airfoils by the erosive action of engine ingested particulate matter is a cause of performance deterioration in commercial aircraft turbine engines. A NASA sponsored JT9D Engine Diagnostics program quantified the problem for the commercial aircraft engine fleet indicating that the performance deterioration of the compressor - and erosion of the compressor airfoils - was related to total engine cycles rather than total engine operating hours. Thus the erosion problem becomes more severe when considering short mission applications where the number of engine operating cycles builds rapidly in relation to total engine operating hours. The appearance of a set of high compressor airfoils operated for approximately 10,000 cycles is shown in Figure 1.

Erosion of turbine engine compressor components has been a serious problem for military helicopter operations. In this application the erosion problem is so severe that factors of ten improvement in erosion resistance are required for any material or coating developed to alleviate the erosion problem. The titanium carbide and titanium diboride coatings that offer this degree of protection also compromise blade fatigue strength to a level not tolerable in commercial turbine engine applications. However, since the erosion problem in commercial engine service is considerably less severe than in military helicopter operations, coating solutions are available that may provide adequate erosion resistance without critically compromising the fatigue strength margin of the airfoils.

In order to evaluate the potential effectiveness of coatings in limiting erosive damage to compressor airfoils, an effort was initiated to evaluate candidate coatings for substrate alloys typically used in commercial engine high compressor blades. Laboratory and rig erosion testing of plasma deposited and diffusion coatings described in this paper has shown the potential of a two-to four-fold improvement in erosion life. The selective application of these coatings to approximately the outer third of the airfoil - the area that is subject to erosion degradation - avoids coating the fatigue critical region of the blade, thus providing erosion resistance potentially without compromising the fatigue strength of the blade. Both the plasma and the diffusion coatings also offer the advantage of low initial cost and a multi-source production base.

---

*The reported work has been performed under NASA Lewis Research Center Contract titled Materials for Advanced Turbine Engine (MATE)(Contract NAS3-20072) P&WA Project 4, Erosion Resistant Compressor Airfoil Coatings.

Coating Selection:

A useful first order classification system for potential erosion resistant coatings identifies three major types of coatings. Specific coatings selected from each class for this study include:

1. Multiphase Overlay Coatings: tungsten carbide-cobalt
2. Diffusion Coatings: chromium-boron
3. Single Phase Hard Compound Overlay Coating: titanium-diboride

The tungsten carbide-cobalt composition is applied by modern plasma spraying. This type of coating has been widely used in the aircraft engine industry principally to minimize contact wear involving galling, fretting and impact. High energy thermal spray processes, the most important of which are plasma spray and detonation gun, have been developed for the application of high integrity coatings. These processes are highly commercialized and supplier facilities capable of producing these coatings exist world wide.

Representing the diffusion coating class is a chromium-boron composition. This type of coating is formed by diffusional interaction of chemical elements with substrate alloys to form erosion resistant phases at the alloy surface. An intensive commercialized technology base exists for the fabrication of diffusion coatings for the turbine engine industry.

Single phase, hard compound overlay coatings such as TiB_2, TiCN and TiC have been demonstrated to provide the greatest degree of erosion resistance, particularly at low particle impingement angles. The two most widely investigated processes for fabrication of these coatings are chemical vapor deposition and fused salt electrolysis. TiB_2 produced by fused salt electrolysis is representative of this type of coating.

Coating Evaluation:

Laboratory Erosion Testing

Coatings were produced on three alloys representing typical materials used in commercial turbine engine compressor airfoils. These alloys are the titanium base alloy Ti-6Al-4V (AMS 4928), a stainless steel alloy (AMS 5616) and a nickel base alloy (IN901). In this paper the laboratory erosion test results are reported for the coatings on stainless steel (AMS 5616). The alloy specimens were coated to a nominal thickness of 50 microns (2 mils).

The laboratory erosion testing was performed using an S.S. White Airbrasive Unit. Aluminum oxide with a nominal 27 micron particle size was used as the abrasive material. The abrasive particles are accelerated to approximately 300m/sec and impinge on the test specimen approximately 1.5cm from the nozzle. Three abrasive impingement angles were tested - 20, 45 and 90 degrees. The erosion resistance was measured by weight and volume change as a function of time, and by the time to erode 25 microns (1 mil) of coating.

Erosion data for the three types of coatings on AMS 5616 at the 20° impingement angle shows considerable improvement in terms of volume loss compared to the uncoated stainless steel alloy (Figure 2). Erosion at this angle is typical of airfoil trailing surfaces.

A comparison of test results at all three abrasive impingement angles is presented as time to erode 25 microns of material. The coatings are particularly effective at the low impingement angles (Figure 3). These data are in general agreement with the literature, with the hard coatings demonstrating greater resistance to erosion at low impingement angles than the baseline uncoated alloys. At the test condition used, the hard compound TiB_2 coating demonstrated improved resistance at a 90° abrasive impingement angle, which is not typical of this type of material in field service engine testing. These laboratory erosion tests are valuable tests to quickly and inexpensively rank coating compositional and processing variations. However, they are inadequate to provide an assessment of the potential life improvement coatings can provide on compressor airfoils.

Rig Erosion Testing

To address the challenge of establishing a test procedure that would simulate relative compressor airfoil life when subject to erosive conditions, a facility was constructed to erosion test actual compressor airfoils. A combustor system was modified to include a particle injection system (Figures 4,5). A holder was designed to place the test airfoil at controlled downstream locations with the airfoil positioned at controlled angles to the particle stream. Airfoil temperatures are monitored using an optical pyrometer. Typically, nominal twenty micron aluminum oxide is used as the erosive agent. The Laser Doppler velocimetry technique was used to determine particle velocity and particle flux in planes at a number of locations from the combustion exit nozzle. These measurements were made as a function of test rig control variables: fuel pressure, air pressure, and particle feedrate. Thus the test rig was calibrated to produce known particle velocities and test airfoil temperatures by varying the rig controls and the airfoil distance from the exit nozzle, providing the capability of simulating the temperature and velocity conditions at each stage of high compressor in gas turbine engines.

To determine the ability of this rig to reproduce erosion patterns seen in field service operated hardware, a group of blades were rig tested. Visual appearance of field service and rig tested blades was similar (Figure 6). Profiles taken at standard planes indicated similar erosion patterns with both types of testing resulting in significant reduction in blade leading and trailing edges as well as thinning of the concave airfoil.

In addition to duplicating the erosion pattern seen on field service operated compressor blades, this rig test has been able to demonstrate the blade leading edge chipping phenomenon seen in field service with titanium diboride coated blades (Figure 7). The blade leading edge blunting is an important effect to determine in screening candidate coatings as the blunt leading edge results in unacceptable aerodynamic penalties and would preclude the use of erosion resistant coatings exhibiting this effect.

Initial rig testing of AMS 5616 compressor blades with approximately 30 micron (1 mil) thick coating of plasma applied tungsten carbide-cobalt and diffusion coated chromium-boron exhibited a three fold improvement in erosion resistance measured by volume loss compared to the uncoated blades (Figure 8). The test conditions used in these tests were a blade temperature of 390°C (730°F) and a particle velocity of 290 m/sec (950 ft/sec). In these tests neither the plasma applied tungsten carbide-cobalt coating nor the diffusion chromium-boron coating eroded in a manner to produce the aerodynamically unacceptable blunted leading edge appearance seen with the titanium diboride coated blades (Figure 9).

These initial results indicate that the plasma applied coatings and the diffusion coatings offer the potential of limiting the erosive damage to high compressor airfoils.

Continuing Activity:

In the next phase of this erosion resistant coating development activity rig erosion resistance data will be generated for selected coatings on a number of airfoil stages chosen to be representative of all stages of modern turbine engine high compressors. In addition fatigue testing of coated blades has been initiated as well as surface treatments to produce blade surface finishes on the order of 20 micron AA.

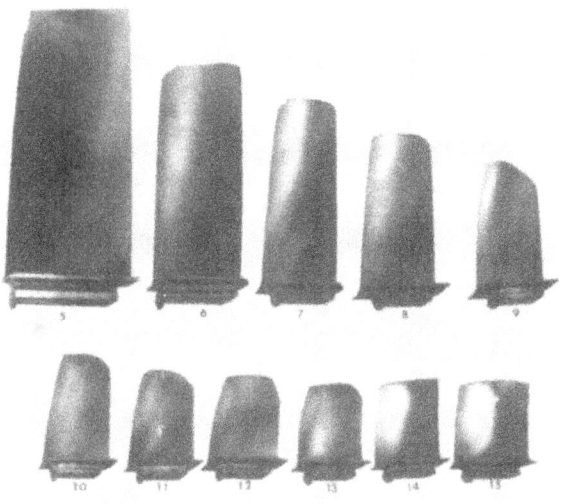

FIGURE 1 COMMERCIAL ENGINE COMPRESSOR
AIRFOILS AFTER SERVICE OPERATION

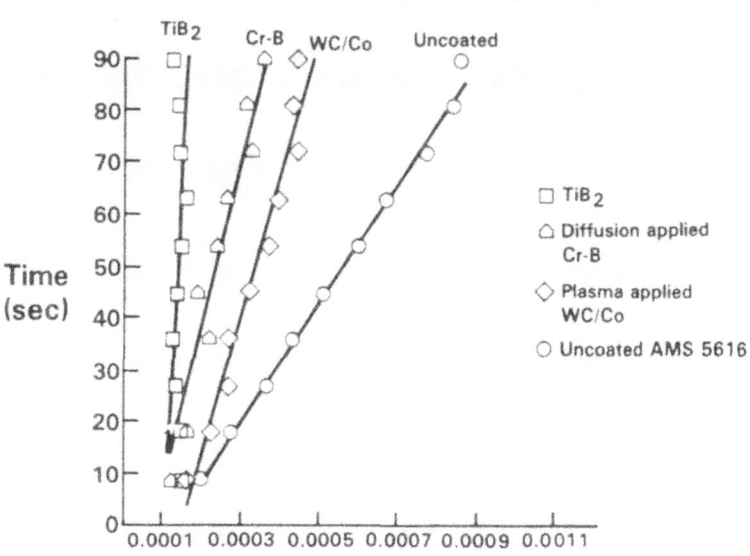

FIGURE 2 LABORATORY EROSION TEST RESULTS
ON COATED AMS 5616 STAINLESS STEEL

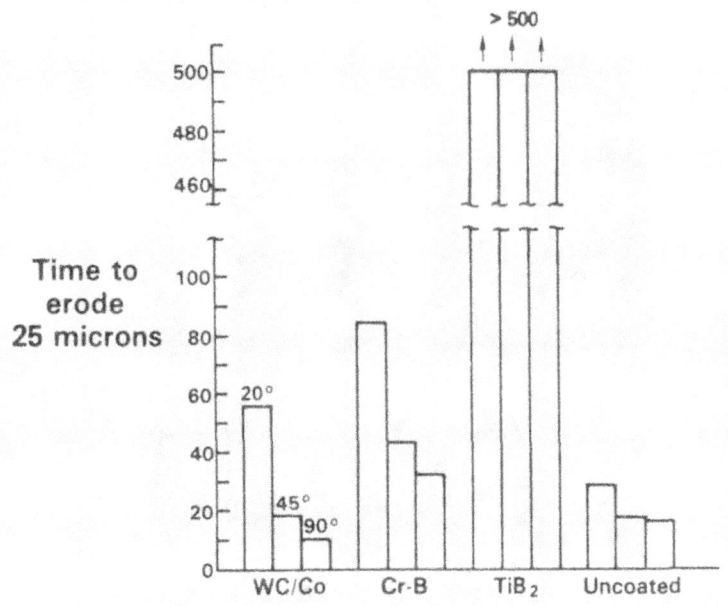

FIGURE 3 EROSION AS A FUNCTION OF ABRASIVE IMPINGEMENT ANGLE

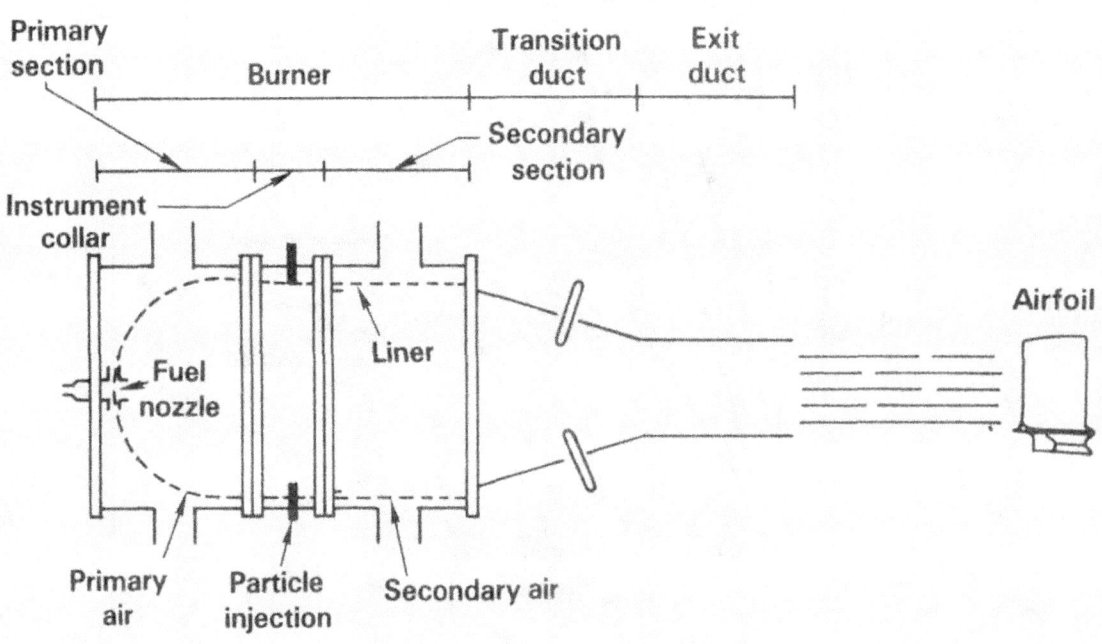

FIGURE 4 SCHEMATIC OF AIRFOIL EROSION FACILITY

FIGURE 5 FACILITY TO TEST EROSION RESISTANCE
OF COMPRESSOR AIRFOILS

Engine service eroded　　　　Rig eroded

FIGURE 6 COMPONENT RIG TEST SIMULATES
ENGINE SERVICE EROSION

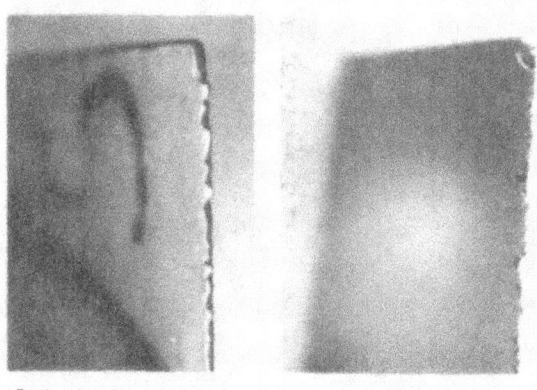

FIGURE 7 COMPONENT RIG TEST SIMULATES ENGINE SERVICE LEADING EDGE CHIPPING EROSION OF TITANIUM DIBORIDE COATED AIRFOILS

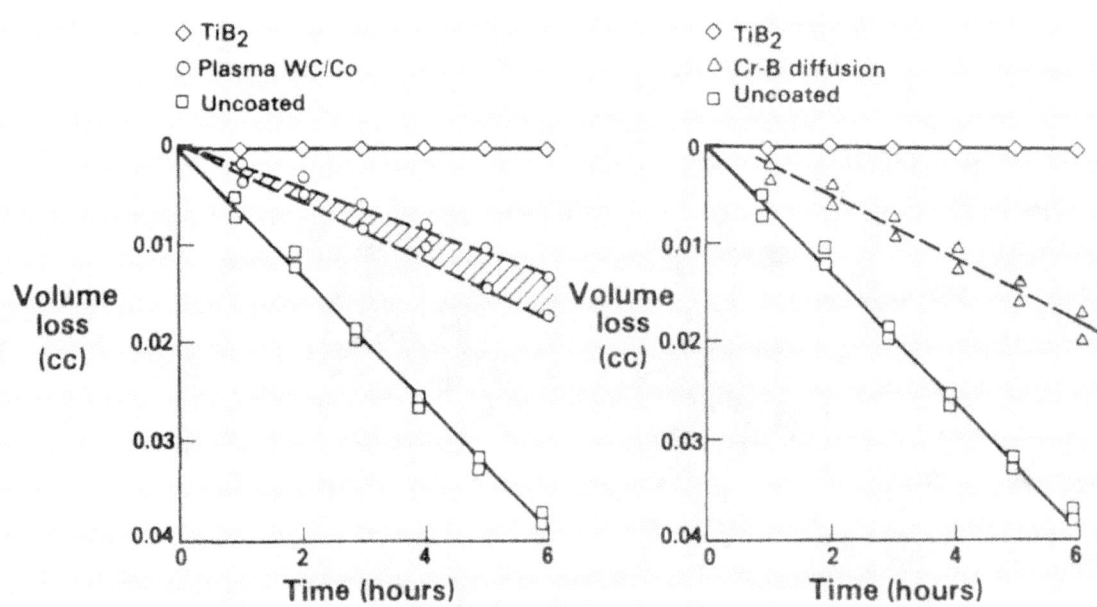

FIGURE 8 EROSION TEST RESULTS OF PLASMA AND DIFFUSION COATINGS ON AMS 5616

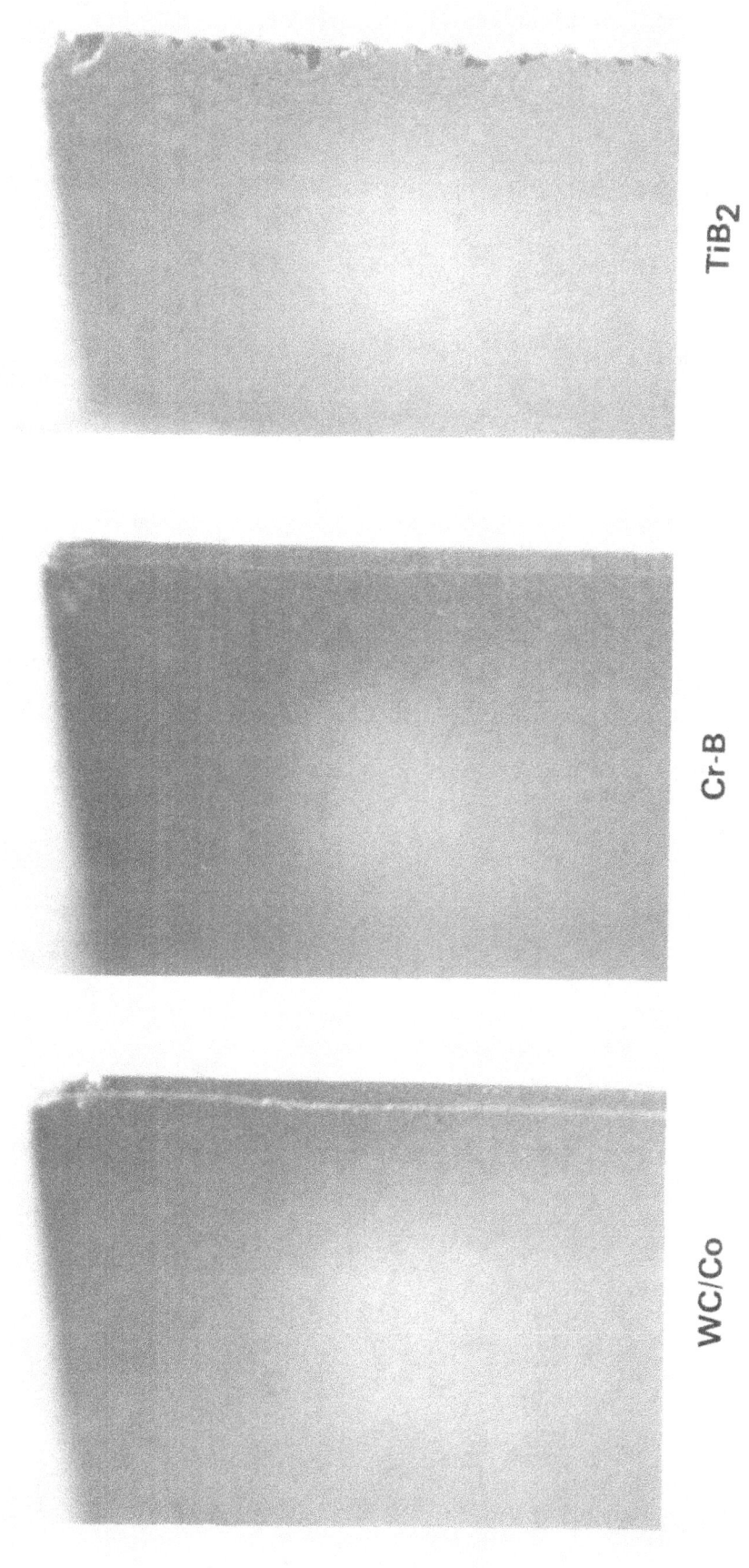

FIGURE 9 PLASMA CARBIDE/METAL AND DIFFUSION Cr-B COATINGS WITHSTAND LEADING EDGE CHIPPING EROSION

Page intentionally left blank

ADVANCED OXIDE DISPERSION STRENGTHENED SHEET ALLOYS
FOR IMPROVED COMBUSTOR DURABILITY*

R. J. Henricks
Pratt & Whitney Aircraft Group

Introduction

Burner durability has become a serious problem in many current generation aircraft gas turbine engines. Advances in structural metal temperature capability and in burner hardware cooling technology have not kept pace with demands for more efficient (higher gas temperature) engine performance. Hastelloy X burners designed for around 871°C (1600°F) metal temperature operation are experiencing hotter streak conditions with heavy penalties to operating life. Both improved burner materials and designs are required to provide the large durability increase essential to future aircraft turbine engine operation and maintenance. A decrease in engine maintenance costs can result both from increased burner life and from reduced turbine section damage caused by burner distortion.

The substitution of advanced oxide dispersion strengthened (ODS) alloy sheet materials with improved creep strength and oxidation resistance compared to Hastelloy X can produce a significant increase in burner durability. Properties of two advanced ODS alloys, Incoloy MA 956 and Haynes Developmental Alloy 8077, compared to Hastelloy X indicate that they exhibit a 167°C (300°F) advantage in creep strength and in cyclic oxidation resistance (Figure 1). However, these ODS materials exhibit low cycle fatigue properties that show no improvement over Hastelloy X.

It is the objective of a NASA/P&WA MATE (Materials for Advanced Turbine Engines) program to evaluate burner design modifications that will take advantage of the improved creep and cyclic oxidation resistance of ODS alloys while accomodating the reduced fatigue properties of these materials. This program will culminate in a JT9D experimental engine test of the selected combustor design and ODS alloy. A status report of this MATE program is the subject of this paper.

Burner Environment

The principal failure modes for louvered combustor liners are creep buckling of the louver lips, oxidation and low cycle fatigue cracking, examples of which are shown in Figure 2. Creep buckling and oxidation are the dominant failure modes in long missions; low cycle fatigue is the dominant failure mode in short missions.

---

* Work performed under NASA-MATE Contract NAS3-20072 with NASA-Lewis Research Center.

Creep buckling of the louver lip results from the strain imposed by the constraint between the thermal growth of the hot louver lip and the relatively cooler weld (knuckle) area over an extended period of time (Figure 3). The louver lip develops a high stress that yields the material in compression. Continued cycling produces circumferential distortions which are sufficiently large to close off the louver gap in local areas (Figure 3). These local closures reduce the cooling air flow for the downstream louver, increasing the local temperature, and accelerating the buckling process on the downstream louver lip. Eventually the severity of the process reaches such a magnitude that rapid oxidation produces a burn-through of the cooled liner. When this point is reached, repairs are necessary.

Low cycle fatigue failures of conventional louver liners are the result of high radial temperature gradients through the liner producing excessive thermal strains. The hot side of the liner is subjected to average temperatures of about 871°C (1600°F) in the area of the welds. The cold side of the liner, in the area of the cooling holes, is subjected to temperatures of about 900°F. The severe thermal strain resulting from this radial temperature gradient is aggravated by unavoidable circumferential variations in hot side temperatures of at least 56°C (100°F).

Burner Design and Structural Analysis for ODS Alloys

Structural analysis of combustor liner cooling geometries was conducted to determine the stress and strain distribution produced by thermal and mechanical loads acting on the liner during an engine flight cycle. Using a finite element analysis, the elastic stress state can be defined with elastic modulus and the coefficient of thermal expansion as a function of temperature. While the thermal analysis and elastic stress are affected by the physical properties of the materials, burner life prediction is dependent upon mechanical properties.

The design system (Figure 4) incorporates the technique of "exhaustion of ductility" for calculating life predictions (ref. 1). The interaction of creep and fatigue modes in a cumulative damage model becomes the failure criterion determined for the design system. For ODS sheet alloys, the ductility determined from tensile data and implied from LCF data is about 5%, whereas the creep ductility measured in creep testing can be 0.1-0.2%. An available ductility of 0.1% for ODS alloys was assumed throughout the design phase. The structural analysis and life predictions are based on the engine operating conditions applicable to an advanced energy efficient engine.

In the design phase of this program, the operating strains for the ODS alloys were minimized in a series of five candidate advanced combustor designs. This was accomplished by designing a series of segments in the circumferential direction to significantly reduce the hoop (circumferential) strains and by eliminating fixity between the hot wall and cold wall shell to reduce radial constraint (strain). Based on the thermal and structural analysis for these five combustor designs the predicted lives were calculated for both ODS sheet alloys.

In addition to the predicted lives for these combustor designs, other factors for assessing the relative benefits of the designs and for selecting the final two were considered; these factors included liner fabricability and engine maintenance and operating costs. The maintenance cost (MC) is based on predicted lifetimes, on initial construction cost and on the type of repair procedure employed for the design. The direct operating cost (DOC) is derived by using the initial fabrication cost, the overall weight and the maintenance cost. Equally important in the determination of the particular designs worth pursuing is the consideration of risk for the construction and repairability of a given design. Based on significant life improvement, lower maintenance and direct operating costs and estimated moderate risk factors for fabrication and repairability, two designs were selected for continued evaluation: 1) a mechanically attached, film cooled segmented louver and 2) a mechanically attached, transpiration cooled segmented twin wall. Schematic diagrams of these two designs are presented in Figures 5 and 6. These two designs were assessed relative to a current commercial engine JT9D-59/70 using the same design selection criteria. Comparison of these designs using ODS alloys to the film cooled JT9D combustor of Hastelloy X in Table I shows a four times improvement in life reflecting the high temperature strength of ODS alloys and the reduced strain range present in the two segmented designs. The lower cooling air levels in the transpiration cooled, twin wall reflects the increased effectiveness in that design. While the initial construction cost and the combustor weight are somewhat higher than for the standard combustor, the maintenance cost and direct operating costs of the advanced designs are significantly lower. It must be pointed out that the relative changes in DOC represent decreases in overall engine operating cost and are not limited solely to combustor cost.

Mechanically Attached, Film Cooled Segmented Louver

A mechanically attached, segmented louver using current film cooling techniques (Figure 5) is attractive and can accomodate the low strain capability of ODS alloys. By mechanically attaching with rivets each ODS segment to a Hastelloy X shell (cold wall) rigidly at only one location and providing room at the other rivet and bushing locations for differential thermal expansion of the hot segments relative to the cold shell, circumferential and radial constraint does not occur under these conditions. The only thermal strains present are those generated from the non-linear temperature variation within each segment. The failure mode in this design is established as initiation of a 0.79mm (1/32") crack at the louver lip.

Calculating the combustor liner life based on the number of cycles of exhaust 0.1% strain as the criterion, the predicted life for MA 956 is in excess of 10,000 cycles, while the predicted life for HDA 8077 is 2000 cycles (Table II).

Mechanically Attached, Transpiration Cooled, Segmented Twin Wall

The "twin wall" transpiration cooled design is an advanced cooling technique with the capability of significantly reducing metal temperature and/or cooling flow and thermal strain. A schematic cutaway view of this transpiration cooled, segmented twin wall combustor is shown in Figure 6. The transpiration cooled panel is attached to the impingement plate with a series of studs so that leakage around the edge of the panel at the operating temperature of 1010°C (1850°F) is less than 10% of the panel cooling air. However, to reduce the strain at maximum temperature, the panel is pre-stressed at room temperature to a predetermined contour duplicating the shape it will assume at the operating temperature of 1010°C (1850°F). An impingement plate which serves as a mandrel employs a contoured edge radius and a centeral positioning stud to impart the desired deflected panel shape.

Analysis shows that during simulated engine operation as the average temperature and the through-thickness gradient increase, the mechanically induced pre-stresses are reduced and become essentially nonexistent at the operating temperature. The only stress at operating temperature is a small cooling air pressure load. The nature of this design is such that the largest stress occurs at 20°C (68°F) where the material strength is the highest. The location of the high stresses is in the center of the edges of the panel at room temperature; during heat-up although some stress redistribution occurs, the maximum stress remains at the center of each edge. Throughout the thermal loading, the maximum stress level remains below the proportional limit so that no plastic damage occurs.

The current method for predicting the effect of hole arrays in a transpiration cooled geometry on the thermal-mechanical fatigue life employs the concept of linear elastic, isotropic fracture mechanics (ref. 2). For transpiration cooled designs, failure is defined as linkup of cracks emanating from adjacent cooling holes. As a result of prestress, operating stress, and thermal cycle, the predicted service lives for the ODS alloys are in excess of 10,000 cycles (Table II). Since there is no thermal interaction between the panel and the studs due to prestressing, at the operating temperatures the cooled studs support only a small radial pressure load.

Alloy Evaluation

Mechanical property testing of the two candidate advanced ODS alloys, Incoloy MA 956 and HDA 8077, was directed towards the selection of one alloy for evaluation in the remainder of the program. The three main criteria for this alloy selection were creep, oxidation and thermal fatigue resistance. Three separate measures of thermal fatigue resistance were defined: creep ductility, isothermal LCF life, and hot spot blister (thermal cycling) cracking.

Incoloy MA 956 alloy sheet of 1.1-1.3mm (0.043-0.051") thickness was supplied by Wiggin, Ltd. of the International Nickel Company and HDA 8077 alloy sheet of 1.1-1.4mm (0.043-0.055") thickness supplied by Cabot Corporation. The nominal chemistry of each alloy is shown in Table III; both alloys are strengthened by a fine dispersion of yttrium oxide (Y_2O_3). The sheet materials were produced by mechanical alloying of powder, powder consolidation and a series of hot and/or cold rolling procedures. These processing techniques result in coarse "pancake" grains in the plane of the sheet and elongated grains through the thickness in both alloys (Figure 7) with fine yttria particles dispersed throughout the structure.

Creep evaluation of MA 956 and HDA 8077 ODS alloys and Hastelloy X, bill-of-material in JT9D combustor liners, in the 871-1093°C (1600-2000°F) range demonstrates significant creep strength advantage for both ODS alloys over Hastelloy X at the higher temperatures (Figure 8); approximately 167°C (300°F) for MA 956 compared to Hastelloy X. These Larson-Miller curves represent an average of time to 0.1% creep strain data generated on these sheet alloys within the present program. Comparison of the ODS alloys shows that HDA 8077 sheet possesses time to 0.1% creep strain and final creep ductility superior to those of MA 956 sheet. Figure 8 compares the high creep ductility of Hastelloy X to the limited ductility of the ODS alloys. The final creep ductility is defined as the last creep extension measurement within two hours of specimen failure. The MA 956 sheet, which is not cross rolled, is anisotropic as exhibited by the difference in creep ductility between the longitudinal (parallel to rolling direction) and transverse (perpendicular to rolling direction) orientations. The average transverse creep ductility of MA 956 is lower than the longitudinal ductility, although the minimum values are similar. There is no difference in time to 0.1% creep strain (creep strength) for these orientations; however, the creep-rupture life for the transverse orientation is lower reflecting the decreased ductility.

Cyclic oxidation testing was conducted at 982°C (1800°F) using a six minute cycle with cooling to 316°C (600°F) using a four minute hold at maximum temperature. Specimens of the the three alloys were tested in a rotating fixture subjected to a JP4-R fuel gas flame for heating and forced air for cooling. Surface attack was determined metallographically on tested specimens. The results of this oxidation testing (Figure 9) show that there was insignificant surface attack <.013mm (0.0005") in MA 956 after 1000 hours and that it is superior to HDA 8077 which had .025-0.051mm (.001-.002") of surface oxidation. Both ODS sheet alloys possess excellent oxidation resistance compared to Hastelloy X, which exhibited 0.36mm (.014") of surface oxidation and spallation after 1000 hours. The relative oxidation resistance of MA 956 and Hastelloy X were verified in a duplicate 1000 hour oxidation test using different heats of material.

Isothermal low cycle fatigue (LCF) testing utilizing strip specimens in a fully reversed bending mode was conducted for the three alloys at 760°C (1400°F) and 871°C (1600°F) with a ± 0.25% strain range (Figure 10). Specimen fatigue life is defined as failure into two pieces. At 760°C (1400°F), the Hastelloy X and HDA 8077 showed similar average failure lives, although the latter exhibited a high degree of data scatter, and both were somewhat higher than the MA 956 life. At 871°C (1600°F), Hastelloy X LCF life was lower than MA 956; however, MA 956 demonstrated approximately a five-fold higher life than HDA 8077 sheet. Limited testing at 982°C (1800°F) of the ODS alloys showed this same 5:1 life advantage of MA 956 over HDA 8077.

A "hot spot blister test" was designed to produce localized thermal fatigue cracking and/or deflection similar to that produced by flame impingement on a burner louver in an engine. Seventy-six millimeter (3.0 inch) diameter disks were subjected to a thermal cycle (5 cycles/min.) from a Tmin of 538°C (1000°F) to a Tmax of 982°C (1800°F) by use of an alternating oxy-acetylene flame and a cool air blast (Figure 11). The Tmin temperature of 538°C (1000°F) was maintained on the edge of the disk using a propane gas burner. Strain analysis of this hot spot blister test shows that compressive strain peaks at maximum temperature and tensile strain at minimum temperature; this type of strain-temperature cycle is typical for current engine combustors. At pre-determined cycle intervals the degree of surface cracking was recorded visually and specimen deflection height was measured. Actual crack depths were determined metallographically on discontinued test samples. As strain range on the cold side of the "hot spot blister" specimen is calculated to be much higher than that of the hot side, initial cracking was observed on the cold side of the sheet. Metallographic determination of cold side cracking shows that MA 956 experiences substantially earlier crack initiation and more severe cracking than HDA 8077 which, in turn, displays more cold side cracking than Hastelloy X as shown in Figure 12. Hastelloy X deflects substantially more than the two ODS alloys, while MA 956 and HDA 8077 exhibit similar deflection heights. The specimen deflections reflect the relative creep strengths of each of the three alloys. Additional testing to a Tmax of 1093°C (2000°F) showed an identical ranking of the alloys in cracking and deflection.

Alloy Selection

Both ODS alloys demonstrated the 167°C (300°F) advantage in creep and oxidation resistance over Hastelloy X. Comparing the ODS alloys, MA 956 is superior in oxidation resistance and isothermal LCF life and HDA 8077 is slightly better in creep strength and ductility and is superior in hot spot blister cracking resistance. Under the MATE program, concurrent to the materials evaluation phase, the two alloy manufacturers were engaged in a sheet reproducibility program. Wiggin, Ltd. of INCO successfully demonstrated MA 956 sheet product reproducibility for both sheet quality and mechanical properties in a second heat of material; Cabot Corporation was unable to reproduce the intial excellent formability and creep properties of HDA 8077 in subsequent sheet product. While neither ODS alloy exhibited an obvious overall superiority in properties, on the basis of product reproducibility and excellent life predictions in both combustor designs, Incoloy MA 956 alloy was selected for evaluation throughout the balance of the MATE program, including component and experimental engine testing.

Low Cycle Fatigue Structural Tests

LCF rig testing of components of the two candidate designs using MA 956 and Hastelloy X alloys is directed toward selection of one design for experimental test in a JT9D engine. Such component rig evaluation allows for testing of the structure of each design under simulated combustor conditions. Feasibility studies were conducted to define the best approaches for LCF structural assessment of the two ODS combustor designs.

For the mechanically attached, film cooled, segmented louver design (riveted louver) a single louver segment of MA 956 or Hastelloy X attached to the inside of a Hastelloy X shell in conjunction with a double return pie-wound induction coil and external cooling air successfully simulated the temperature profile along the louver as shown in Figure 13. The test consisted of a 45 second heating cycle to the desired temperature profile with a maximum temperature of 1010°C (1850°F) at the lip, a two minute hold at this condition and a 30 second cooling cycle to a louver lip temperature of 538°C (1000°F).

A total of six (three each) of MA 956 and Hastelloy X riveted louver segments were tested in this induction heated rig; the results are reported in Table IV and Figure 14. MA 956 segments exhibited considerably more dimensional stability (less distortion) than did the Hastelloy X segments. While there was a significant degree of test scatter in both materials, a comparison on the basis of cycles per millimeter points up the greater resistance to buckling for the MA 956 alloy. Typical bow of the panels removed from the Hastelloy X shell are shown in the photograph in Figure 14.

No crack indications were evident by Post Emulsion Fluorescent Penetrant inspection on any of the test segments. The Hastelloy X tests were discontinued when the bow in the segments became excessive and preluded maintaining the axial temperature profile on the bow or distortion; specifically after 66, 242 and 1,500 cycles. The MA 956 tests were discontinued after 4000, 5000 and 6000 cycles. These test results demonstrate the excellent creep resistance of MA 956 compared to Hastelloy X in this component test and the low strain ranges achieved as evident by no cracking in the segmented louvers.

A second component test of the riveted louver design was defined and conducted in a thermal cycle rig (Figure 15). The axial temperature profile established in the louver at transient conditions approximated the steady state profile of the induction heated rig test. Rotating gas burners impinge on the lip of the segmented louvers during heat-up and an air manifold directs cooling air onto the louver lips during the cool-down for a total cycle time of 60 seconds. The lip is cycled between 954°C (1750°F) and 593°C (1100°F), while the Hastelloy X shell is cycled between 538°C (1000°F) and 399°C (750°F). In this test, alternating segments of MA 956 and Hastelloy X were installed around the ID circumference of the Hastelloy X shell. To date, the first test component has achieved 4000 cycles with no distress evident in the attachment rivets or bushings of any of the louver segments. Additional thermal cycle testing of this component and of a duplicate is scheduled.

For LCF structural testing of the transpiration cooled, twin wall design the test rig is shown in Figure 15. A stationary gas burner heats the hot side of the pre-stressed panel/impingement plate assembly which rests on a box providing a plenum of cooling air. The hot side of the panel reaches a maximum temperature of 927°C (1700°F) and an average through-thickness gradient of 22C° (40F°) and is lowered from the flame, applying increased cooling air flow until it cools to 649°C (1200°F) at which time it is raised back into the gas flame and the airflow is reduced. The entire test cycle is 30 seconds in length (22 sec. heating and 8 sec. cooling). An initial MA 956 panel tested for 2000 cycles contained laser drilled cooling holes; all subsequent panels were electrochemical machined (ECM) because of improved cooling hole integrity in the MA 956 panels. The film from the transpiration holes is an effective cooling mechanism and useful in obtaining maximum life of the segment. Since the holes are ECM drilled at an acute angle, one edge of the panel in the rig is void of film. (In actual engine use, a layer of film would be established to provide insulation until the transpiration film became established.) This region of low film in the rig tested panel exhibited numerous hot and cold side cracks between the edge and the middle rivet (Figure 16). The cracks extended from acute corners of the cooling corners of the cooling holes in this high strain region of the panel (edge center); several cracks linked up to form a larger cracks. The laser holes contained pre-existing .08-.10mm (.003-.004") cracks resulting in earlier crack growth than would be expected with ECM holes. MA 956 and Hastelloy X pre-stressed panels have each achieved 10,000 cycles without any evidence of crack initiation at the ECM holes.

Summary

A NASA-sponsored MATE project for ODS alloy combustor liners is in progress; a summary of the program to date follows:

1. Five advanced combustor designs were evaluated based on preliminary analysis and life predictions, on construction and repair feasibility and on maintenance and direct operating costs. Two designs - the film cooled, segmented louver and the transpiration cooled, segmented twin wall - were selected for LCF component testing.

2. Detailed thermal and structural analysis of these designs established the strain range and temperature at critical locations resulting in predicted lives of 10,000 cycles for MA 956 alloy.

3. ODS alloys, MA 956 and HDA 8077, creep strength and oxidation resistance demonstrated a 167°C (300°F) temperature advantage over Hastelloy X alloy. MA 956 alloy was selected for mechanical property and component test evaluations.

4. MA 956 was superior to Hastelloy X in LCF component testing of the film cooled, segmented louver design.

5. Thermal cycle testing of the riveted louver design and LCF structural testing of the twin wall design are in progress.

References

1. Polhemus, J.F., Spaeth, C.E.., and Vogel, W.H., "Ductility Exhaustion Model for Prediction of Thermal Fatigue and Creep Interaction", <u>Fatigue at Elevated Temperatures</u>, ASTM STP 20, American Society for Testing Materials, 1973, Pages 625-636.

2. Gemma, A.E. and Phillips, J.S., "The Application of Fracture Mechanics to Life Predictions of Cooling Hole Configurations in Thermal-Mechanical Fatigue", <u>Engineering Fracture Mechanics</u>, 1977, Vol. 9, Pergammon Press, Great Britain.

LIFE/COST COMPARISON OF DESIGNS
(MA 956)

	Cooling air % W_{AB}	Total strain range (%)	Life cycles/hrs	Cost $ K	Weight lbs	MC $/hr	DOC %
JT9D base	45	~0.40	1.0	1.0	1.0	1.0	Base
Film cooled, segmented louver	45	0.145	4X	1.26x	1.06x	0.63x	−0.21
Segmented twin wall	33	0.225	4x	1.48x	1.03x	0.65x	−0.21

Table I

ODS ALLOY COMPARISON

	MA 956		HDA 8077	
Design	Total strain range (%)	Life (cycles)	Total strain range (%)	Life (cycles)
Film cooled, segmented louver	0.145	>10,000	0.185	2,000
Transpiration cooled, segmented twin wall	0.225	>10,000	0.245	>10,000

- Strain range differences are associated with thermal expansion characteristics

Table II

CANDIDATE ODS ALLOYS

	Fe	Ni	Cr	Al	Ti	Y_2O_3
Incoloy MA 956	Bal	—	20.0	4.5	0.5	0.5
HDA 8077	—	Bal	16.0	4.0	—	0.8

Table III

PROPERTIES OF CANDIDATE ODS ALLOYS

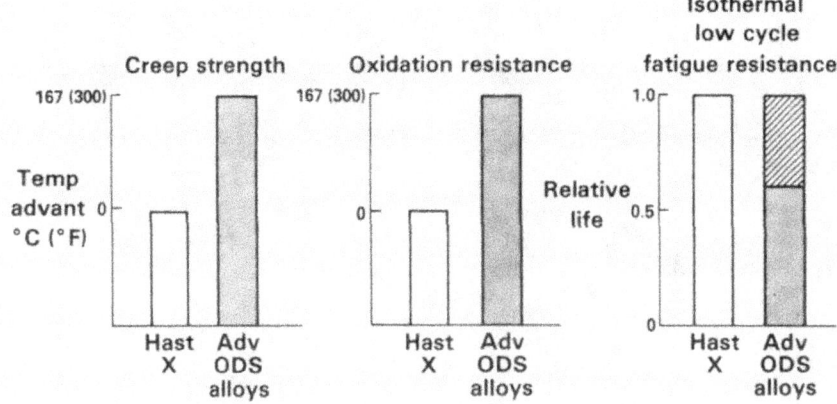

Figure 1

COMBUSTOR FAILURE MODES

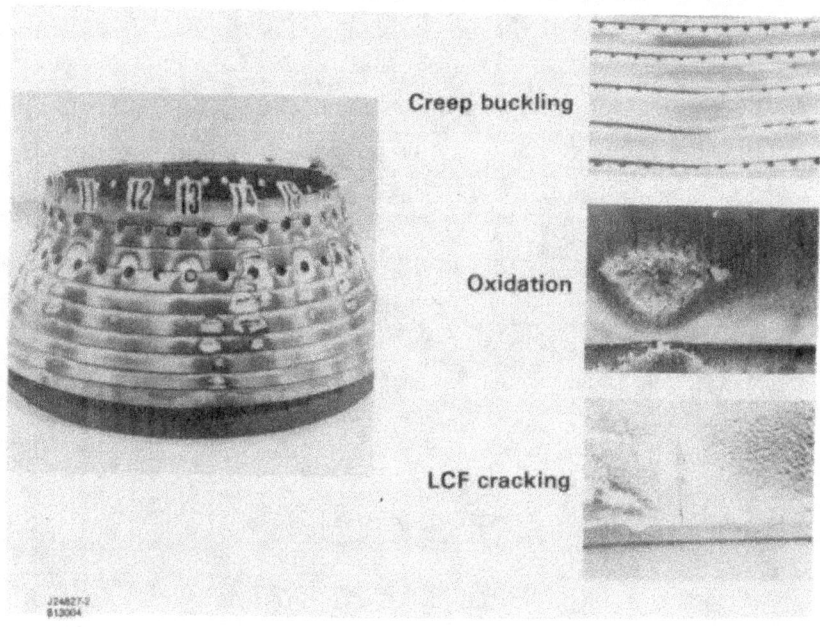

Figure 2

CONVENTIONAL FILM COOLED LOUVER COMBUSTOR

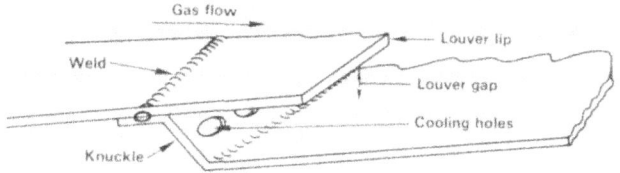

Figure 3

DESIGN SYSTEM

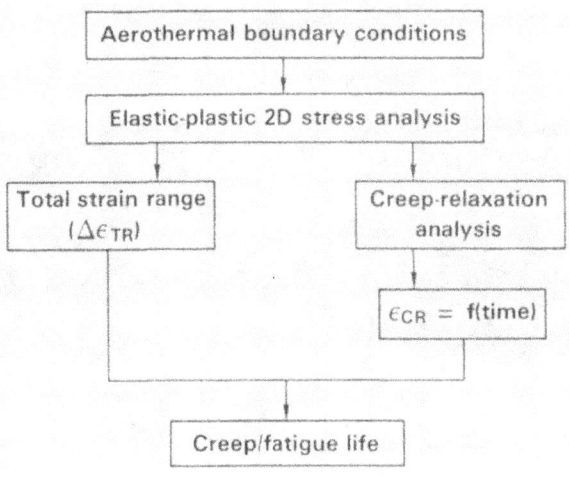

Figure 4

MECHANICALLY ATTACHED, FILM COOLED SEGMENTED LOUVER

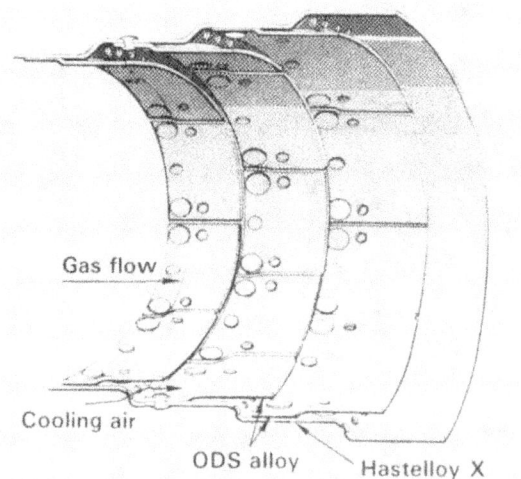

Figure 5

TRANSPIRATION COOLED, SEGMENTED TWIN-WALL DESIGN

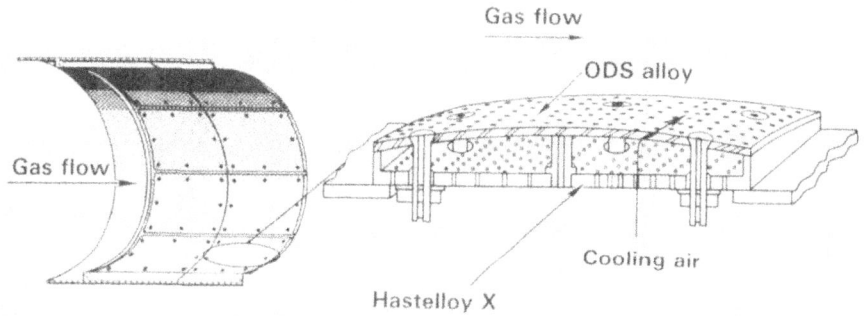

Figure 6

TYPICAL MICROSTRUCTURES OF CANDIDATE ODS SHEET ALLOYS

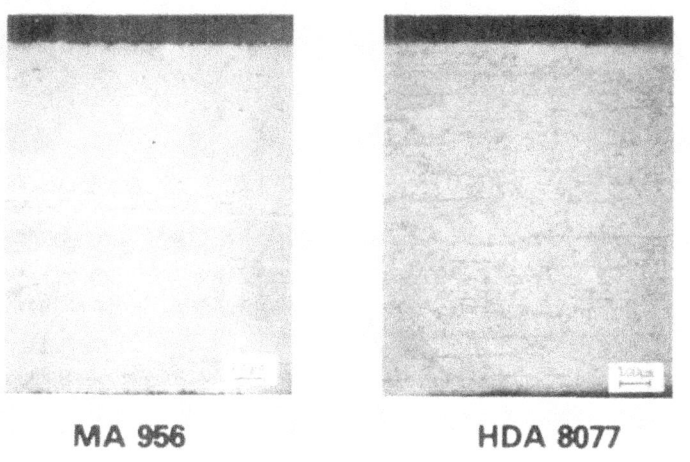

MA 956   HDA 8077

Figure 7

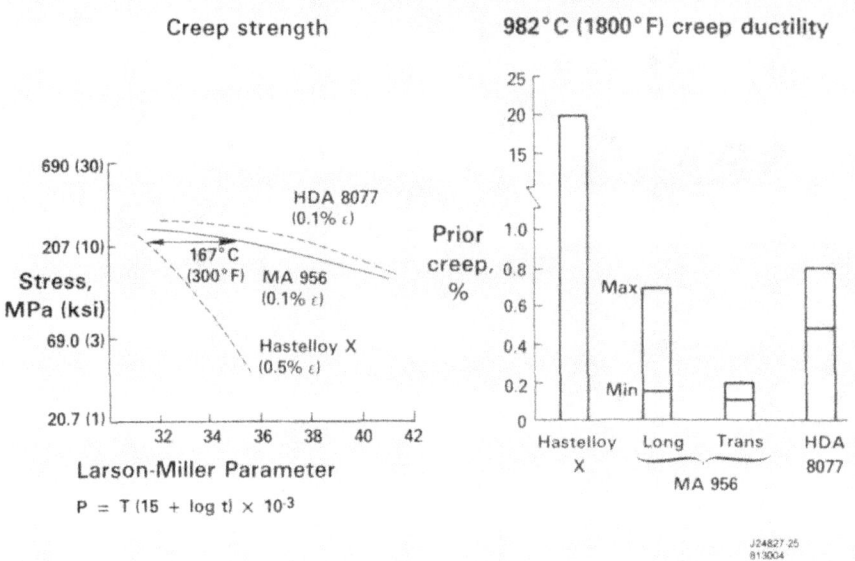

Figure 8

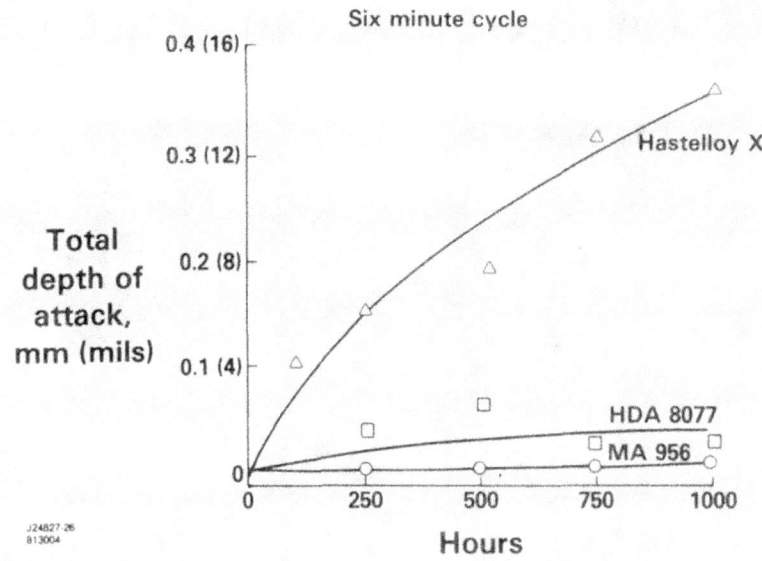

Figure 9

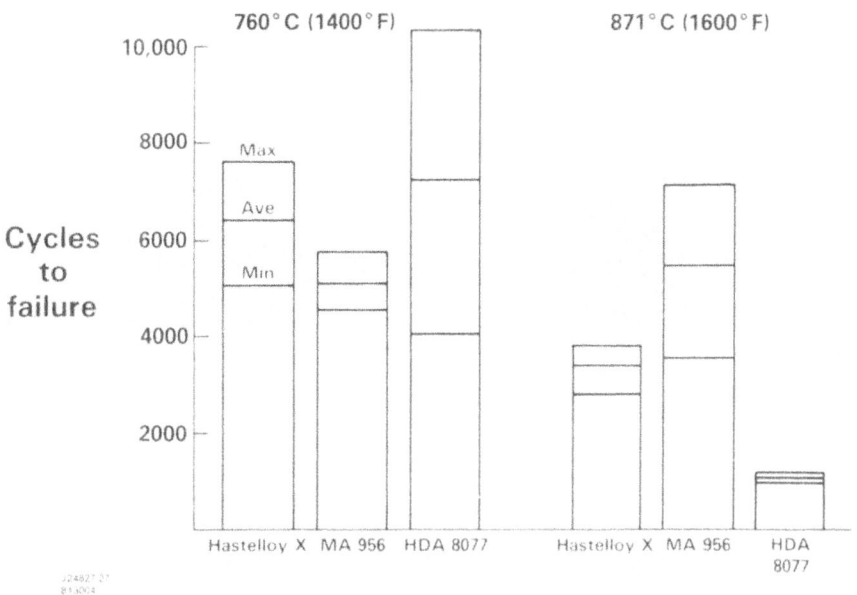

Figure 10

THERMAL FATIGUE
Hot spot blister test

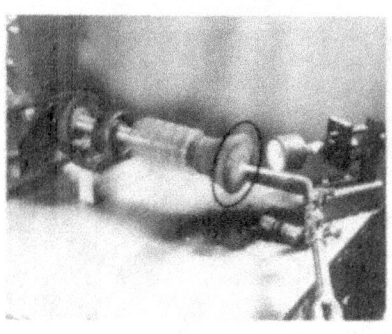
Test rig

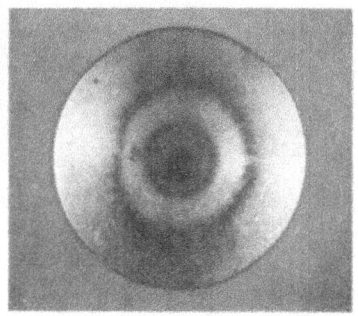
Test specimen

Figure 11

HOT SPOT BLISTER TEST RESULTS

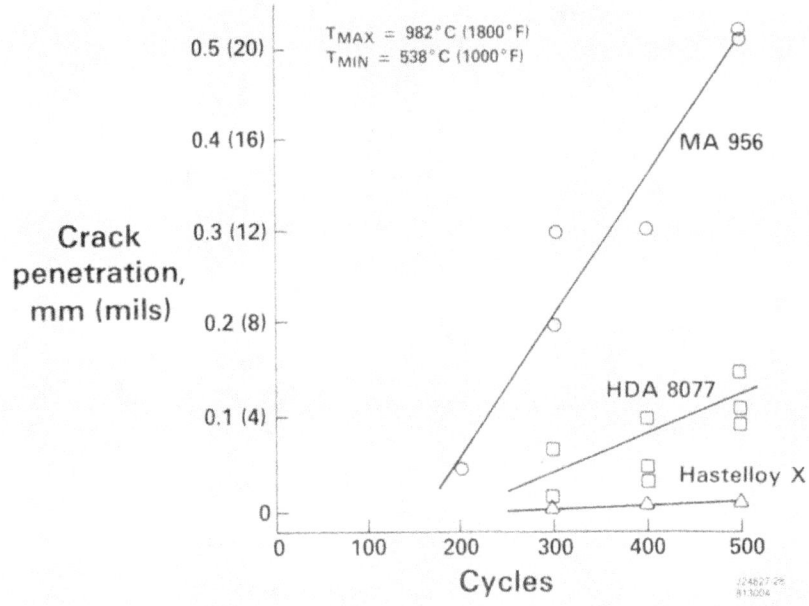

Figure 12

RIVETED LOUVER LCF STRUCTURAL TEST

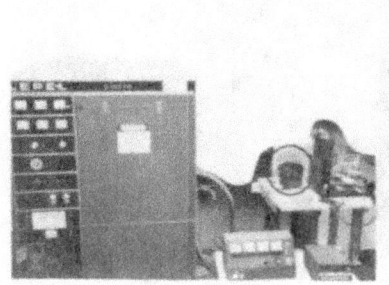
Test rig

Test specimen

Figure 13

RIVETED LOUVER LCF TEST RESULTS

- No edge cracking
- Dimensional stability of MA 956 superior to Hastelloy X

Conclusion: MA 956 demonstrates excellent creep resistance with no LCF cracking

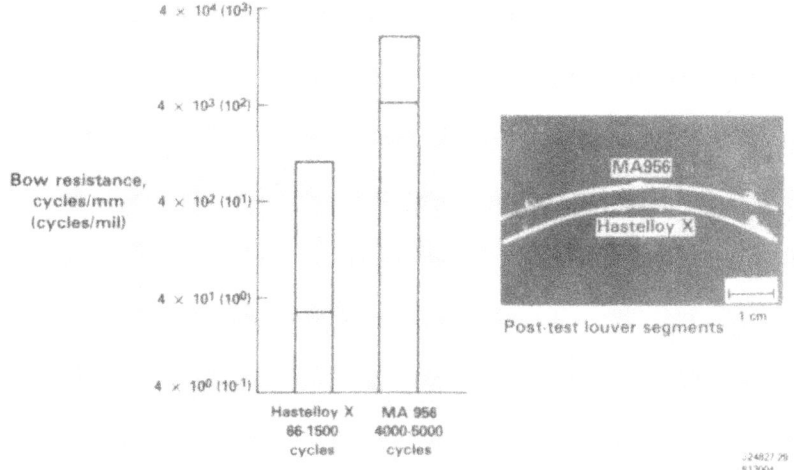

Figure 14

COMPONENT RIG TESTS

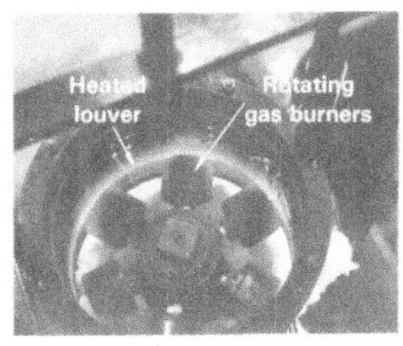

Riveted louver
thermal cycle test

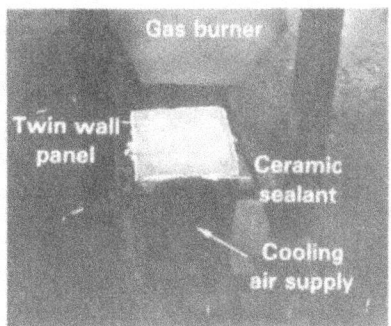

Twin wall
LCF test

Figure 15

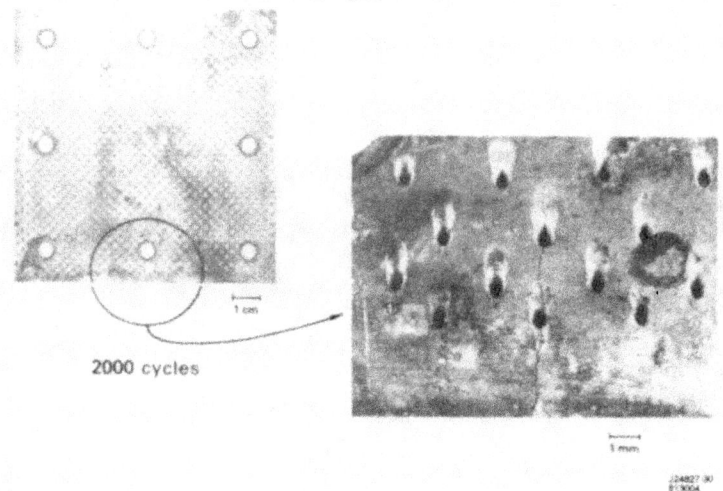

Figure 16

ADVANCED TURBINE BLADE TIP SEAL SYSTEM

J. W. Zelahy
General Electric Company

In axial-flow gas turbines, the turbine is designed to minimize the radial clearance between the blade tips and mating shroud segments. This helps to maximize aerodynamic efficiency. In spite of the designers' best intentions, the shroud assembly may go out-of-round, and/or the rotor and shroud may be slightly eccentric resulting in potential interference between the blade tips and the shrouds. Any interference which occurs generally removes material from the blade tips (Figure 1a) in preference to the stationary shroud, creating a larger annular clearance between the rotor and stator than if the blade tips had remained unaffected and the shroud material had been removed. Furthermore, the blade tip may be damaged, reducing useful blade life, and/or requiring expensive repair operations. At best, any rub on the bucket tip removes the environmental coating, thus making the blade vulnerable to both oxidation and hot corrosion (Figure 1b).

A NASA-sponsored (MATE Project 3) program is being conducted to establish and demonstrate the payoff of an advanced blade/shroud system designed to maintain close clearance between blade tips and turbine shrouds and at the same time, be resistant to environmental effects including high-temperature oxidation, hot corrosion and thermal cycling.

The target goal of this project is to demonstrate the increased efficiency and increased blade life attainable by using the advanced blade tip seal system. Increased efficiency results from the improved clearance control when blade tips preferentially wear the shrouds. Increased blade life results from the superior single-crystal superalloy tip.

The project will establish tip design, joint location, characterize the single-crystal tip alloy, finalize the abrasive tip treatment, fabricate blades, component test and engine test. The project will also establish quality control plans and define the total manufacturing cycle required to fully process the blades.

The turbine blade tip is of a multicomponent construction consisting of an Activated Diffusion Bonded (ADB) oxidation/hot corrosion resistant single-crystal superalloy squealer capable of withstanding thermal cycling, combined with a thin layer of alumina (Al_2O_3) abrasive particles held in place by an oxidation/corrosion resistant matrix (Figure 2). The shroud materials investigated included the current CF6 shroud (Bradelloy) and two advanced shroud materials, Genaseal and Vacuum Plasma Deposited (VPD) CoNiCrAlY.

The project is structured toward the successful engine demonstration of an improved efficiency, long life turbine blade tip system. The technical effort is divided into nine principal tasks.

Initial blade tip design work established the joint design and location, optimum squealer thickness and single-crystal orientation (Figures 3 and 4). The design that was established allows the single-crystal tip-to-blade bonding to be accomplished very early in the manufacturing cycle (possibly at the casting vendor) thereby not appreciably altering the standard manufacturing sequence. The tip design eliminated inside contour mismatch, located the joint in a low stress region and had total manufacturing acceptance. Using property data of both the single-crystal tip material and the bond joint, an economic benefit analysis (payoff) was subsequently performed by CF6-50 engineering on the single-crystal/abrasive tip system. The analysis predicted a minimum 2X increase in blade life via the superior tip material and a 0.013" tip clearance improvement (0.43% Specific Fuel Consumption (SFC) reduction) as the result of the abrasive tip treatment.

Since the 2X blade life goal was totally dependent upon both the increased environmental resistance of the single-crystal blade tip and the strength of the activated diffusion bonding (ADB) tip attachment process, a comprehensive evaluation of the mecahnical and physical properties of both the Normalloy (single-crystal tip material) and the Normalloy-to-Rene'80 (blade material) was conducted. The evaluation included elevated temperature, tensile, rupture, oxidation, corrosion and simulated engine thermal shock (SETS) testing. The results of the testing (Figures 5-8) confirmed that the properties exceeded those required for safe engine operation and would be expected to achieve the goal of 2X tip life.

The SFC reduction attainable with the advanced tip system is the direct result of the capability of the abrasive-tipped turbine blade (Figure 9) to resist wear during rub interactions with the shroud material. Several factors including particle size, particle type, particle relief, incursion rate, tip speed, test temperature and to a large degree shroud material have been shown to affect the wear characteristics of the abrasive system. Variations in particle size and type, degree of particle relief and rub incursion rate were evaluated. Test temperature (2000F) and tip speed (1400 ft/sec) were held constant. Three shroud materials: Bradelloy, Genaseal, and VPD CoNiCrAlY were evaluated. The particle types included various grades of aluminum oxide (Al_2O_3) and Borazon (Cubic Boronitride). In all cases, the method of abrasive application was the electroplate encapsulation process.

All wear testing was conducted at the Solar Research Laboratory (division of International Harvester) in San Diego, CA. Solar's facility has the capability of 1400 ft/sec. tip speed, 2000F shroud temperature, and direct readout/record of all vital functions including chamber temperature, shroud temperature, rotor speed and incursion rate. Measurements of both the blade specimens and shroud specimens were made before and after wear testing to establish the total wear of each. In addition, thermocouples were placed at the surface and 0.050" into the shroud specimens to record surface temperature and shroud temperature rise (and rate) as the result of the incursion. After each test, the blade specimens were evaluated visually, dimensionally, microstructurally and in some cases, by SEM analysis to establish both the total amount of blade and shroud wear and the wear mechanism (i.e., machining, compaction, melting, etc.) of each (Figure 10). Throughout the program over 50 wear tests were conducted.

The results of the testing showed that in all cases the abrasive tips resisted wear when rubbed into the Genaseal (both new and preoxidized) and the VPD CoNiCrAlY (Figures 11 and 12). The new Bradelloy was shown to be moderately abradable. The oxidized Bradelloy, however, was extremely difficult to "cut" and in most cases, after a small incursion into oxidized Bradelloy, the abrasive tips were rapidly consumed (Figure 13). The results of all of the wear testing are summarized in Figure 14. With respect to particle type, with the exception of Borazon, all particles behaved similarly. The Borazon system, in virtually all instances, abraded the shroud materials to a greater degree; even the oxidized Bradelloy was abraded more effectively by the Borazon particles. In addition, neither increased size nor relief significantly affected the abrasive characteristics of any particular system. The only test variable shown to appreciably effect abrasiveness was incursion rate. Slow incursions, i.e., 0.001 inch per sec or less, were shown to generate higher shroud temperatures and resulted in greater tip wear than at the 0.002 and 0.004 inch/sec tests (typical incursion rates in engines have been estimated at 0.002 inches/second or greater).

The results of the above wear testing have tentatively indicated that:

1. A large allowable latitude in abrasive system variables exists, i.e., particle type, particle size, relief, and environmental coating can be varied considerably without decreasing the abrasive characteristics of the tip treatment.

2. An oxidation resistant shroud material (e.g. Genaseal or VPD CoNiCrAlY) should be used to achieve full benefit of the abrasive system.

3. Alundum 38X, 100 grit aluminum oxide/NiCr electroplate with a Codep aluminide coating is the best all-around tip system.

4. Slower incursion rates (i.e., $\leq$ 0.001 in./sec.) are more detrimental to the abrasive system than faster incursion rates (0.002 to 0.004 inch/sec).

The abrasive tip system designated for component and engine testing is defined below.

- particle type: 38X alundum (Al_2O_3)

- particle size: 0.005" - 0.007" diameter

- matrix: 0.006" Ni, 0.001" Cr Diffusion H.T. with aluminide coating

- relief: matrix plated "flush" with particles

- shroud: either Genaseal or VPD CoNiCrAlY

Using both simulated and actual hardware, the environmental resistance and abrasive capability of the environmental/abrasive tip/shroud system was verified. Wear testing was conducted on Solar wear specimens that were modified with single-crystal/abrasive tips (Figure 15). The wear testing of the simulated tip system specimens indicated the tip system was capable of withstanding the rigors of severe shroud rub with no deleterious affects on either the single-crystal tip material or the ADB joint. The single-crystal-to-Rene'80 joint sustained very severe rub loading, particularly in the case of one bare bladed rub where $\approx$ 0.050" of tip was removed and no joint degradation was evident. Although minimal success was achieved in rubs of abrasive tipped blades into Bradelloy, successful rubs were made into Genaseal and CoNiCrAlY shrouds without loss of abrasives.

The environmental testing (i.e., oxidation, corrosion), impact and thermal shock testing will be conducted on actual hardware (scrap "fall-out" from fabrication task). This testing is currently in progress.

An integrated quality control plan including control over the tip material, the attachment process, the abrasive treatment and all related blade processing operations is currently being prepared. Temporary specifications have been issued and will be revised and updated as needed. Drawings for the single-crystal tip have been issued defining crystallographic orientation and tip configuration. Tooling for inspection of joint thickness has shown dimensional accuracy of $\pm$ 0.0005" and has been used to inspect all fabricated blades to date.

Each of the separate processing steps established in earlier tasks were formulated into an integrated processing sequence for the manufacture of turbine blades with the advanced tip system. The sequence of operations allowed the single-crystal tip bonding to be accomplished without any appreciable changes in normal blade processing (Figure 16). The blades were removed from the production airfoil operation immediately prior to tip cap cavity EDM operation and ground to a specified length. The single-crystal tips were bonded to the blades and the blades were re-introduced to the airfoil operation for the tip cap cavity EDM operation. The EDM operation provided a smooth tip squealer/blade internal wall surface and eliminated any need for internal tip/blade "blending" operations. This task is also still in progress and when completed will fully define the blade casting configuration, tip preparation and heat treatment, the single-crystal tip configuration, orientation and processing, the bonding process operations, fixturing and inspection, the abrasive tip treatment and all nonstandard operations associated with the blade manufacture. A total processing plan, including step-by-step sequence, will be provided.

A total of 171 blades were subsequently fabricated using the manufacturing sequence defined earlier. Tips were bonded in "dead weight" load fixtures in a cold-wall high vacuum furnace. The activated diffusion bonding (ADB) alloy was D15 (Rene-80-BASBD Chemistry) and was applied as 0.003" foil. Of the 171 parts that were bonded only 2 failed inspection (joint thickness measurement). Approximately 150 blades are fully manufactured (Figures 17-20) and are either undergoing or awaiting factory engine test evaluation (Figure 20). The remaining blades will undergo exhaustive destructive evaluation to further assess process reliability and reproducibility.

Two engine tests are planned to fully evaluate the payoff of the advanced tip system. The first engine test will evaluate the benefits of the single-crystal tip via "C-cycle" (simulated flight cycle) endurance testing (1000 cycles minimum). The second engine test will evaluate the abrasive capability of the system via performance testing under closely controlled clearance and engine operating conditions. The second test will be of short duration and is designed to "push" the abrasive tip system to the "limit" to fully establish maximum abrasive capability.

The results of the engine tests will be evaluated and analyzed to assess the effectiveness of the entire system to achieve the program goals.

Successful completion of the program can provide engine manufacturers a viable approach to increase blade life and reduce fuel consumption.

(a) TIP WEAR

(b) OXIDATION/CORROSION/CRACKING

FIGURE 1. TYPES OF TIP DETERIORATION

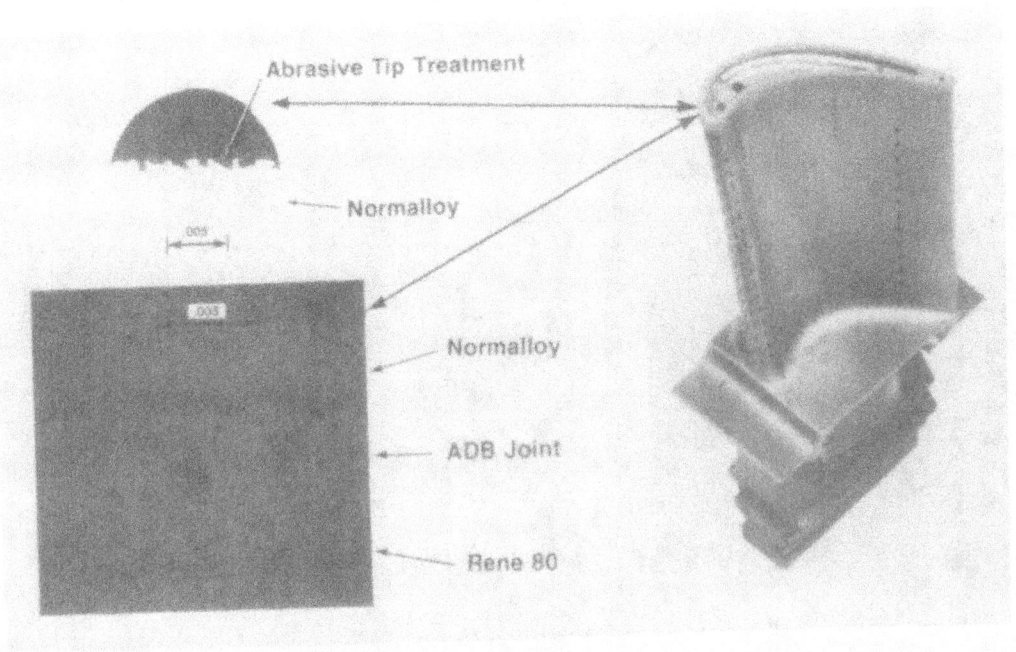

FIGURE 2. CF6-50 STAGE 1 HPT BLADE WITH ADVANCED BLADE TIP SYSTEM

Task I — Blade Tip Seal System Design
Joint Design Location

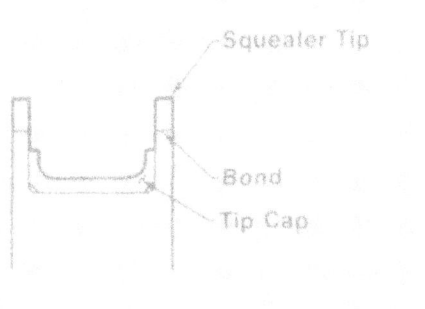

- Low Stress/High Reliability
- Ease of Manufacture
- Consistent With Current Blade Processing

FIGURE 3. JOINT DESIGN LOCATION

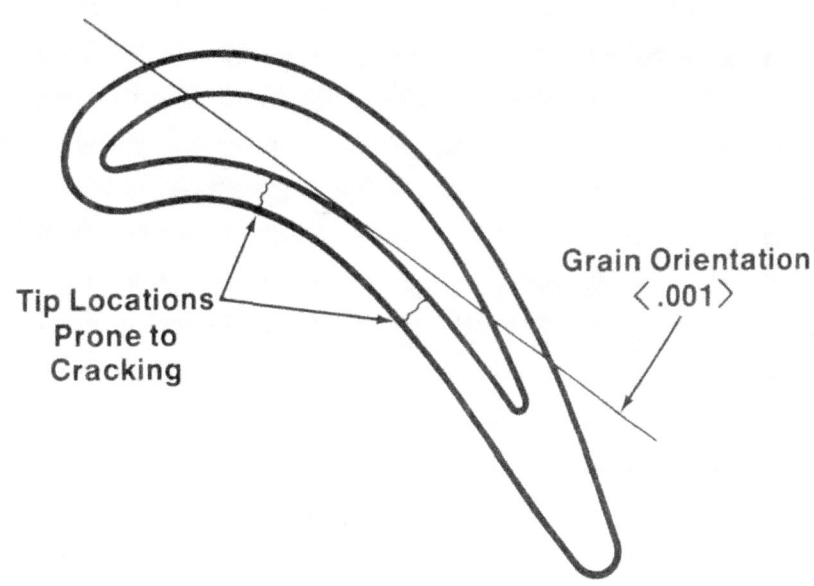

FIGURE 4. MONOCRYSTAL ORIENTATION

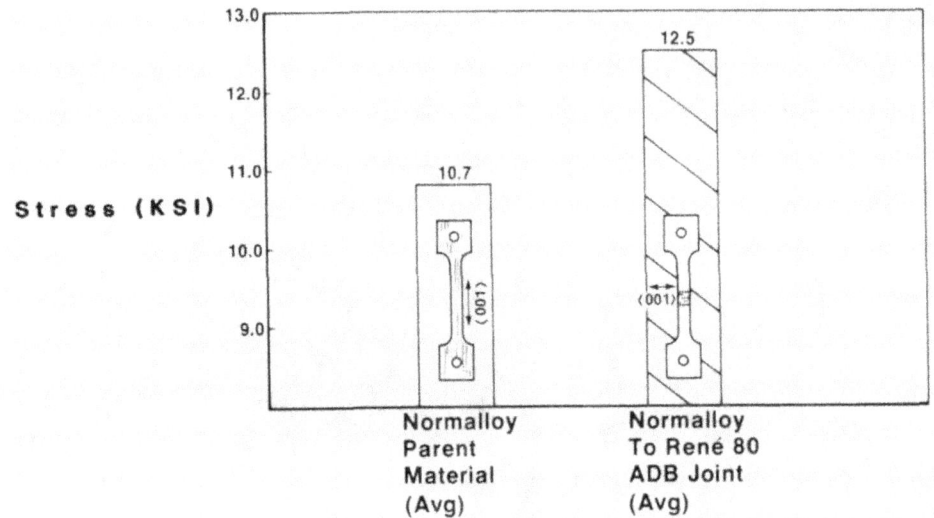

FIGURE 5. 2000F TENSILE PROPERTIES

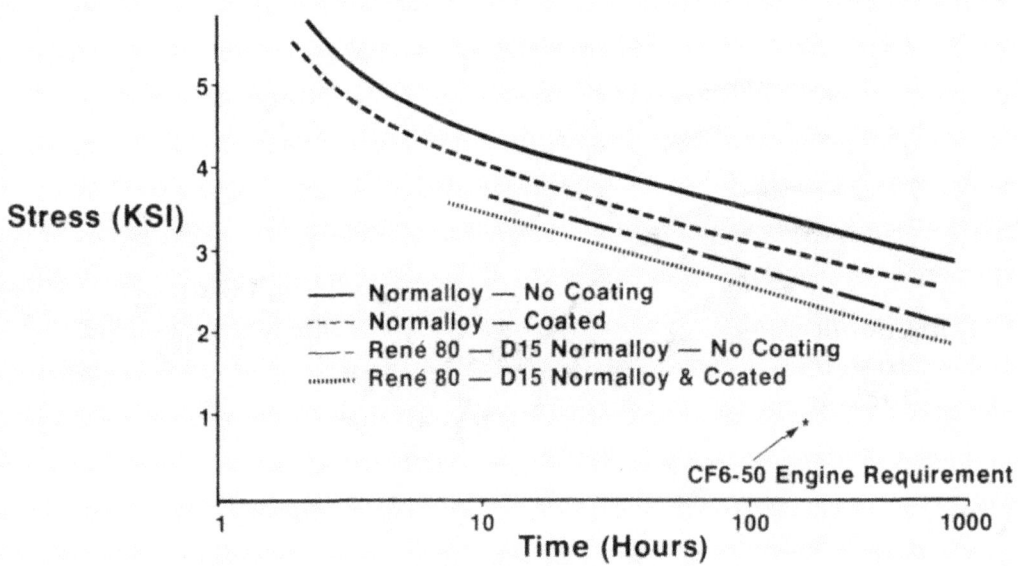

FIGURE 6. 2000F STRESS RUPTURE (AVERAGE)

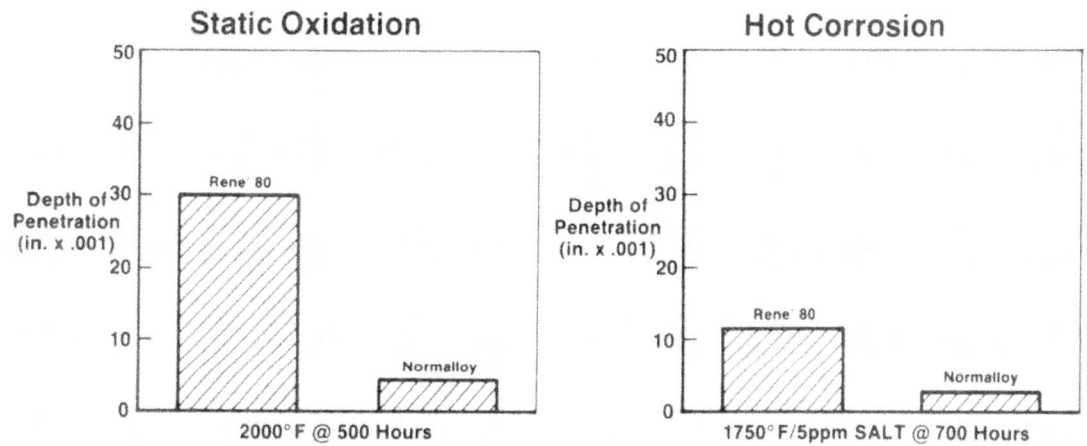

FIGURE 7. ELEVATED TEMPERATURE OXIDATION & CORROSION TESTING

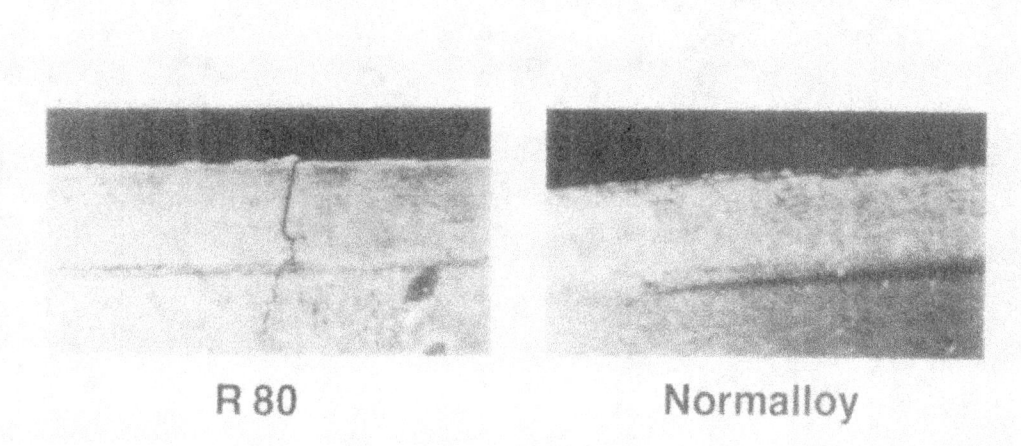

FIGURE 8. THERMAL FATIGUE RERISTANCE (2000 THERMAL CYCLES)

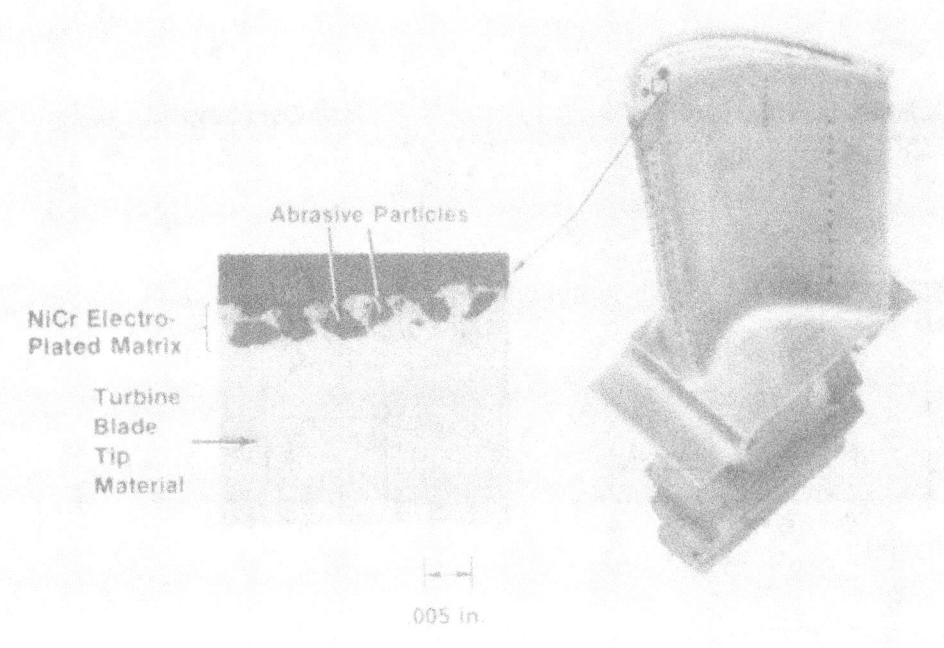

FIGURE 9. ABRASIVE TIPPED TURBINE BLADES

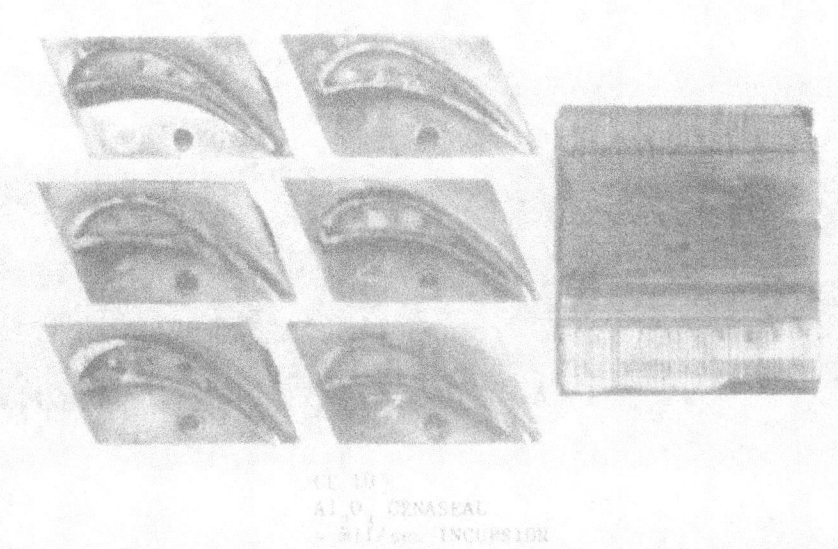

FIGURE 10. SOLAR WEAR TEST SPECIMENS - AFTER TEST

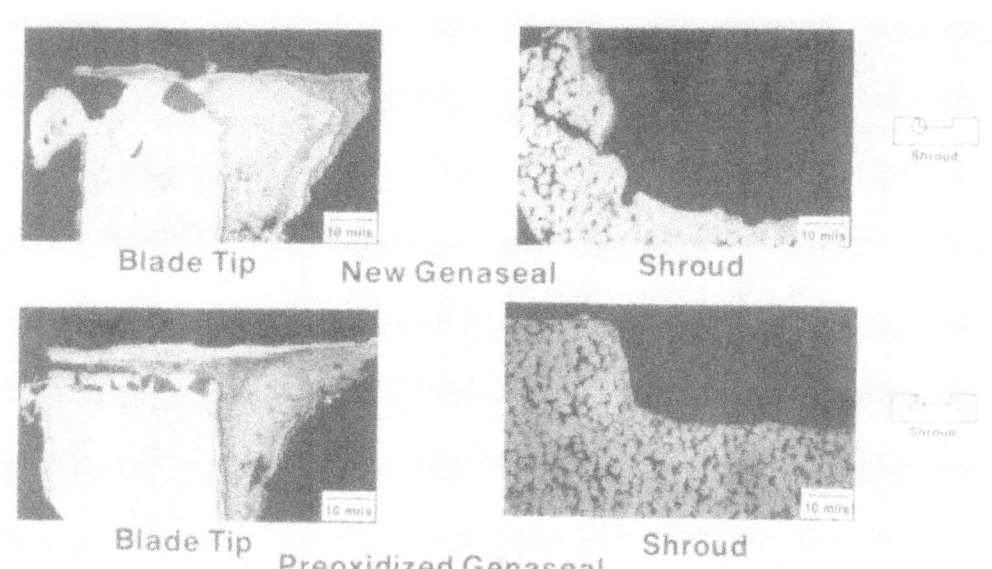

FIGURE 11. WEAR TESTING - ABRASIVE TIP INTO NEW & PREOXIDIZED GENASEAL

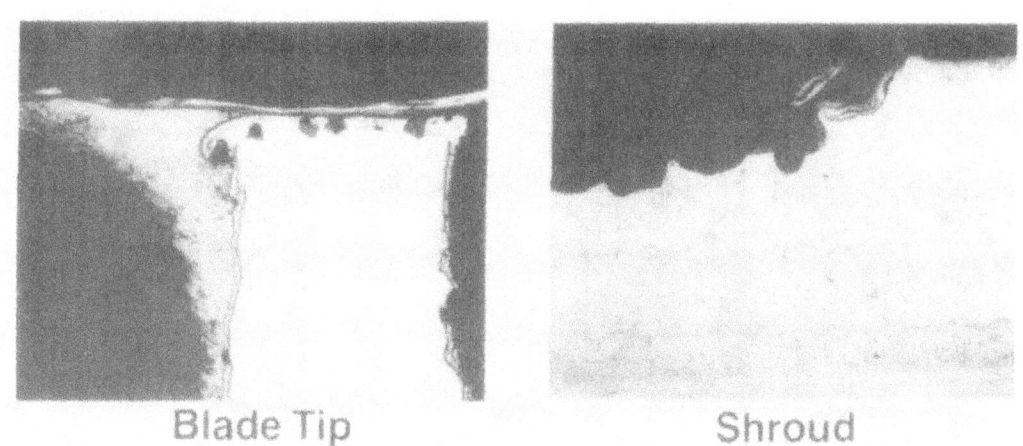

FIGURE 12. WEAR TESTING - ABRASIVE TIP INTO NEW VPD CoNiCrAlY

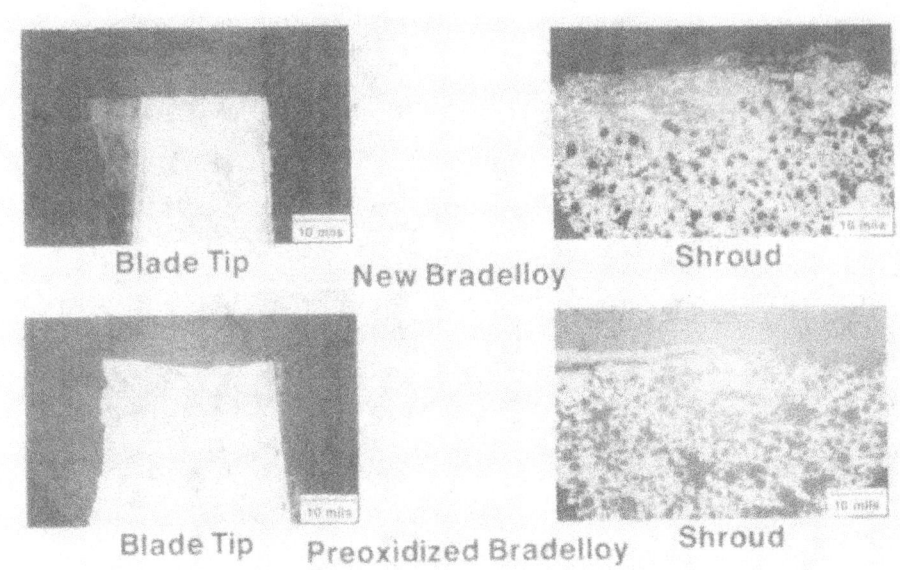

FIGURE 13. WEAR TESTING - ABRASIVE TIP INTO NEW & PREOXIDIZED BRADELLOY

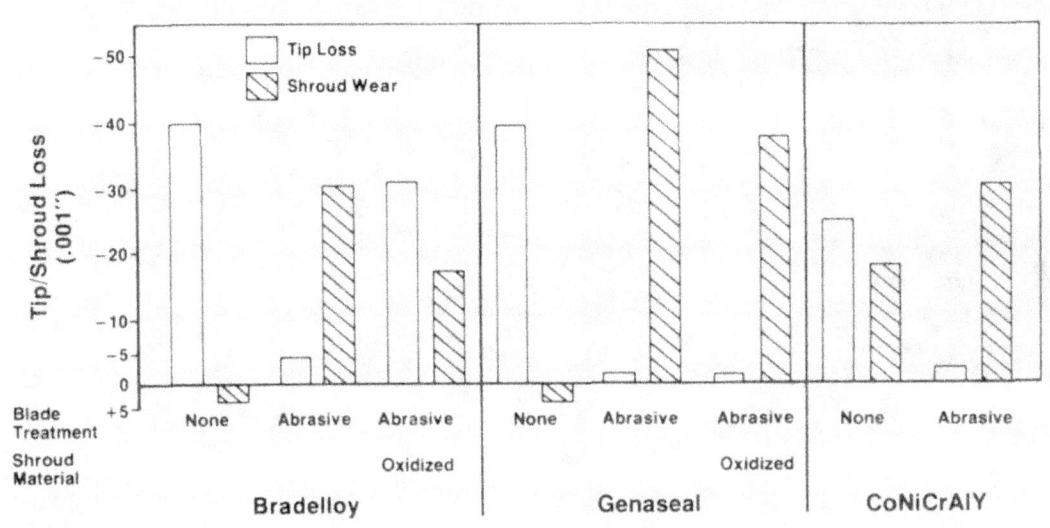

FIGURE 14. WEAR TEST RESULTS (2000F, 1400 ft/sec)

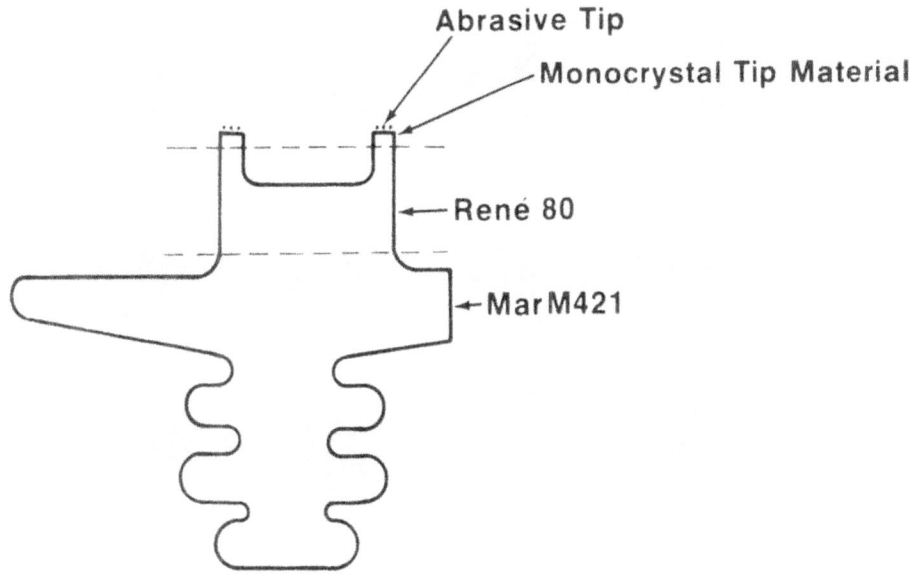

FIGURE 15. WEAR TEST SPECIMEN

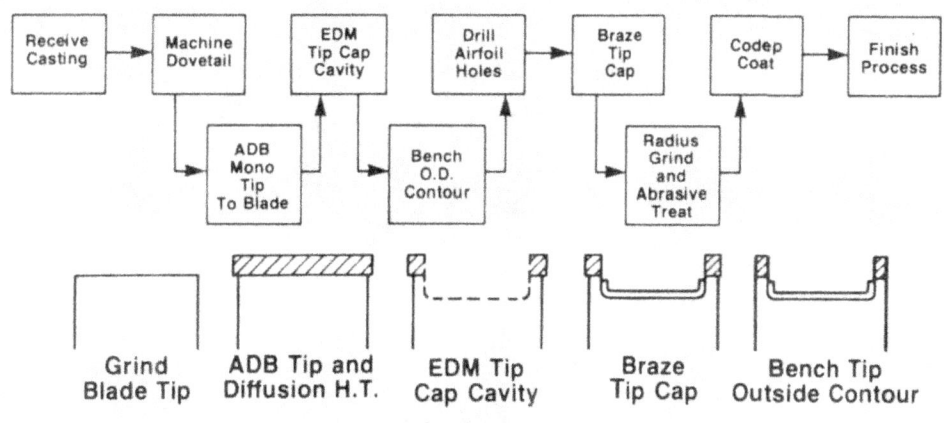

Finishing Steps:
- Finish Process as Standard Blade
- Radius Grind Tip
- Apply Abrasive and Diffusion H.T.
- Coded Coat and Age

FIGURE 16. MANUFACTURING PROCESS PLANS

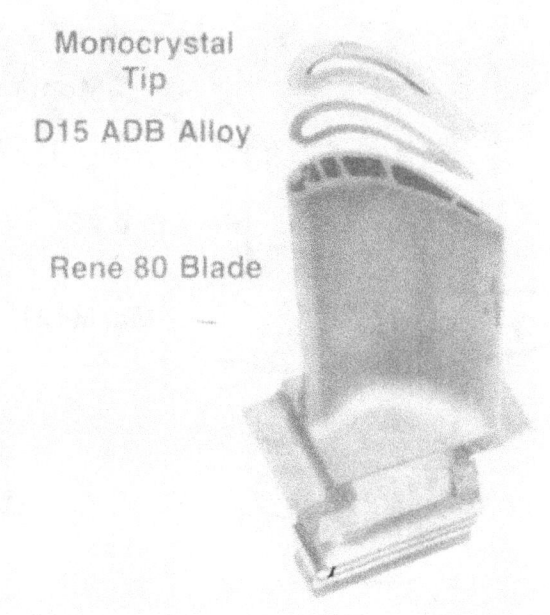

FIGURE 17. BLADE/TIP/ADB ALLOY ASSEMBLY

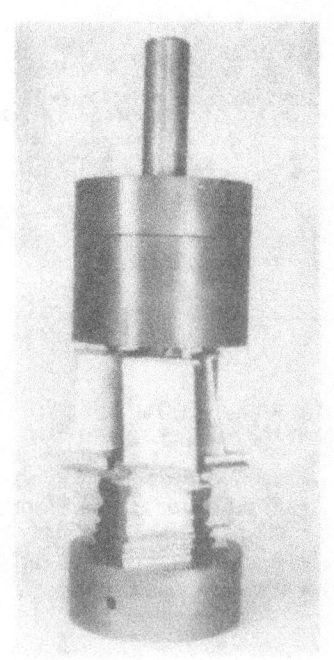

FIGURE 18. BLADE SUBCOMPONENTS & FIXTURE

FIGURE 19. BLADE WITH MONOCRYSTAL TIP AFTER BONDING

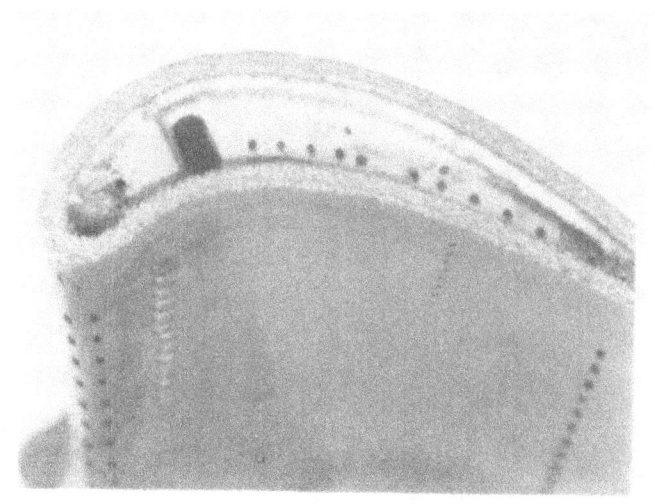

FIGURE 20. ABRASIVE TIP APPLIED & FULLY PROCESSED

FIGURE 21. STAGE 1 HPT BLADES WITH ADVANCED TIP SYSTEM ASSEMBLED IN TURBINE ROTOR PRIOR TO ENGINE TEST

AN INTRODUCTION TO NASA'S TURBINE ENGINE HOT SECTION TECHNOLOGY (HOST) PROJECT

Daniel J. Gauntner and C. Robert Ensign
NASA Lewis Research Center

INTRODUCTION

Today's modern gas turbine engines with their high thrust to weight ratio and low specific fuel consumption are comprised of many sophisticated components utilizing the latest high strength materials and technology. This is especially true in the hot section components of the combustor and turbine where high temperature superalloys and protective coatings are necessary in an environment where gas temperatures are well above the melting point of the materials. Current hot section components must endure higher temperatures, higher stresses, and more severe thermal transients than ever before. The durability and efficiency goals of the hot section components operating in this adverse environment will be difficult to achieve. Any shortfalls in achieving these goals could have significant effects on the overall operating cost of the modern gas turbine engine. Early in 1978, NASA began to plan a major project of turbine engine hot section research. Plans called for in-house and contract research to develop and improve the accuracy of current analysis methods so that increased durability could be designed into future engines. This paper is an overview of the new NASA Turbine Engine Hot Section Technology (HOST) project that began officially in January, 1981.

The HOST project was formulated around a simple, yet basic premise. Specifically, present analysis methods for designing combustor and turbine components need improvements in accuracy and applicability before increases in life can be attained during the initial design process of advanced turbine engines. The improved accuracy in life prediction can be attained by conducting focused and directed research efforts in each of the areas involved in component design, including description of the thermal and aerodynamic environments, the material's mechanical response, and the interactions between environmental and structural response. Verification of the more accurate predictions will be a necessary element of the HOST project and it will require high temperature instrumentation capable of measuring near-engine environment effects. The achievement of these improvements will require a rigorous and systematic research effort, beginning with evaluations of current predictive methods by comparing their predictions to benchmark data from special component tests, followed by supporting research to improve the modeling of the physical phenomena, and concluding with tests to verify the improved models in each of the pertinent discipline areas. These areas include structural analysis, surface protection, combustion, turbine heat transfer, and instrumentation.

TURBINE ENGINE HOT SECTION

The hot section components of an advanced turbine engine include the combustor and the turbine. A schematic of a typical turbine engine hot section is shown in figure 1. The contoured shaded areas represent an annular flow combustor connected to an axial flow turbine. The combustor liner and turbine airfoil outlines are represented in the figure. The arrows on the schematic represent the flow of the hot section cooling air around the components and through the turbine disk cavities. Because the liner of the combustor and the airfoils of the turbine are the hot section parts exposed to the highest temperatures and consequently suffer a large degree of damage, the research efforts in HOST will be concentrated on improving the analysis methods used to design these three parts. Typically the hot section has twenty percent of the engine weight but accounts for almost sixty percent of the maintenance costs. The consequences of combustor liner failures are generally more economic than operational and result only in a slow, general deterioration of the engine. It is included with the turbine airfoils as part of the HOST project primarily because of the combustor's close coupling and direct effect on the turbine durability.

A knowledge of the basic functions of the combustor and the turbine is necessary if the impact and importance of hot section durability problems is to be understood. In the combustor, the basic release of energy to the core airflow takes place with the burning of the turbine engine fuel. Involved in this energy release are many phenomena, including flow mixing, combustion kinetics, turbulence, flame radiation, soot formation and consumption, liner heat transfer, and gradual acceleration of the high temperature combustor airflow into the turbine. The control of these phenomena by suitable design factors will determine the temperature distribution in the combustor liner and the exit temperature profiles of the airflow leaving the combustor and entering the turbine. In the turbine, this entering airflow is channeled through a set of inlet guide vanes to properly align the flow vectors for optimum and efficient transfer of momentum and energy to the rotating blades of the turbine. The efficiency of the turbine, which contributes greatly to the overall performance and fuel efficiency of the engine, is directly related to the gas flow and temperature distributions. Besides the gas temperature distribution, the gas flow behavior is also needed. Any flow disturbance that inhibits uniform circumferential temperatures or proper radial temperature distribution imposes a penalty on engine performance. This is particularly true near a turbine hub, where large secondary flow vortices are often generated. The extent of these vortices depends upon the quality of the flow entering the turbine. They can be partially controlled by the radial gradients of the energy extracted from a turbine. The temperature and flow phenomena must be better understood and predicted with greater accuracy if life prediction methods are to be improved.

HOT SECTION DURABILITY PROBLEMS

The durability of hot section components is highly dependent on such factors as the type of aircraft mission flown, the geographical location of the operating base, and pilot operation. All of these factors affect the temperature and pressure environment and the cyclic load history of the parts in the hot section. During a typical turbine engine design, the type of aircraft mission expected to be flown is expressed in terms of engine cycle information. Design life predictions are made for accumulating levels of repetitive, and somewhat simplified engine cycles. The engine's hot section temperatures and pressures from the engine cycle information are used in these predictions. Variation in conditions due to geographical locations and weather conditions are accounted for in non-standard day test conditions. Variations due to individual pilots, however, can not be treated deterministically. The design assumes that the engine operates along a worse case cycle.

The incorporation of these real-life variants is beyond the scope of the HOST project. What is possible is to look at factors which affect the durability of the individual components in the gas path of the hot section. Other programs have gathered experimental and field service data regarding the actual and probable modes of failure for combustor liners and turbine airfoils. Examples of durability problems in components are shown in figure 2. Typically, air-cooled combustion chambers experience large, thermally induced strains that exceed elastic limits of materials at points of maximum stress and/or temperature. Creep-low cycle fatigue interactions and louver lip collapse have been established as primary burner liner failure modes. Oxidation/erosion modes tend to be secondary failures, usually caused by some other damage mechanism. For turbine airfoils (vanes and blades) creep-fatigue cracking and oxidation/corrosion tend to be dominant failure modes. But for the airfoils, these modes are of more importance and usually necessitate engine removal when detected to prevent further damage such as blade or vane rupture. These modes of failure for the combustor liner and the turbine airfoils have been selected as pertinent examples, but they are not all inclusive.

RESEARCH EFFORTS

Approach

The HOST project will support research to improve the accuracy of analysis methods, which can be used during engine design to increase component durability levels, thereby reducing maintenance and operating costs of the turbine engines. Research will be funded in the areas of structural analysis, surface protection, combustion, turbine heat transfer, and instrumentation. The overall approach of HOST in each of these areas will be to: (1) evaluate existing models; (2) quantify their strengths and weaknesses; (3) conduct new experimental and analytical research to more accurately model the physical phenomena; (4) use the new models in predictive analyses and verify their improvements in accuracy; and, (5) conduct a sensitivity study to assess the improvements in overall hot section durability to be achieved by use of a combination of these new methods.

The HOST project emphasizes the coordination of the research activities (fig. 3) to provide a system of more accurate analysis methods. The use of these improved methods will lead not only to enhanced durability, but also to lower maintenance costs for the hot section, freedom for more innovative design and checkout of new ideas, the ability to perform more accurate trade-off studies between performance and durability plus high reliability in future engines.

The specific elements of research that will be supported and coordinated in HOST are shown in the work breakdown structure in figure 4. The technical aspects of the activities in the six columns of figure 4 will be managed by staff members from four different divisions at the Lewis Research Center. This delegation of technical responsibility is illustrative of the matrix management concept that will be used for the HOST project. Descriptions of the planned research are presented in the following paragraphs.

Structural Analysis

Some typical damage observed on one component from the hot section of a turbine engine is shown in figure 5. This section of a combustor liner shows thermal fatigue cracking. To approach such a problem, the structural analyst must have sophisticated tools for accurate analysis. These include a knowledge of the thermal and mechanical loads, inelastic methods of analysis such as nonlinear finite element computer codes, cyclic constitutive (stress-strain) relationships, and the capability to determine the effects of the creep-fatigue interactions on crack initiation.

The structural analysis efforts under HOST will pursue areas such as thermal loading prediction methods, specialized vane and liner geometric and structural analysis models, methods and procedures to determine time and temperature structural response characteristics, and improved methods to describe time dependent and time independent inelastic material behavior. In addition, material constitutive relationships will be improved for predicting material behavior response to cyclic variations in stress, strain, and temperature with time. Also, life prediction methods will be developed for crack initiation models. The existing methods for such problems will be improved and automated to reduce the required manpower and computer time. For instance, the thermal analysis methods will be integrated with the structural analysis codes, so that the relatively coarse thermal map of a component becomes the input to the more detailed finite element program. Also, the methods will include self-adaptive solution strategies that use substructuring to examine the inelastic regions of a component with an overall elastic behavior.

The specific elements under HOST in the area of structural analysis are listed in figures 6(a) and 6(b) along with the expected results. The bars show the expected starting times and durations of each effort, in terms of fiscal years, which run from October through September.

The first element listed under structural analysis in figure 6(a) represents a planned effort to develop a computerized method to transfer the thermal loads that a burner liner might experience to a structural analysis model. The method will automatically integrate information from a thermal analysis computer code with an advanced nonlinear structural analysis code. The next element, shown by the bar extending from FY82 to 86, extends the application of the first method to prediction of loads that are component related, and time dependent for other hot section components for various engine mission cycles. It also will include effects of local hot streaks, cooling holes, and thermal anisotropy, as found in turbine blades and vanes. The structural analysis methods of the future will require improved versions of today's computer capabilities such as 3-D nonlinear finite element methods that can handle plasticity, creep, strain concentration, and unsymmetrical thermal effects found in hot section components. For effective structural analysis of the hot section, the codes will handle all these interacting inelastic effects. HOST will develop these capabilities and determine strategies and self-adaptive algorithms for solution of such complex analysis problems. After these automated modeling and solution strategies are completed, they will be verified by comparison with data from tests of specific engine components subjected to typical thermal and mechanical forces from appropriate mission cycles. The final program element in figure 6(a) will include component specific models and verification of the above efforts.

Within the computerized structural analysis methods are equations which model the behavior of the material when it is subjected to various loads and temperatures. These equations represent different theories and engineering models that attempt to describe the physical phenomena taking place. The theories, and hence these constitutive equations, must describe the response of a material subjected to both mechanical and thermally induced stresses and strains. For low temperatures, when the material is in the elastic range, the theories are quite adequate. But in the hot section of the turbine, most parts are well into inelastic behavior, and the modeling becomes very complex. Many theories have been proposed to describe this behavior. Several elements of HOST in figure 6(b) will evaluate the various theories and models to understand and improve upon the constitutive equations.

The first of these elements in figure 6(b) will determine the best model to represent the cyclic behavior of isotropic materials. The model will include the complexities of creep-plasticity, multi-axial stress and strain, plus the effects of long-time exposure of surfaces. The second element will develop and verify similar constitutive models for anisotropic materials, such as those used in the manufacturing of directionally solidified vanes or blades. The final product of these efforts will be sets of equations that represent the inelastic response of hot section components with greater accuracy than today's methods.

The second aspect of the HOST project in figure 6(b) is the prediction of the life of hot section component parts using an understanding of the synergistic effects of creep, fatigue, and environment on crack initiation behavior. Existing models that are explicit in the primary variables of stress, strain, temperature, environment, and time will be screened. The first element under Life Prediction Methods will select and develop a model for a specific isotropic material/coating combination that is typical for a liner or vane. The effects of mission loading, multi-axial stress and thermal cycling will be included. A second and parallel program element of HOST will develop a similar life prediction method for an anisotropic material/coating combination for a liner or blade. Both of these efforts will consist of a concentrated effort of laboratory testing resulting in modifications to life prediction methods. After sufficient validation, a second material/component combination will be examined in each of these efforts.

Surface Protection

Significant work has been done to further the science and technology of coatings. Figure 7 shows micrographs of a NiCrAlY coating before and after soaking for a long time at 1366 K. The coating is degraded not only by the hostile environment, but also by its diffusion at the substrate boundary. HOST will concentrate on analytical methods to account for each of the effects of environment, corrosion/erosion, oxidation/diffusion, and metallic coatings to be able to predict the time to crack initiation of coatings and coated hot section parts.

The HOST effort will concentrate on modeling the effects of environmental attack and coatings on crack initiation, the location and rate of erosive particle impact and corrosive salt depositions on airfoils, and also the coating degradation on blades, vanes, and combustors to provide coating life predictions. The various surface protection elements under HOST will study the phenomenological effects and interactions, and will produce analytical models for different types of components (i.e., turbine blades, vanes, and combustor liners). All of these models will be evaluated and verified using either real data from engine field failure experience or laboratory data from erosion/corrosion burner rigs.

The first surface protection element of HOST in figure 8(a) will model the effects of environment and coatings on the creep-fatigue crack initiation of isotropic materials used for liners and vanes. Later, another element will produce similar models, but for anisotropic materials such as directionally solidified vanes and blades. These two elements will be combined with the Life Prediction Methods of figure 6(b), as represented by the dashed lines of figure 8(a). As another element, the behavior of sheet materials coated for use as combustor liners will also be obtained during cyclic testing in a suitable test rig.

Other research efforts in the area of Surface Protection are shown in figure 8(b). The effects of corrosion and erosion, and their interaction, will be modeled and then used as part of a more comprehensive coating life model. To assist in developing a corrosion/erosion model, research elements investigating mass deposition on airfoils and the location and rates of erosion on airfoils will be conducted. The third element of the corrosion/erosion model effort will include rig burner tests of the combined corrosion/erosion mode to verify the deposition and erosion models.

Under the coating life model of figure 8(b), the first research element will collect engine field failure experience data for coatings to provide a real environment data base. The effects of oxidation and corrosion will be investigated and then modeled in inhouse tests to verify the effects of this dual cycle mode of attack. Next, the corrosion/erosion and dual cycle models will be combined. Life predictions will be verified in test rigs. Finally, a test program will obtain correlations of rig test effects and engine test effects on coating life.

Combustion

Present turbine engine combustors exhibit very complex flow conditions and high levels of heat transfer by radiation (fig. 9). These conditions make the prediction of gas and metal temperatures very difficult. To aid in this task, the combustion research will be conducted in the areas of aerothermal modeling and liner cyclic testing. To support this analytical work, a test program will be developed to study gas flow and mixing phenomena and flame radiation effects. Also, plans call for the design of a test rig that can obtain accelerated low life data for liner segments subjected to thermal cycling. The design will be difficult to obtain, since the thermal loads on the liner segment must simulate the real loads on a full circular liner, if the accelerated life data is to be useful.

The first combustion element in figure 10 will assess the existing aerodynamic and thermodynamic models to determine their capabilities, deficiencies, and the priority of areas requiring improvement. Research and model refinements will then be made in areas such as internal flow and exit temperature pattern factor, as well as in the mathematical routines used in computer solutions (e.g., faster solutions of the Navier Stokes equations). An experimental study of the penetration and thermal mixing characteristics that result when secondary (dilution) jets are used in combustors will be conducted. This work will add the empirical relations to explain the effect of dilution jet parameters on the exit temperature profile. Another experimental program will provide comprehensive luminous flame radiation and liner heat flux data for varying gas flow conditions. The effects of pressures up to 40 atmospheres on the luminous flame radiation will be included in the test program, and the results modeled.

The final aspect of the aerothermal modeling activities under HOST will begin in 1984. This "integration" phase will put together all of the submodels and routines that will be developed in the model refinement and testing activities. It will also include an assessment of the improvements made.

The low cycle fatigue life of combustor liners will be studied by running thermal cyclic tests on segments of liners. This data will be used with the life prediction methods described earlier under structural analysis. The effects of hot streaking on the life of combustor liners will also be investigated.

Turbine Heat Transfer

The turbine heat transfer research to be conducted in this area includes research into gas path analysis, gas side heat transfer, coolant side heat transfer, and metal temperature prediction. Advanced turbine engine design requires accurate predictions or knowledge of the local metal temperatures of the various static and rotating parts. For the turbine, as exemplified by the schematic cutaway in figure 11, these analyses must consider the characteristics of the gas flow at the entry, including its temperature, pressure and turbulence levels. The extremely complex flow field around the blades and along the walls must be understood and modeled, before the temperature of static and dynamic airfoils can be calculated. If the gas temperature and heat flux conditions for each row through the turbine are known, the heat transfer coefficients for blades, vanes and endwalls can be calculated. The coefficients can then be used to calculate the operating temperatures of these parts for transient as well as steady-state conditions. Finally, the information can be used to analytically optimize the design (and durability) of the components for various materials and geometries.

The efforts under HOST will be both experimental and analytical. They will establish benchmark quality data, model the complex heat transfer mechanisms, and, finally, provide the methodology for determining temperatures and heat transfer coefficients, which can then be input to structural analyses routines. The first two turbine elements in figure 12 will evaluate the effect on flow transition of variables such as Reynold's number, turbulence, geometry, and temperature ratios for vanes with and without the effects of film cooling. Viscous 3-D analyses to predict heat transfer and gas flow for stator and rotor cascades, including side and endwall effects, will be undertaken as part of HOST. In the next element, the heat transfer and flow characteristics will be determined for various geometries of multiple jet impingement arrays. The influence of rotational (Coriolis) forces and entrance geometry on the prediction of coolant-side heat transfer coefficients will be studied in another element of HOST. Steady-state and transient metal temperature prediction codes will be improved, and interfaced with structural analysis codes, by using the improved flow and heat transfer models above.

Research to measure local heat transfer coefficients over a stator vane, and a rotating blade will be included to verify the development of the above models. Measurements using improved instrumentation will be made to help evaluate the accuracy of codes for predicting gas-side heat transfer coefficients, metal temperatures and static strains in the materials.

Instrumentation Development

Crucial to the experimental effort of HOST are accurate measurements of the temperature, pressure, strain, and heat flux in the hot gas flow stream of the turbine engine. These measurements will be made to provide the benchmark quality data required for verification of the models developed in the other areas. Many of the measurements will require instruments that extend the present state-of-the-art. Fortunately, new techniques and computerized instrumentation (fig. 13) offer promising solutions and exciting extensions of current technology.

The first element in figure 14 will make use of new thin film sputtering techniques to develop a miniature heat flux sensor that is applied directly on blades and vanes. Also to be developed is a method for measuring the radiation portion of the total heat flux to sections of the combustor. Current static strain gages can operate at temperatures of 650 K (or 920 K for a few hours). By using thin film or powder metallurgy techniques, HOST will develop new static strain gages and installation methods for temperatures up to 1250 K. The third element of figure 14 will be the development of a viewing system that is needed for observation of the hot section components during operation at near engine condition temperatures and pressures. For instance, inside the combustor, the edges of the liner could be viewed to see if they are buckling or closing. Also, the interactions of the swirling flow of gases and fuel spray could be carefully studied. The increasing of the clarity of the view of these phenomena within the hot section will be a major part of this effort.

An automated laser anemometer system to measure the three components of average and fluctuating velocities will be developed. The final effort in instrumentation under HOST will produce a probe to measure dynamic gas temperatures up to 1000 Hz. Present temperature probes with fine wire thermocouples having electronic compensation are limited to a frequency response of about 30 Hz. The compensation depends on the gas stream flow properties (Mach number, density, etc.) but these vary during a test, so the compensation must also be dynamic, following these parameters in real time.

CONCLUDING REMARKS

The Turbine Engine Hot Section Technology (HOST) project, discussed above, will utilize current models and conduct new research to develop improved and more accurate analysis methods for the design of advanced turbine engine components. The research in the five areas of structural analysis, surface protection, combustion, turbine heat transfer, and instrumentation will be focused so that problems in hot section component durability can be understood and overcome. Current plans for the research call for eighty percent of the work to be done by engine manufacturers and other competent research institutions. The remaining twenty percent of the HOST effort will be accomplished inhouse by NASA Lewis Research Center technical personnel.

Although the HOST project includes research efforts in a number of separate technical disciplines, its organization is that of a systems technology project. As such, it has identifiable schedules with intermediate milestones and project end dates. The specific project products, as defined above in the text and in figures 6 to 14, are the key part of a systems technology project. While current plans and thinking are presented, it must be recognized that a certain amount of risk exists that some of the project's products may prove to be too far beyond the state of the art or not achievable by the end of the HOST project in fiscal year 1986. Individual research efforts will be monitored and appropriate plan changes made, if required, to ensure that the HOST project attains its objective.

The products of all of the HOST-supported research, excluding the instrumentation development, will be presented in the form of individual models, or in some cases, as computer modules, that can be acquired separately and utilized by engine manufacturers in analyzing designs of advanced turbine engine components. No attempt will be made during the HOST project to integrate the individual models into one overall model. All improved models, benchmark data bases, and any programmed computer modules will be disseminated to the U.S. domestic aerospace industry through formal reports and at suitable workshops and meetings. Thus, the U.S. engine manufacturers will be able to improve the durability of hot section components in their advanced turbine engine designs. This enhanced ability will enable the U.S. aerospace industry to maintain its favorable position in an increasingly competitive world aerospace market.

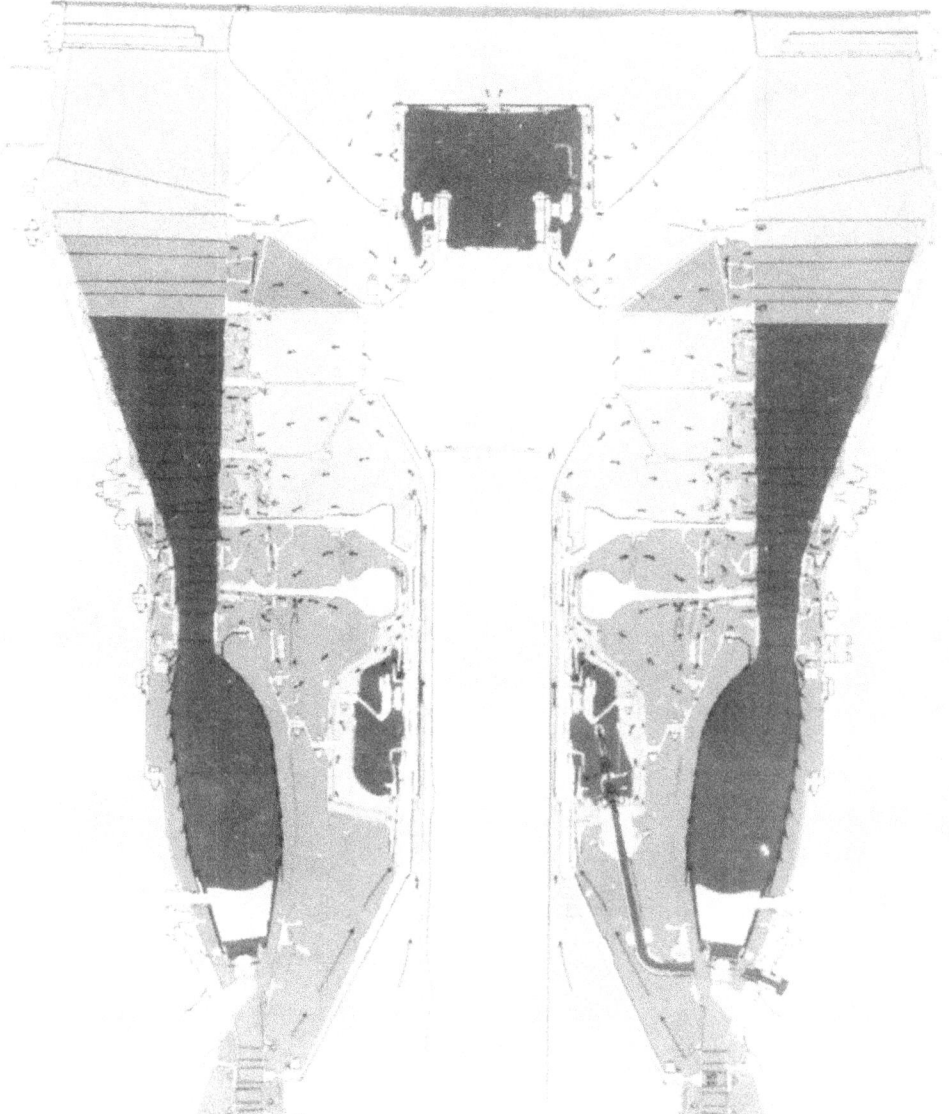

Figure 1. - Turbine engine hot section.

Figure 2. - Hot section components.

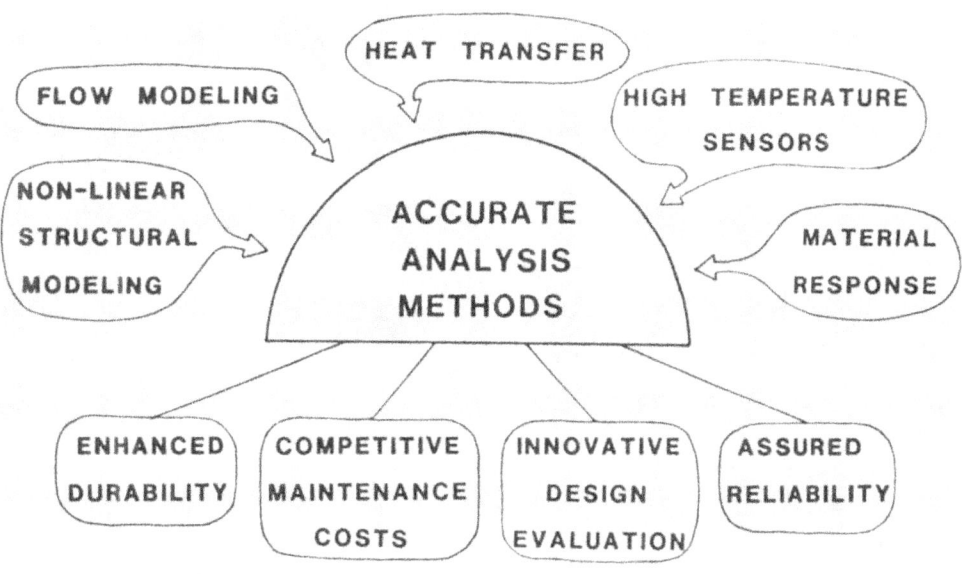

Figure 3. – Coordinated research activities.

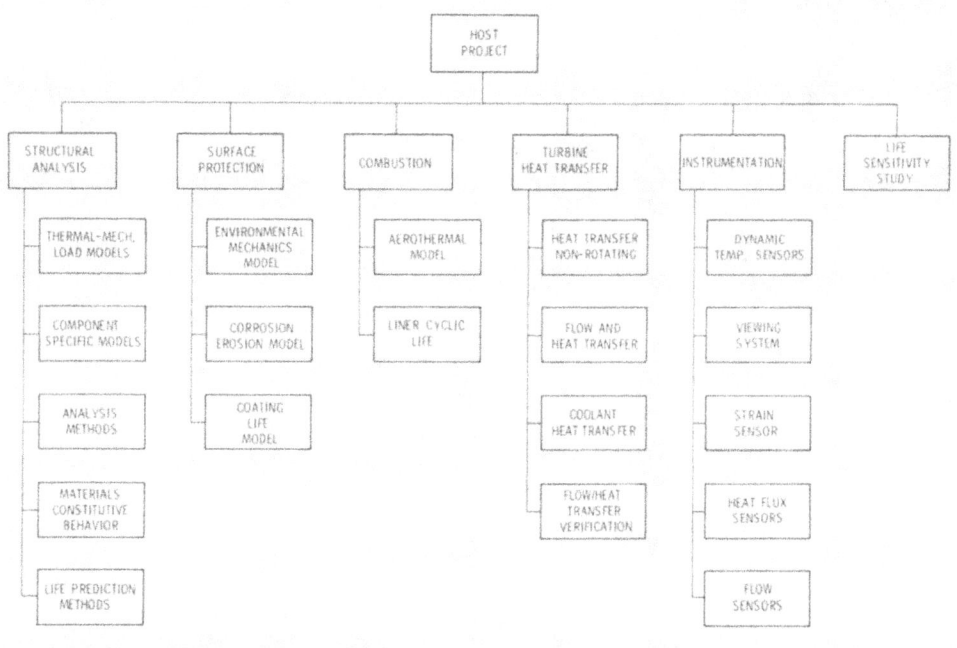

Figure 4. – Work breakdown structure.

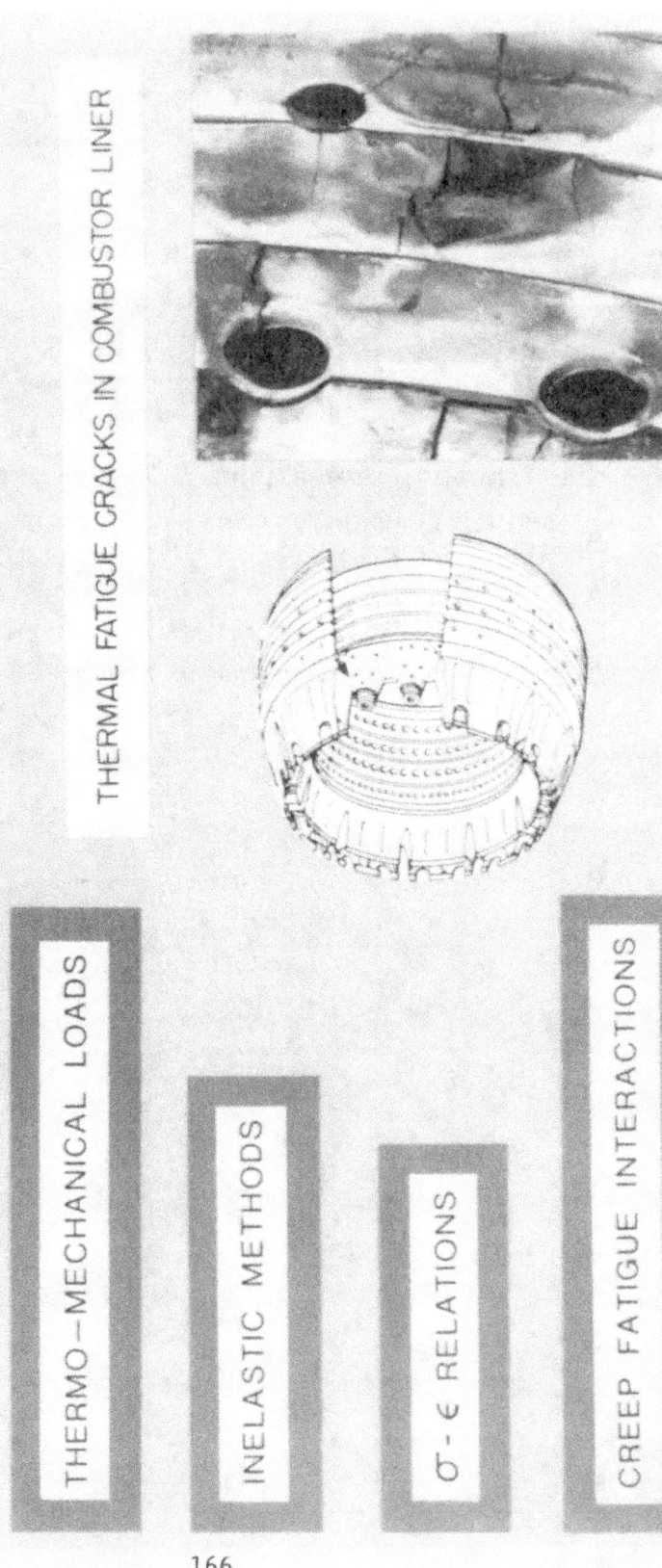

Figure 5. - Structural analysis research.

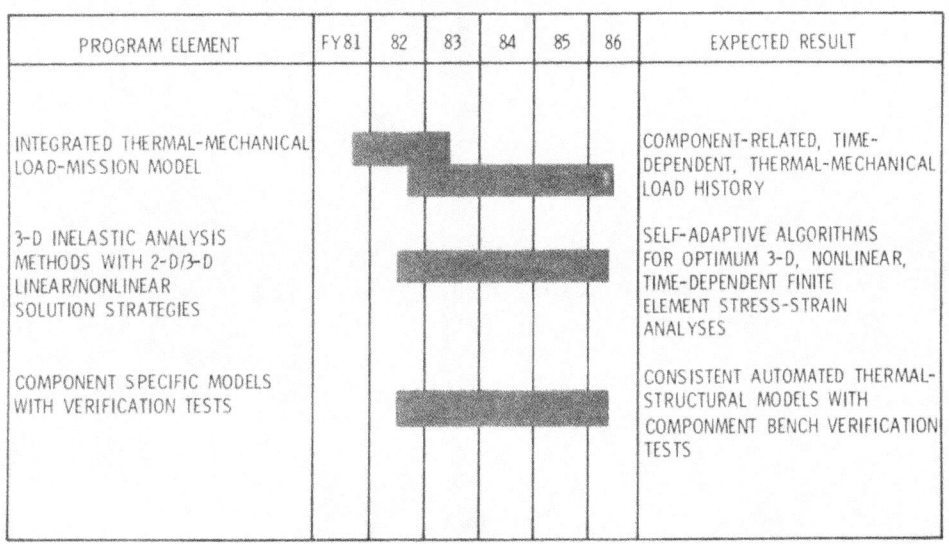

(a) Analysis methods.

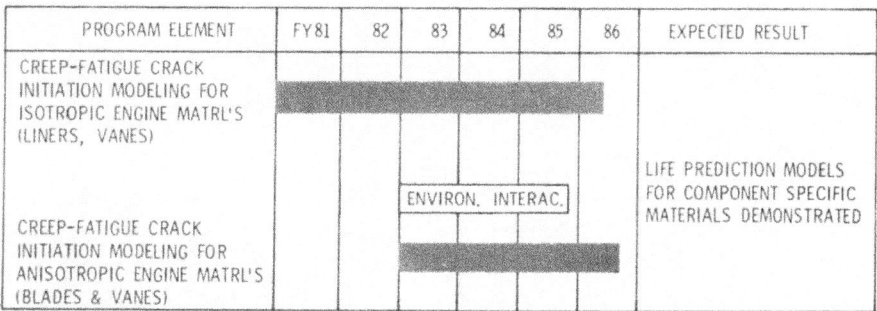

(b) Materials constitutive relations and life prediction methods.

Figure 6. – Structural analysis.

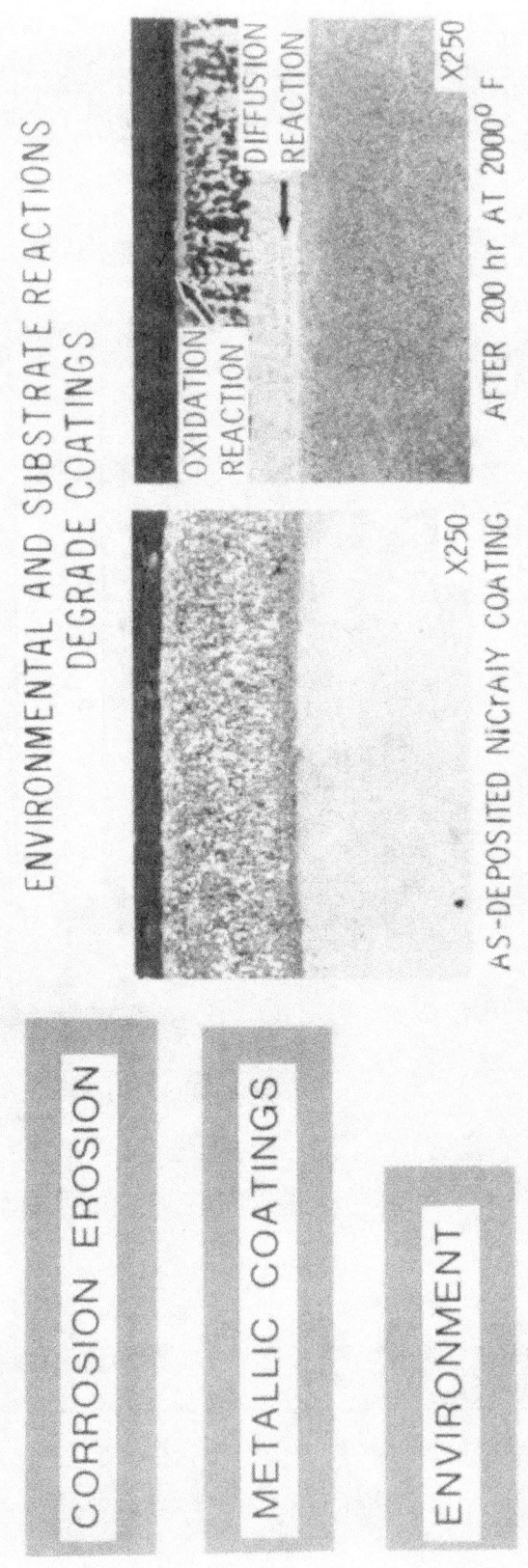

Figure 7. - Surface protection research.

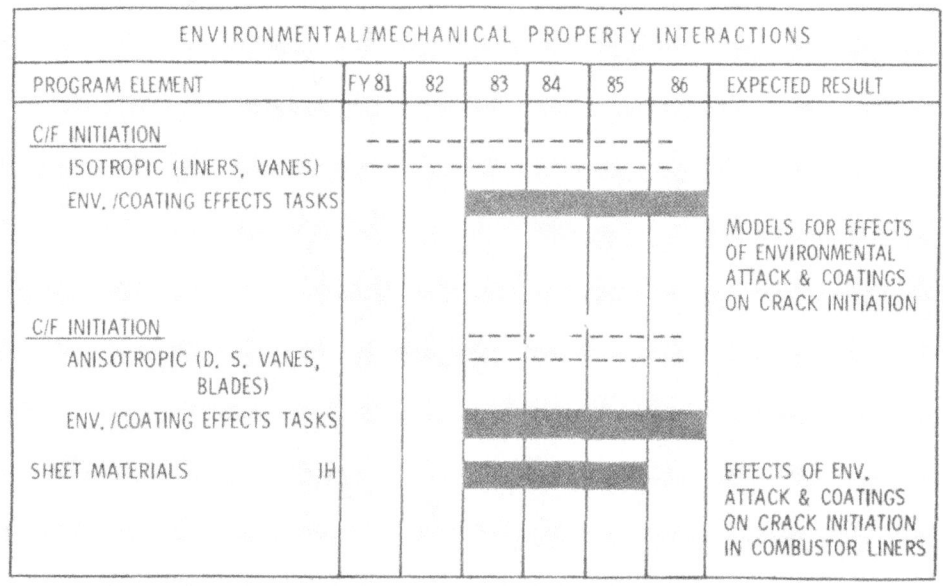

(a) Environmental/mechanical property interactions.

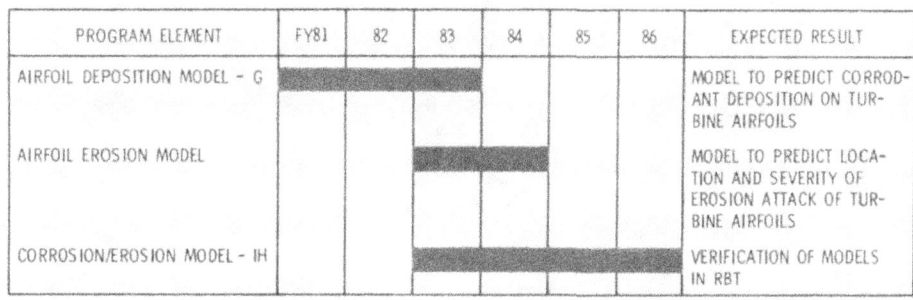

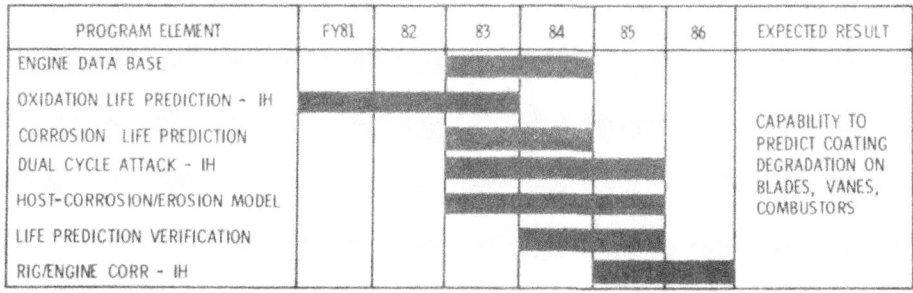

(b) Corrosion/erosion and coating life models.

Figure 8. – Surface protection.

COMBUSTION

AEROTHERMAL MODELING

LINER PANEL SEGMENT

THERMAL CYCLIC DATA

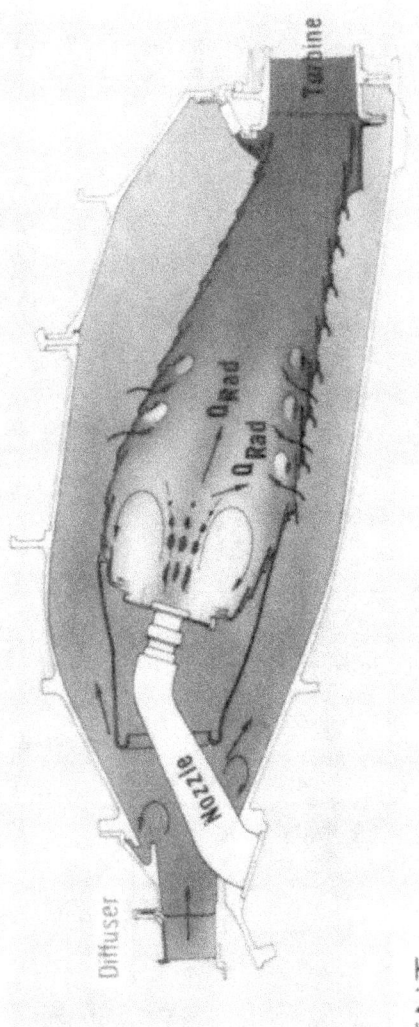

Figure 9. - Combustion research.

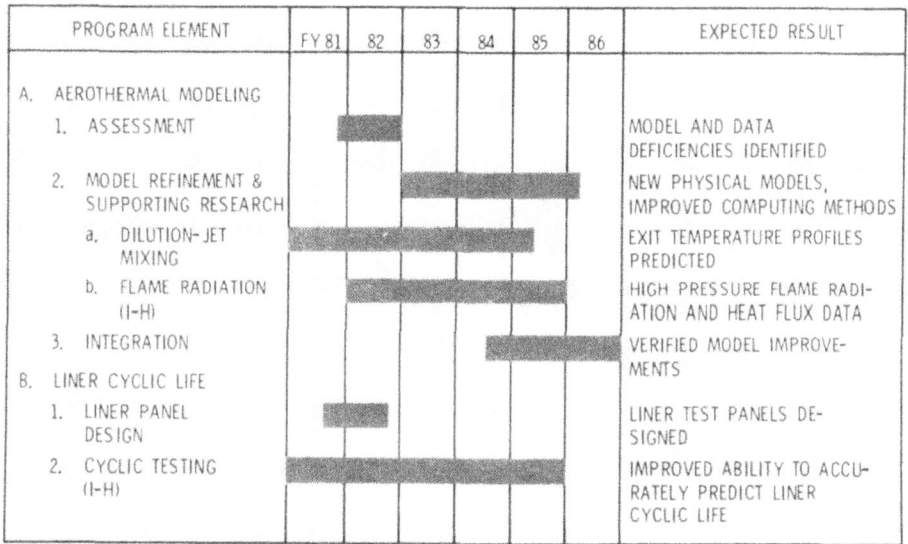

Figure 10. - Combustion analysis.

- ENTRY GAS FLOW CONDITIONS
    - TEMPERATURE, PRESSURE, & TURBULENCE CHARACTERISTICS
- GAS CONDITIONS FOR EACH ROW
- HEAT TRANSFER COEFFICIENTS
    - BLADES, VANES, END WALLS
- TEMPERATURE OF COMPONENTS
    - STEADY STATE & TRANSIENT
- OPTIMIZATION TO IMPROVE DURABILITY
    - METAL TEMPERATURES & GEOMETRY

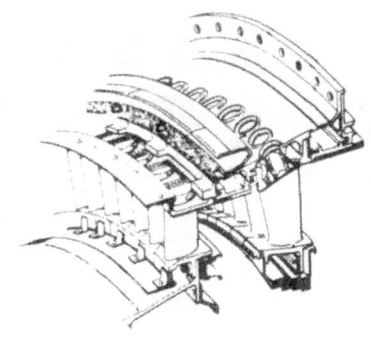

Figure 11. - Turbine heat transfer research.

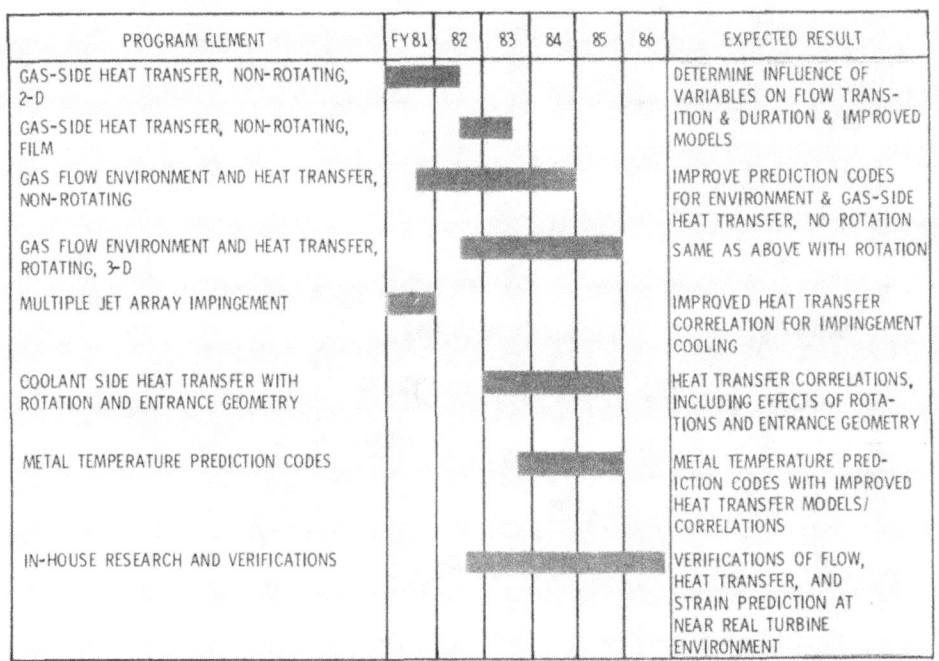

Figure 12. - Turbine heat transfer.

- SENSORS: HEAT FLUX, STRAIN & TEMPERATURE
- LASER ANEMOMETER

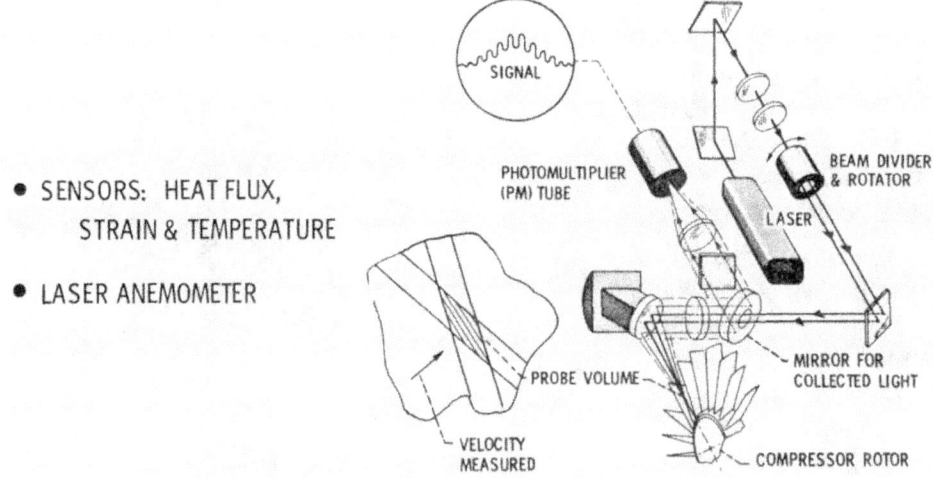

Figure 13. - Instrumentation development.

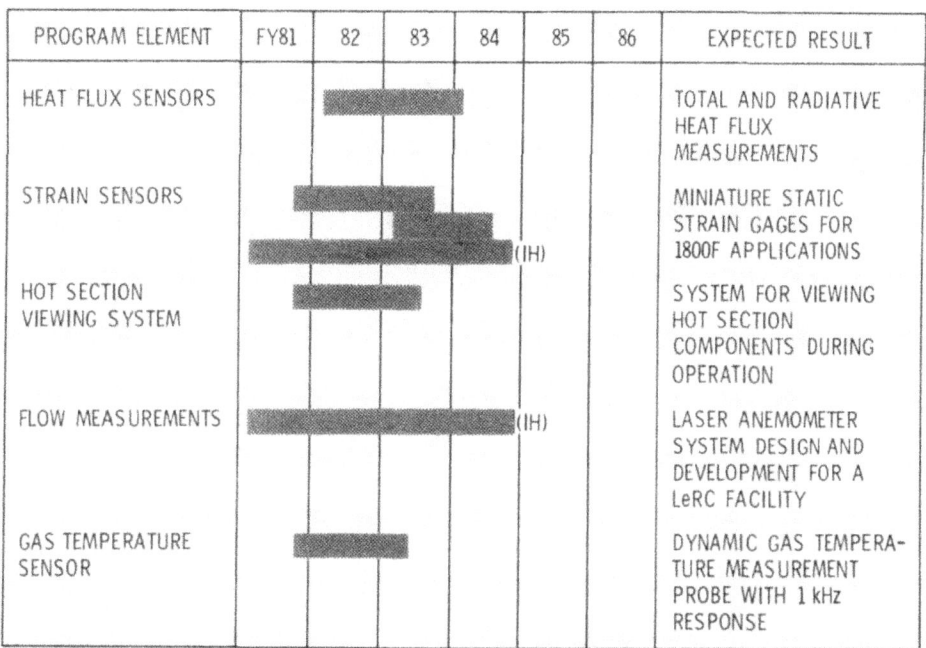

Figure 14. - Instrumentation.

THE NATURE OF OPERATING FLIGHT LOADS AND
THEIR EFFECT ON PROPULSION SYSTEM STRUCTURES

Kenneth H. Dickenson
Richard L. Martin
Boeing Commercial Airplane Company

ABSTRACT

Past diagnostics studies revealed the primary causes of performance deterioration of high by-pass turbofan engines to be flight loads, erosion and thermal distortion. This paper examines the various types of airplane loads that are imposed on the engine throughout the lifetime of an airplane. These include flight loads from gusts and maneuvers and ground loads from take-off, landing and taxi conditions. Clarification is made in definitions of the airframer's limit and ultimate design loads and the engine manufacturer's operating design loads. Finally, the influence of these loads on the propulsion system structures is discussed.

INTRODUCTION

The traditional transport airplane structures analyst's treatment of an engine is very simple: The engine is a "concrete block" whose properties are entirely inertial, and the main concerns are that it should not fall off the strut and that it is located properly from a wing flutter viewpoint (figure 1). In the era of the turbojet and the low bypass ratio turbofan, such treatment was acceptable because it was nearly correct. Engines and their inlets were so compact and rigid (figure 2) that the internal structural problems could be left to the engine manufacturers who needed only to be advised of the accelerations to be applied at the mount locations.

In the late 1960's the advent of the high bypass ratio turbofan engine with it's large fan case relative to its core brought change. Early in the Boeing 747 program, for example, it was found that thrust forces caused "ovalization" of the engine case because the engine's combination of large diameter (figure 3) and high thrust imposed a substantial couple at the engine mounts. This problem was alleviated by adding a "thrust yoke" that transferred thrust directly to the strut and reduced distortion of the case due to thrust.

The high bypass ratio turbofan engines have large inlet airflows relative to the engine core size. Thus, a large momentum change is required to align the airflow with the engine at high angles of attack. Since inlets are usually bolted to the front flange of the fan case, the inlet aerodynamic loads associated with this momentum change induce bending and distortion into the smaller diameter engine core case. These case distortions may cause rubbing between the rotors and the static case structure while the desire for higher overall pressure ratio requires better control of tip clearances. The aerodynamically induced operating loads that act on the inlet are modest in comparison to overall airframe design loads. It remained for the 1973 oil embargo and the ensuing dramatic rise in fuel prices to motivate a deeper investigation of airframe and engine structural interaction.

Attention was focused on the causes of engine fuel consumption deterioration in the NASA sponsored Pratt & Whitney Aircraft JT9D Diagnostics Program. In this effort, several different probable deterioration mechanisms were identified and evaluated analytically. Prominent among them was rubbing between the rotor tips (of the fan, compressors, and turbines) and the engine case caused by flexing of the engine under operating loads. The result of rubbing was increased clearance between the rotor and case since material was worn from the "rub strips" and the blade tips. Increased clearance reduced component efficiency and increased specific fuel consumption.

AIRPLANE LOADS

Design Loads

Due to the overriding importance of safety, airplane design loads have been studied intensively for many years and are the subject of a large body of doctrine and practice developed by airframe manufacturers and enforced by the Federal Aviation Administration (FAA). A "limit design load" is determined for airframe structure as the maximum load that the structure can be expected to encounter during the entire life of the airplane fleet. At the limit design load, the structure is not permitted to suffer permanent deformation; i.e., the maximum stress may not exceed the elastic limit. To provide an added degree of safety, an "ultimate design load" is also specified, usually as 1.5 times the limit load value. Up to the ultimate load, the structure is permitted to suffer "permanent set", but it must not fail.

The design loads are determined by analyzing the airplane in a variety of load conditions that are contained within the envelope of the "V - n diagram" (a plot of the accelerations that the airplane must withstand versus airspeed). Three main types of load conditions are considered. The first is maneuver. Transport category airplane limit loads are determined in 2.5 g turns or pull-ups with flaps retracted and in 2.0 g turns with flaps down. The second type of load condition is due to atmospheric turbulence. It is assumed that gusts of a defined shape and velocity will be encountered by the airplane at speeds specified in relation to the design operating speed limits chosen by the manufacturer. The airplane's response is determined by analyzing the aerodynamic, elastic, and inertial characteristics in detail. The third loading category is associated with the ground. These loading conditions include take-off, landing and taxi. Analytically determined design loads are corroborated by extensive flight load surveys using accelerometers, strain gages and pressure transducers.

The oustanding structural safety records of today's commercial air fleets demonstrate that the design loads issues are very well understood. However, the loads that cause day-to-day TSFC deterioration are less severe than design loads and are not so well understood. They may be termed "operating" loads, and a different approach is needed to understand them.

Operating Loads

The parameters for determining operating loads are the same as those for design loads; i.e., airplane aerodynamic, elastic, and inertial properties on the one hand, and flight conditions (maneuvers, turbulence, etc.) on the other. To a considerable degree, the problem resembles that of analyzing structural fatigue. Service life, maintenance, and economy are the main considerations, with statistical descriptions of the operating environment being the scenario, as opposed to a set of extreme conditions.

Fatigue damage is assessed from the cumulative occurences of different stress levels. The part of TSFC degradation due to clearance change, on the other hand, depends on the probable time period (or number of flights) until any given load level is exceeded once. Figure 4 shows the analysis sequence. The starting point is the set of mission profiles that typify the airplane's utilization (upper left corner). Mission length is important because it determines how many "ground-air-ground" (GAG) cycles are flown per hour of operation. The altitude and speed profiles determine the frequency and severity of gusts.

Load exceedance probability per flight can be inferred from airplane characteristics and the mission profile. The sketch at the upper right of figure 4 refers to inertia load exceedances, but a similar plot can relate to airloads. When the probable loads are known, the probable tip clearance changes can be inferred from the elastic properties of the engine itself. These may be obtained by analyses varying from simple beam representations to finite-element models containing thousands of elements. Recent experience supports the need for the more complex finite-element approach. When tip clearances become negative, rubs are indicated, and TSFC deterioration can be expected.

In addition to revenue service missions, other flight profiles must be considered, such as crew training. A significant mission that occurs only once on each airplane is the "acceptance flight" (figure 5). All transport airplanes are checked for satisfactory flight characteristics and functioning of warning systems before delivery to the customer airline. In such flights, the airplane is not tested to the limit loads of the flight design envelope but to more normal operating conditions, such as maximum airspeed (dynamic pressure), maximum Mach number, and minimum airspeed (stall warning) where warning devices such as stick shakers automatically alert the pilot to the situation. Since such a flight always occurs first in an airplane's history, it establishes a starting set of rubs and clearances for subsequent exceedance studies.

Statistical descriptions of the inertia load environment have been obtained by accumulation of a great many speed/acceleration/altitude ("VGH") recordings made in actual airline service (figure 6). Airspeed (V), normal acceleration (g), and altitude (H) are recorded continuously. The recordings are later analyzed by counting acceleration peaks. The number of "occurrences" of a particular acceleration level is defined as the number of peaks found over some time period that fall between the upper and lower bounds of that level. The number of "exceedances" of that level is the sum of the occurrences of that level and all higher levels as shown in figure 7.

Histograms (figure 8) can then be constructed which show occurrences and exceedances per flight hour or per flight versus load level, and plots such as shown in figure 9, depicting probable nacelle inertia load exceedances, can be drawn. This figure, incidentally, shows a characteristic feature of the environment of wing-mounted engines. The motions and accelerations of the nacelles are larger than those at the airplane center of gravity because of the wing aeroelastic response to dynamic loads such as gusts and landing impacts. In additon to accelerations, gyroscopic loads caused by airplane angular motions must also be considered.

Under normal conditions, the most severe engine aerodynamic loading occurs at takeoff when the maximum engine thrust produces a high mass flow rate through the inlet combined with a high angle of attack. These effects are illustrated in figure 10 which shows low pressure caused by suction on the lower inside lip of the inlet. This is associated with high local velocity as the flow turns sharply. Supersonic flow usually occurs in this region, followed by shocks and sometimes by local flow separation.

Of lesser importance, but significant because it is a condition that creates a load reversal, is the maximum dynamic pressure condition shown in figure 11. This condition involves a negative local inlet angle of attack and an inlet pitching moment acting downward. This moment is, however, of much smaller magnitude than the nose-up moment at takeoff.

Neither the maximum dynamic pressure condition nor the one shown in figure 12 -- stall warning at 10° flaps -- are normal revenue service flight conditions. Both of these, however, are currently flown in the flight acceptance test of every new airplane.

One of the more uncertain assumptions regarding inlet pressures has been the circumferential distribution. A simple, one wave cosine distribution, illustrated in figure 13, rotated to account for non-symmetric effects has been customary. This assumption awaits validation by the results of the Nacelle Aerodynamic and Inertia Loads (NAIL) flight test program.

To illustrate the joint efforts in the JT9D Diagnostics Program and the interdependence of the engine and airframe manufacturer in the propulsion interface, Figure 14 shows the mathematical model used to analyze the 747 propulsion system. Government and industry foresight several years ago provided the NASTRAN finite element program giving wide availability to this technology. Air breathing propulsion structures are a relatively late application of this technology, and there is currently a large effort being made toward test and analysis correlations to enhance this application.

Currently, Boeing and the three major jet engine manufacturers are utilizing the type of models illustrated here to calculate deflections, clearance changes, internal loads, and vibration behavior on the 767, 757 and 737-300 new airplane programs and on future powerplant installations for the 747 airplane program.

The airframe and engine manufacturer must each conduct their own analyses for their specific needs. An exchange of data files provides each with this capability and the integrated model as illustrated in figure 15. Recent trends in nacelle design have resulted in much closer structural coupling between the engine, nacelle, and strut. For example, in the 767 design the front mounting system has been placed to minimize thrust bending moment, and the thrust reverser and fan exhaust cowling are hinged from the strut and clamp onto the engine through circumferential V-grooves. This not only simplifies engine removal and maintenance but also serves as a dual load path with the mounts which provides the redundancy required for fail safe design in case of mount failure. An extra benefit from this dual load path is a reduction of engine loading which enhances engine performance through reduced clearance changes under flight operating loads. The close coupling inherent in this type of design necessitates use of detailed nacelle-engine-strut finite element models to define interface loads accurately.

A characteristic of engine structure relative to conventional airframe structure is its inherently greater stiffness and complexity. This obviously must be the case in order to maintain the dimensional constraints so important to engine performance. The maximum engine bending deformations are typically one to two orders of magnitude less than maximum strut deflection as exhibited in figure 16 for a "g" loading condition. The attainment of accuracy in the engine and nacelle math model comparable to conventional airframe structures , therefore, requires a great deal of experience and effort and should rely heavily on accurately measured data when available.

To illustrate this point, figure 17 shows typical calculated clearance change contour lines for a normal takeoff condition. This plot is for the inboard side of the number three engine on the 747 airplane. Clearance closure is denoted by the shaded regions. The information shown here is used by the engine manufacturer in a separate post processor program that calculates blade rubs, stage-by-stage clearance increases, and TSFC deterioration.

The ultimate goal in the diagnostics effort is to provide adequate data for taking actions toward eliminating performance deterioration. Much of the required data has been generated in the early tasks. The flight loads portion of the JT9D Diagnostics Program in which engine clearance changes are measured in flight and the concurrent NAIL flight loads program will complete the data. The task ahead is the application of this data and the appropriate use of design tools in concerted efforts between the engine and airframe manufacturers to reduce performance deterioration. A considerable effort has evolved in the area of integrated engine-nacelle design studies aimed at stiffening current power plant installations. More important is the use of the diagnostics data in systems currently under design and development that recognize and build deterioration prevention into the initial designs.

DESIGN LOAD PHILOSOPHY

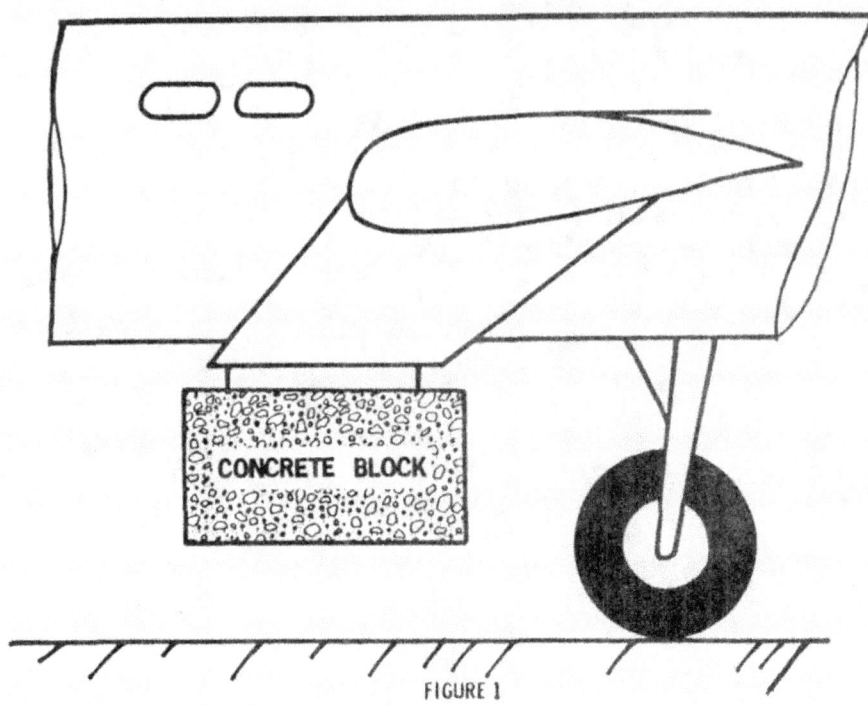

FIGURE 1

EARLY TURBOJET INSTALLATION

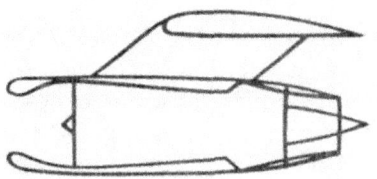

FIGURE 2

MODERN ENGINE INSTALLATION

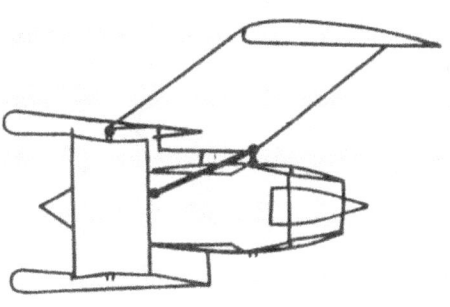

FIGURE 3

RELATION OF AIRPLANE FLIGHT LOAD EXCEEDANCES TO BLADE TIP CLEARANCE

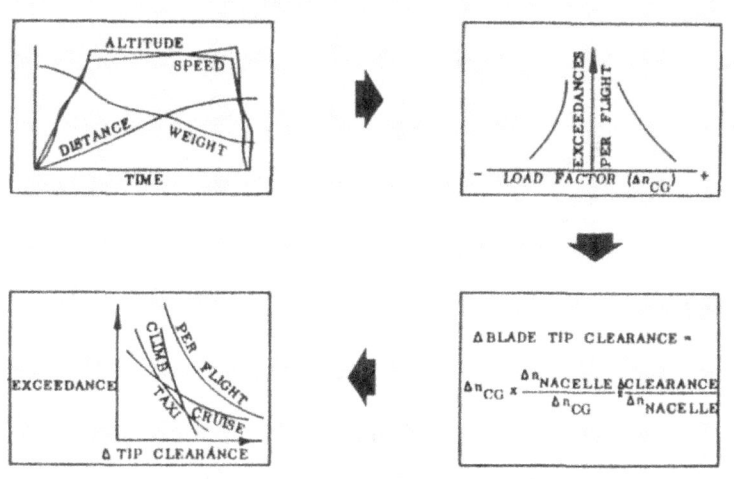

FIGURE 4

ACCEPTANCE FLIGHT PROFILE

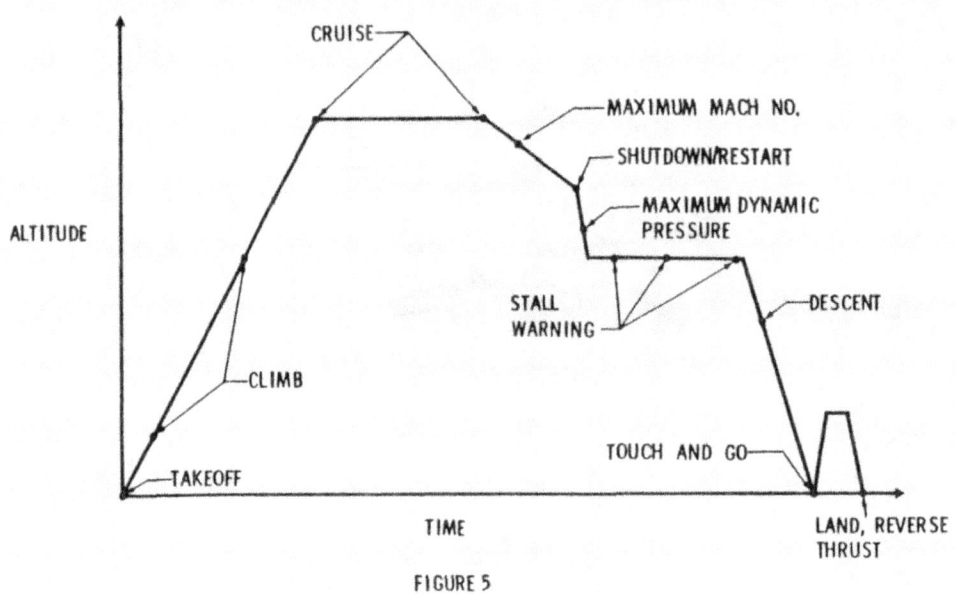

FIGURE 5

ILLUSTRATIVE VGH RECORDING

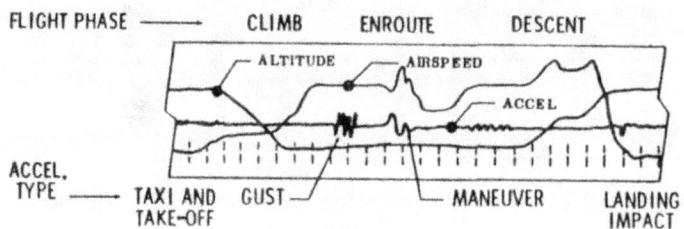

FIGURE 6

ANALYSIS OF LOAD FACTOR TIME HISTORY FROM VGH RECORDER

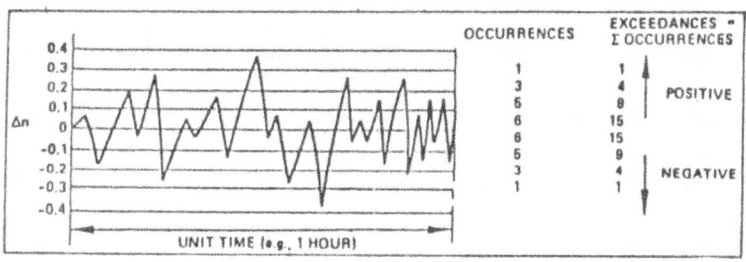

FIGURE 7

HISTOGRAM AND PLOT OF OCCURRANCES

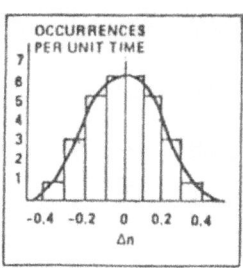

FIGURE 8

NACELLE LOAD EXCEEDANCE DATA

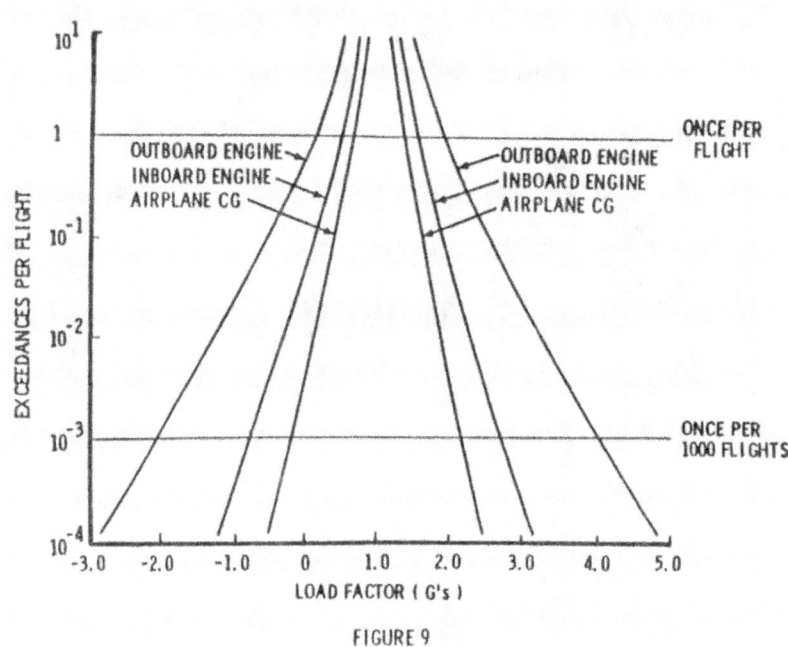

FIGURE 9

AIRLOADS
TAKEOFF ROTATION

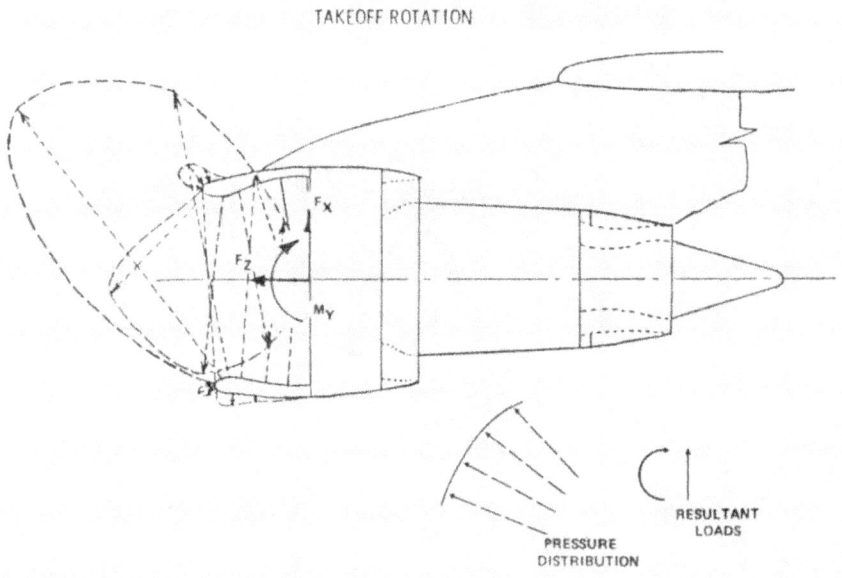

FIGURE 10

AIRLOADS
MAXIMUM DYNAMIC PRESSURE

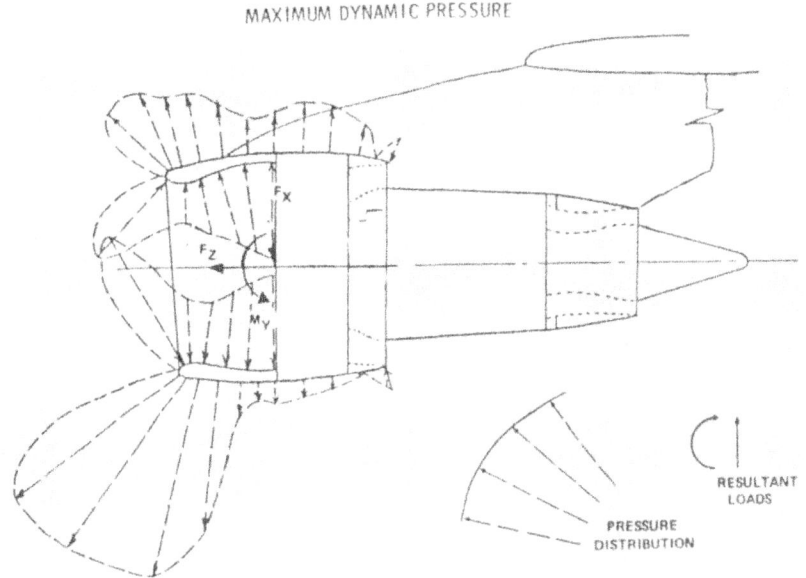

FIGURE 11

AIRLOADS
STALL WARNING, $10°$ FLAPS

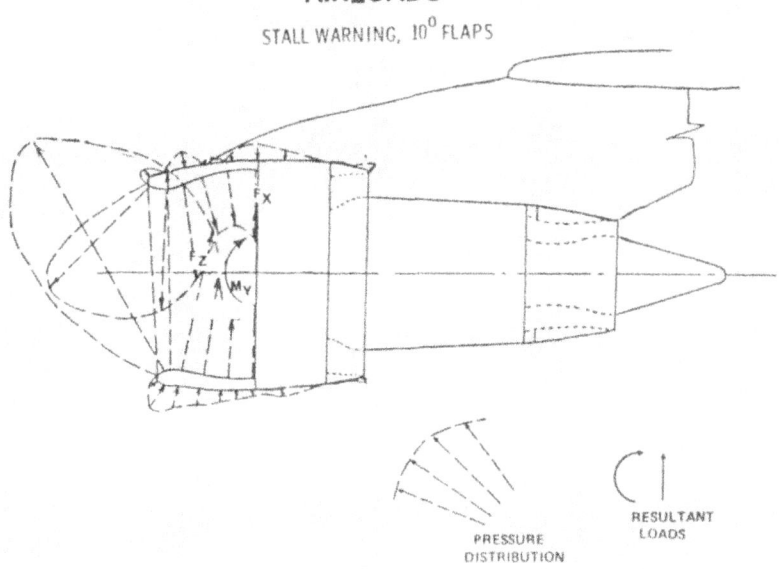

FIGURE 12

INLET CIRCUMFERENTIAL PRESSURE DISTRIBUTION

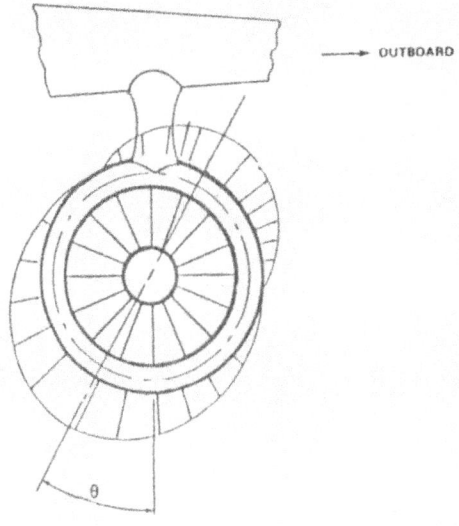

FIGURE 13

PROPULSION SYSTEM SUBSTRUCTURES AND RESPONSIBILITIES

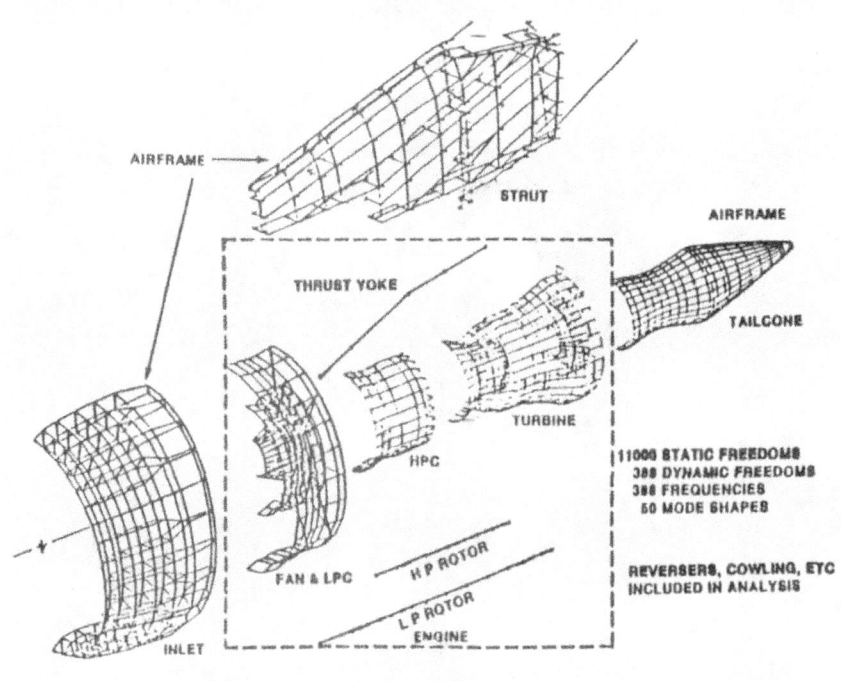

FIGURE 14

FINITE ELEMENT MODEL

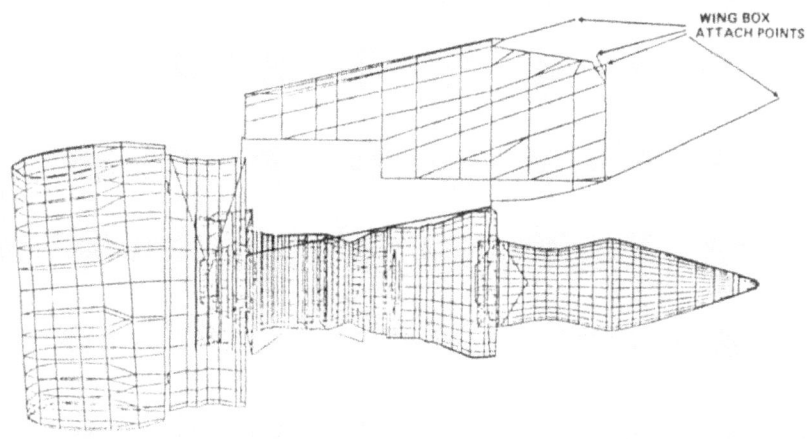

FIGURE 15

STRUCTURAL DEFLECTION

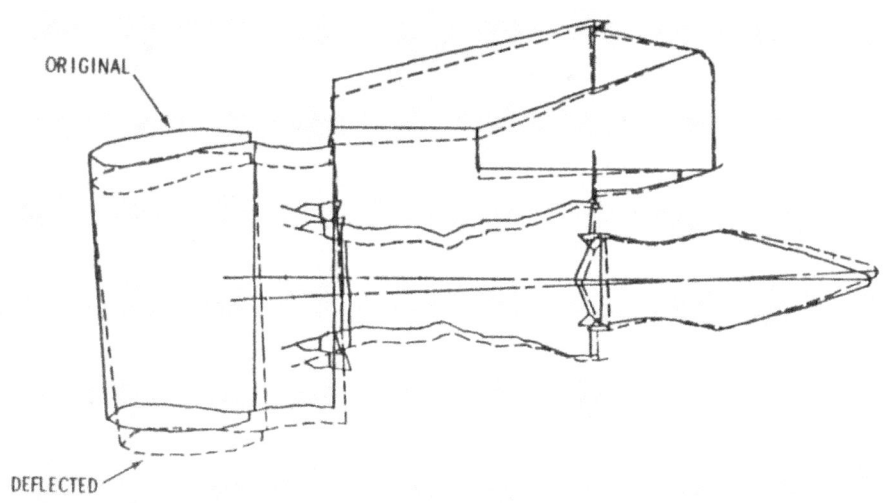

FIGURE 16

CALCULATED CLEARANCE CHANGES (INCHES)

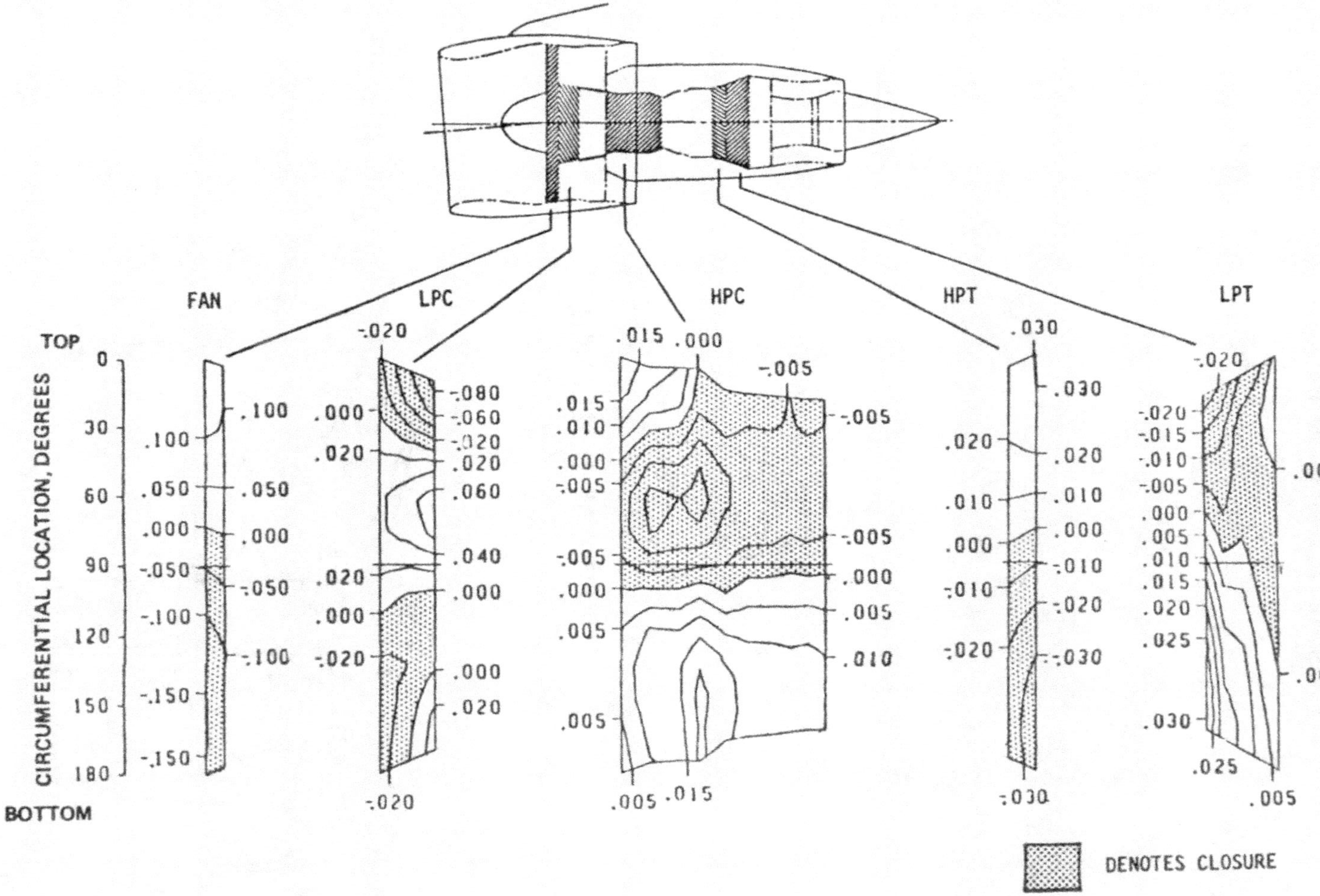

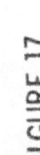

FIGURE 17

CONSERVATION OF STRATEGIC AEROSPACE MATERIALS (COSAM)

Joseph R. Stephens
NASA Lewis Research Center

SUMMARY

NASA has undertaken several projects directed at conserving strategic materials used in the aerospace industry. Research efforts involving universities and industry as well as in-house activities at the NASA Lewis Research Center comprise the current "Conservation of Strategic Aerospace Materials" COSAM effort. The primary objective of COSAM is to help reduce the dependence of the United States aerospace industry on strategic metals, such as cobalt (Co), columbium (Cb), tantalum (Ta), and chromium (Cr), by providing the materials technology needed to minimize the strategic metal content of critical aerospace components for gas turbine engines. Thrusts in three technology areas are appropriate for COSAM. These include near-term activities in the area of strategic element substitution; intermediate-range activities in the area of materials processing; and long-term, high-risk activities in the area of "new classes" of high temperature metallic materials. This paper describes in some detail the efforts currently underway and the initial results generated to date. Initial emphasis has been placed in the area of strategic element substitution. Specifically, the role of cobalt in nickel-base and cobalt-base superalloys vital to the aerospace industry is being examined in great detail by means of cooperative university-industry-government research efforts. Investigations are also underway in the area of "new classes" of alloys. Specifically, a study has been undertaken to investigate the mechanical and physical properties of intermetallics that will contain a minimum of the strategic metals. Current plans for COSAM are presented in this paper also.

INTRODUCTION

The United States relies heavily upon foreign sources for the supply of most strategic metals required by our aerospace industry. With the exception of molybdenum, iron, magnesium, and the rare earths, the United States imports from 50 to 100 percent of such aerospace metals as Co, Cb, Ta, Cr, And Mn (ref 1). However, the potential for foreign cartels, political unrest, and production limitation is great and is intensified by steadily declining known reserves. Thus, the United States can expect to be faced with supply shortages and price escalation for many strategic metals. Since these metals are vital to the welfare of the nation's economy, their continued availability at a reasonable cost is a national issue which requires cooperative action between the aerospace industry and appropriate government agencies.

The aerospace industry is currently a major factor in the positive inflow of funds from U.S. exports (ref. 2). This industry, and within it the aircraft engine industry in particular, relies heavily upon imports for several key strategic metals including cobalt, columbium, tantalum, and chromium. In order to offset or minimize future disruptions in supply, efforts to develop viable options must begin now, since a new material can take from 5 to 10 years of research and development efforts before qualifying for aerospace service.

NASA currently has plans to address the aerospace industry's needs to minimize the use of strategic metals for advanced aerospace systems. COSAM has as its broad objective the reduction of the dependence of the U.S. aerospace industry on strategic metals. This objective can be accomplished by providing the materials technology options needed to allow individual companies to trade-off the material properties of critical components versus cost and availability of their strategic metal content. This paper summarizes NASA's current activities in this area and broadly outlines the plans for COSAM.

STRATEGIC MATERIALS

A definition of strategic materials as used in this paper is given in figure 1. Strategic materials are those predominantly or wholly imported elements contained in the metallic alloys used in aerospace components which are essential to the strategic economical health of the U.S. aerospace industry. As the basis for what are considered strategic metals, we will focus on the aircraft engine industry's needs. Based on a survey of the ASME Gas Turbine Panel and a subsequent survey of a number of aerospace companies, the elements listed in figure 1 were considered to be the most strategic with respect to the aerospace industry. As a result of prioritizing by NASA's COSAM planning team supplemented by further discussions with several aircraft engine manufacturers, four elements emerged that were of particular concern. The alloys used to build the critical high temperature components for aircraft propulsion systems require the use of the four metals - cobalt, columbium, tantalum, and chromium. These metals are contained in steels, stainless steels, and superalloys that are used in engine manufacturing. Figure 2 lists these four elements in the high priority category with a brief rationale for this ranking. The remaining five strategic elements evolving from our surveys were given a lower priority and figure 2 also contains a short explanation for this ranking.

The location of these metals in aircraft engine compressors, turbines, and combustors is illustrated in figure 3. The need for such metals has increased as the demands have grown for higher durability plus high performance, fuel efficient aircraft turbine engines. Based on the essential nature of these metals and for the U.S. aircraft industry to maintain its competitive position, it is necessary that supplies be readily available at a reasonably stable cost. To achieve these requirements, domestic sources of key metals are desirable.

Today, we are almost totally dependent on foreign sources for these metals as shown in figure 4. In several of the countries listed in figure 4, recent political disturbances have led to supply interruptions. Therefore, the U.S. aircraft engine industry can be seen to be highly vulnerable to supply instabilities of the essential metals for engine manufacturing. Accompanying supply disruptions or increased demand is an accelerated price increase. Escalated prices during the recent few years are evident for tantalum, columbium, cobalt, and to a lesser degree for chromium, as shown in figure 5. These rapid price increases illustrate the additional vulnerability of the U.S. aircraft engine industry to cost fluctuations. The essential nature of cobalt, columbium, tantalum, and chromium along with their vulnerability to supply instabilities and cost fluctuations combine to cause these metals to be classified as strategic aerospace metals.

The portion of these four metals used in superalloys for the aerospace industry compared to all other U.S. uses is shown in figures 6 to 9. The he use of these metals in superalloys as compared to total U.S. consumption in 1979 was: cobalt - 30 percent, columbium - 28 percent, tantalum - 5 percent, and chromium - 3.4 percent. These data reveal that superalloys comprise the largest single use of both cobalt and columbium.

OVERVIEW OF COSAM

COSAM has as its primary objective the reduction of the dependence of the U.S. aerospace industry on strategic metals. COSAM can also provide the industry with some options for making their own property versus availability/cost trade-offs when selecting aerospace alloys. These objectives will be achieved by providing the technology needed to minimize the strategic metal content of critical components in aerospace structures. Initial emphasis will be placed on the aircraft engine industry. COSAM initially is focused on conservation of the strategic metals cobalt, columbium, tantalum, and chromium. Strategic metals such as titanium, the precious metals, tungsten, and others may be brought into COSAM as it progresses.

Along with prioritizing the strategic elements that were identified, the role that the NASA's COSAM effort should encompass was also evaluated. Options that were considered are listed in figure 10. All of these options could contribute to the conservation of strategic materials and minimization of U.S. aerospace industry vulnerability. However, within the scope of our program a decision was reached based on Lewis' traditional roles and expertise to focus on the three areas noted in figure 10. These areas consist of strategic element substitution, process technology, and alternate materials. Contributions to the other areas may benefit from COSAM through cooperative programs with other governmental agencies such as in the area of scrap reclamation or through cooperation with technical societies in establishing a critical material index. Having selected a list of four high priority strategic elements and having defined the areas of emphasis for COSAM and specific objectives, a technology approach was adopted as shown in figure 11. Conservation, as well as reduced dependence on strategic metals, will be achieved in the area of strategic element substitution by systematically examining the effects of replacing cobalt, columbium, and tantalum with less strategic elements in current, high use engine alloys. This will help guide future material specifications if one or more of these metals becomes in short supply. Conservation through process technology will be achieved by advancements in those net-shape and tailored-structure processes that minimize strategic material input requirements. This will lower total usage. And in the longer term, development of alternate materials that replace most strategic metals with those highly available in the U.S. could lead to a substantial reduction in the U.S. dependence on foreign sources. Both of the later two technology areas will help conserve the four strategic metals Co, Cb, Ta, and Cr.

EARLY COSAM ACTIVITIES

COSAM efforts began in FY'80. Efforts on planning and organizing are still underway. In addition to the planning activities, several small research activities have been initiated. These research activities focus on two of the three major thrusts of COSAM - strategic element substitution and development of alternate materials. Special emphasis of these initial efforts is on developing a fundamental understanding of the role of strategic elements in current aircraft engine alloys so that effective alloying element substitution can be conducted. Similarly, in the development of alternate materials, a basic understanding of materials properties and alloying concepts is being emphasized. Consequently, university grants play a major part in COSAM. In addition, cooperative programs with industry augmented by in-house research at the NASA Lewis Research Center comprise the approach used in these initial projects. This cooperative approach will continue to be followed in COSAM and industry, universities, and government in-house research will each play a key role. The subsequent paragraphs will describe in some detail early COSAM research efforts.

Strategic Element Substitution

Four metals were mentioned previously as being classified as high priority strategic metals. Cobalt was selected from these four metals for the early COSAM strategic element substitution research. The basis for selecting cobalt was twofold. First, the largest single use of cobalt in the U.S. is in superalloys for jet engine applications as was shown in figure 9 (ref. 3). Many of the other applications indicated in figure 9 are also important to the nation's economy and security as well. Secondly, the specific roles that cobalt plays in nickel-base superalloy fabrication and performance has not been clearly established. Most superalloys currently in use were developed at a time when cobalt was plentiful and inexpensive. Literature results (Ref. 4) are conflicting as to the role that cobalt plays in nickel-base superalloys in important areas such as phase stability, γ' partitioning, strength, fabricability, and oxidation and hot corrosion resistance. Because of these uncertainties, there exists a strong possibility that the strategic element cobalt can be substantially reduced or possibly eliminated from several superalloys without sacrifice of the key properties for which these alloys were selected for engine service.

Four nickel-base and one cobalt-base superalloys were selected for this investigation. The five alloys are listed in figure 12 along with their typical applications in the aircraft engine industry, the forms in which the alloys are used, and remarks as to why they were selected for this activity. Applications include turbine disks, turbine blades, and combustors. A variety of product forms are represented by the applications of the five alloys as noted in figure 12. The selection of the five alloys was based primarily upon the considerations given in this figure. Waspaloy* was selected because it represents the highest tonnage of cobalt now in commercial aircraft engines. Selection of Udimet-700* was based on the fact that this alloy is used in the as-cast, as-wrought ingot, as-wrought powder, and as-HIP powder metallurgy fabricated conditions. The potential for determining the impact of cobalt on both conventionally cast as well as on single crystal turbine blades was the reason for selecting Mar-M247*. Rene' 150* was chosen because it is one of the most advanced directionally solidified alloys. The wrought, sheet alloy HA-188* was selected because it represents one of the largest uses of a cobalt-base alloy in aircraft engines.

The primary purpose of the cobalt strategic element substitution research is to determine the fundamental role of cobalt in a wide variety of nickel-base superalloys and in a high-use cobalt-base superalloy. A secondary purpose is to develop the methodology to explore the roles of other strategic elements in similarly chosen alloys so as to have maximum impact on a wide range of users.

Figure 13 shows the participants in this COSAM effort on cobalt strategic element substitution. These initial research efforts are planned for a three-year period and consist of cooperative programs involving universities, industry, and NASA Lewis Research Center. Nominal compositions of the five alloys given in figure 13 indicate that cobalt content ranges from 10 percent in Mar-M247 to 39 percent in HA-188. In addition, the γ' phase ranges from 20 percent in Waspaloy to 65 percent in Rene' 150. The first phase in each research effort will involve substituting the less strategic element, nickel, for cobalt in incremental steps to a zero cobalt content. The effect of this substitution on properties and phases present, such as γ', will make up the major portion of the research effort in the first year of each program element. Efforts in subsequent years will be directed at identifying and optimizing alloying elements as substitutes for cobalt in the five alloys so as to maintain the key properties of these alloys.

The cooperative nature of the research being conducted on Waspaloy and Udimet-700 is illustrated in figure 14. The role of industry as represented by Special Metals Corporation is outlined. Their primary role is to characterize and optimize fabrication and heat treating procedures for the reduced

---
*Trademarks
Waspaloy   United Technologies Corporation
Udimet   Special Metals Corporation
Mar-M   Martin Marietta Corporation
Rene'   General Electric Corporation
HA   Cabot Corporation

cobalt Waspaloy and Udimet-700 alloys. The university role in this effort is also shown in figure 14. Columbia University will be involved with mechanical property characterization, structural stability, microstructural features, and theoretical formulations to identify future alloy modifications if required for the second phase of the project. Purdue University will be primarily responsible for microstructural and microchemistry characterization of the reduced cobalt content alloys. To round out the program, NASA Lewis Research Center will be involved in further mechanical and physical metallurgy characterization of the alloys as shown in figure 14. The output of this cooperative effort is expected to be a clearer understanding of the role of cobalt in nickel-base superalloys.

Some preliminary results on the effects of reducing cobalt in Waspaloy, a 13 percent cobalt alloy, are shown in figure 15 (ref. 5). Tensile strength appears to be insensitive to the amounts of cobalt in the alloy. However, rupture life decreased with decreasing amount of cobalt in Waspaloy. Further testing will be required to better characterize this apparent effect.

Similar effects of cobalt on the rupture life of Mar-M247 have been determined as shown in figure 16 (ref. 6). A possible contributing cause to this reduction in rupture life is the decrease in amount of γ' in this alloy with decreasing cobalt content as shown in figure 17. Also shown in figure 17 is the change in γ' composition. As cobalt is removed from the alloy, the largest change in the composition of γ' is the increase in tungsten content. Further studies are underway to clarify the role of cobalt in this alloy.

The research efforts on Udimet-700 and Rene' 150 parallel the previously described efforts on Waspaloy and Mar-M247. It is anticipated that these studies will lead to an understanding of the fundamental role of cobalt in a variety of conventional and directional nickel-base superalloys. These results should provide an improved technical base to develop modified superalloys in later stages of COSAM, as illustrated in figure 18.

Alternate Materials

Research in this area must be considered to be high risk and long range, but it has the potential of a high payoff in terms of significantly reducing the nation's dependence on strategic materials. As an example of alternate materials, intermetallic compounds are currently being investigated for possible structural applications. Initial efforts are centered on nickel and iron aluminides. Successful development of this type of alternate material offers the possibility of partially or totally replacing all the strategic materials in components where intermetallic compounds can be utilized.

Intermetallic compounds are of interest because of their potential high temperature strength as shown in figure 19 (ref. 7). It can be seen in this figure that nickel aluminides have the strength capability of competing with current nickel-base alloys. However, a possible disadvantage of this type of material is that simple binary aluminide compounds have shown a lack of room temperature ductility (fig. 20). The factors which influence the high ductileto-brittle transition temperature of nickel aluminide (~600 + °C) are currently being investigated. A NASA grant with Dartmouth University is aimed at understanding the fundamental deformation mechanisms in nickel aluminide. From these investigations, methods of improving the low temperature ductility of nickel aluminide may be suggested. An accompanying in-house research project at NASA Lewis Research Center is focusing on the high temperature mechanical properties of aluminides. These studies can provide a fundamental basis for more extensive research to develop these nonstrategic, alternate materials as shown in figure 21.

COSAM PLANS

Future COSAM efforts can build on the fundamental understanding from the early research for cobalt substitution, as was shown in figure 18. Major efforts will be devoted to developing, and if warranted, to scaling-up low or no-cobalt nickel-base superalloys for fabrication into various components. Demonstration of continued promise could also lead to verification in engine tests by major engine producers. Similar efforts will also be conducted for other strategic metals such as columbium and tantalum.

In the area of alternate materials, much more work will be required to develop materials such as intermetallic compounds. As was shown in figure 21, initial efforts will focus on fundamental studies aimed at improving low temperature ducitility and high temperature strength of FeAl and NiAl intermetallics. Complete property characterization will follow on more promising compositions. Reiterations of these basic steps will be required to further optimize the alternate materials and make them viable candidates as structural materials for aircraft engines. Scale-up and rig testing of promising compositions for blades and vanes will follow. The development of alternate materials will help conserve the strategic metals Co, Cb, Ta, and Cr.

A third area of the COSAM consideration involves conservation through improved materials processing technology. Although none of these activities have been initiated, plans have been made for investigating processing technology in such areas as advanced melting techniques, tailored fabrication, advanced coatings, joining techniques, and fabrication efficient processes. A reduction in strategic material usage should result from these processing technologies. For example, early efforts on near-net-shape fabrication of a turbine disk (ref. 8) have been shown to be able to reduce input material weight compared to conventional casting/forging practice and further gains appear possible. Improved processing technology will also help conserve the strategic metals Co, Cb, Ta, and Cr.

CONCLUDING REMARKS

This paper has presented NASA's COSAM efforts and planning. The primary points are summarized below:

1. Advancements in materials technologies are needed to provide the aerospace industry with alternative materials options in the event of future strategic metal shortages or excessive price increases.

2. The primary role of NASA's COSAM efforts will be to address strategic material problems within the aerospace industry. COSAM should make contributions to a national data base that will benefit many other domestic industries as well.

3. COSAM was designed to involve cooperative research efforts with industry (alloy producers, component fabricators, and engine manufacturers), with universities, and with government research facilities (primarily the Lewis Research Center).

REFERENCES

1. Mineral Commodity Summaries 1979. Bureau of Mines, U.S. Department of the Interior, 1979

2. "Aerospace Review and Forecast 1979/80", Aerospace, vol. 18, no.1, Winter, 1980, pp. 2-4.

3. Mineral Industry Surveys. Bureau of Mines, U.S. Department of the Interior, Dec. 1979.

4. Tien, J. K.; et al.: Cobalt Availability and Superalloys. J. Metals, vol. 32, no. 10, 1980, pp. 12-20.

5. Maurer, G. E.; Jackman, L. A.; and Domingue, J. A.: Role of Cobalt in Waspaloy. Superalloys 1980, J. K. Tien, ed., American Society for Metals, 1980, pp. 43-52.

6. Nathal, M.: The Role of Cobalt in a Nickel Base Superalloy. Presented at the AIME 110th Annual Meeting, Chicago, Il., Feb. 22-26, 1981.

7. Schulson, E. M.: The Ductility of Polycrystalline NiAl. Oral Presentation at NASA Lewis Research Center, Sept. 1979.

8. Dreshfield, R. L.; and Miner, R. V., Jr.; Application of Superalloy Powder Metallurgy for Aircraft Engines. NASA TM-81466, 1980.

STRATEGIC MATERIALS

- DEFINITION: THOSE PREDOMINANTLY OR WHOLLY IMPORTED ELEMENTS CONTAINED IN THE METALLIC ALLOYS USED IN AEROSPACE COMPONENTS WHICH ARE ESSENTIAL TO THE STRATEGIC ECONOMIC HEALTH OF THE U.S. AEROSPACE INDUSTRY

- SURVEY RESULTS

	STRATEGIC ELEMENTS IDENTIFIED
A.S.M.E. GAS TURBINE PANEL SURVEY	Cb, Co, Cr, Ta, AND W
NASA AEROSPACE COMPANY SURVEY (ADDITIONAL ELEMENTS)	Mn, Pd, Pt, Sn

Figure 1. - COSAM background.

HIGH PRIORITY

Co	SUPERALLOYS LARGEST SINGLE USER (30% OF TOTAL)
Cb	SUPERALLOYS LARGEST SINGLE USER (28% OF TOTAL)
Cr	CRITICAL TO ENVIRONMENTAL RESISTANCE OF ENGINE COMPONENTS
Ta	CRITICAL TO ADVANCED ENGINE ALLOYS

LOWER PRIORITY

W	NEW U.S. MINES ON STREAM, PROJECTED SELF-SUFFICIENCY BY 1985
Mn	WIDELY USED IN STEEL INDUSTRY, HOWEVER POTENTIAL LOW COST ALTERNATIVE TO Ni
Pd, Pt	USED FOR ELECTRICAL/ELECTRONIC APPLICATIONS
Sn	USED FOR Al AIRFRAME ALLOYS AND IN SOME Ti ENGINE ALLOYS

Figure 2. - Strategic element focus.

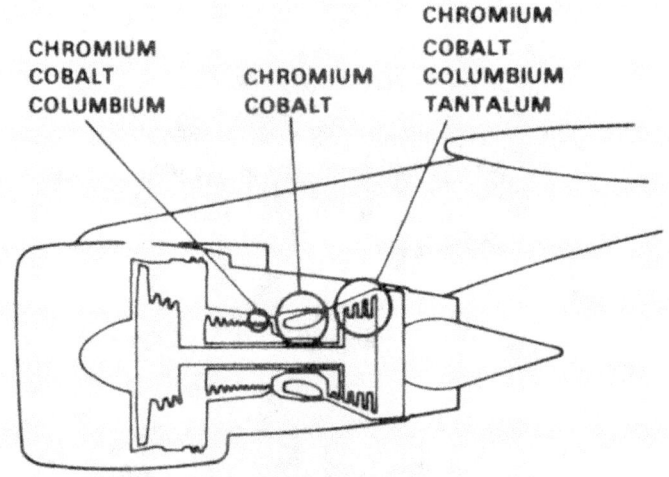

Figure 3. - Strategic metals are critical to turbine engines.

METAL	% IMPORTED	MAJOR FOREIGN SOURCE
COBALT	97	ZAIRE
COLUMBIUM	100	BRAZIL
TANTALUM	97	THAILAND
CHROMIUM	91	SOUTH AFRICA, ZIMBABWE

Figure 4. - U. S. aerospace is vulnerable to supply instabilities.

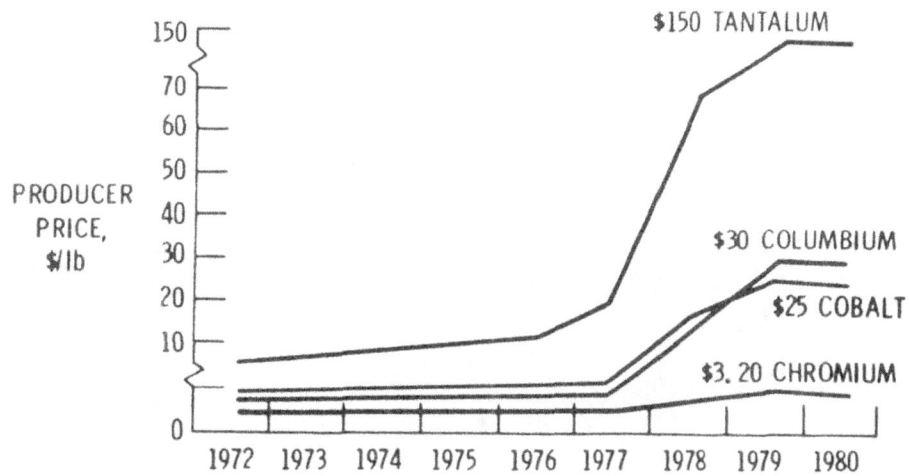

Figure 5. - U. S. aerospace is vulnerable to cost fluctuations.

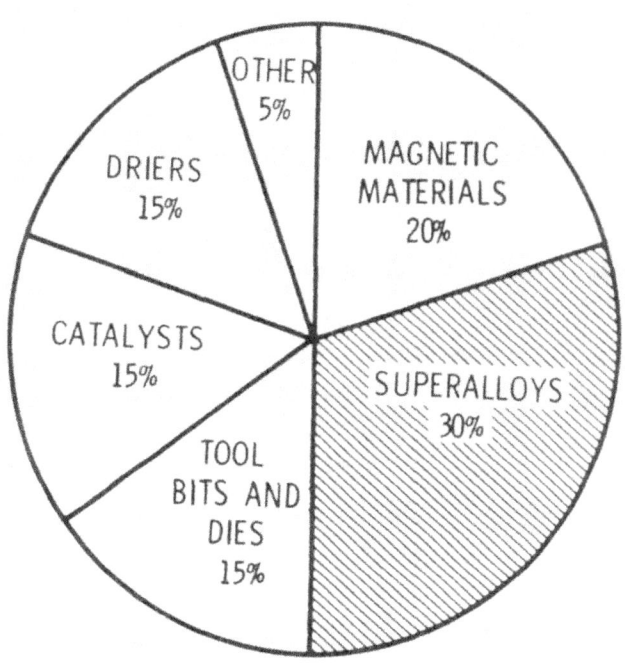

Figure 6. - U. S. consumption of cobalt in 1979 (Total pounds, 20.3 million).

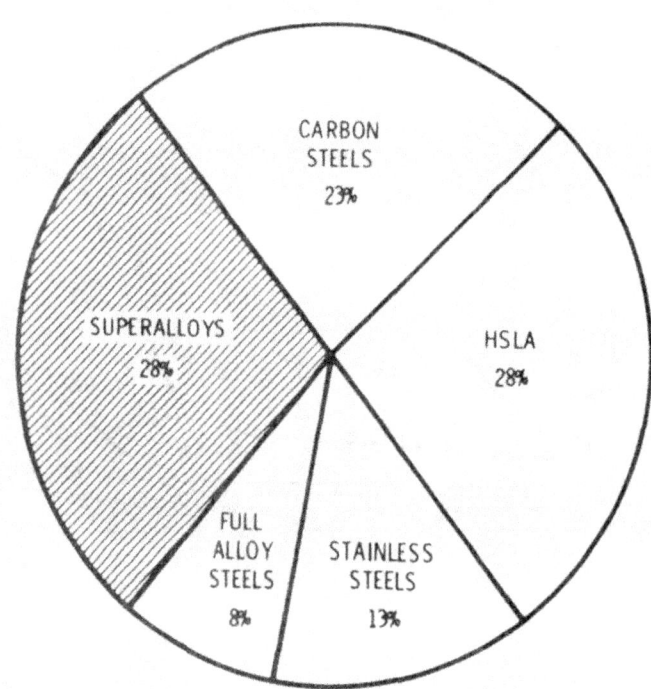

Figure 7. - U. S. consumption of columbium in 1979 (Total pounds, 6.3 million).

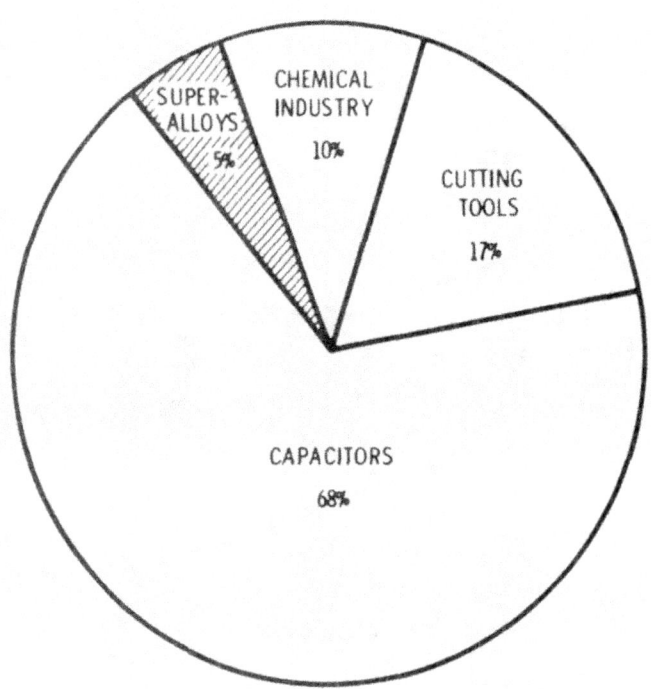

Figure 8. - U. S. consumption of tantalum in 1979 (Total pounds, 1.7 million).

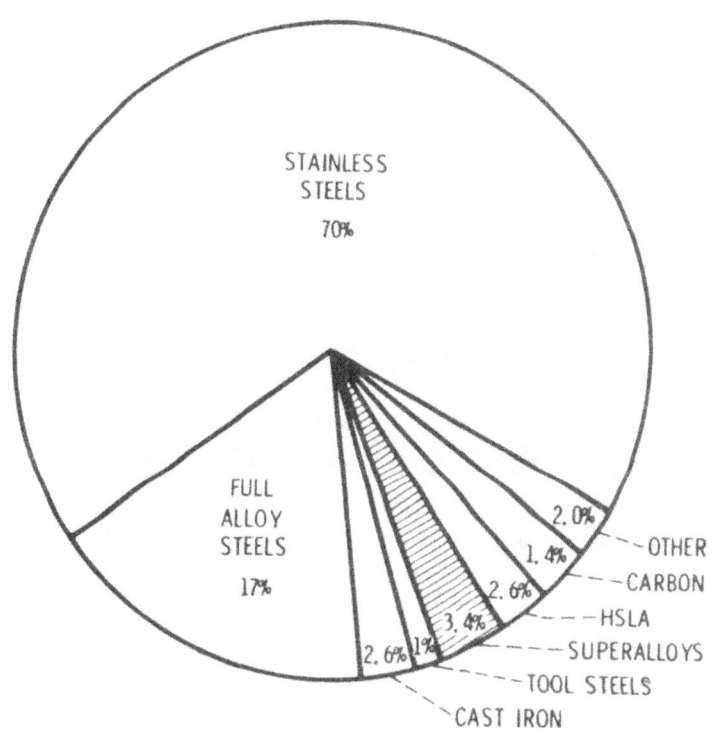

Figure 9. - U. S. consumption of chromium in 1979 (Total pounds chromium ferroalloys, 1.0×10^9).

OPTIONS CONSIDERED IN PREPARATION FOR
STRATEGIC MATERIALS SHORTAGE

- EXPAND EXPLORATION
- IMPROVED RECOVERY
- STRATEGIC MATERIAL SUBSTITUTION ✓
- SCRAP RECLAMATION
- REDUCED WASTE IN PROCESSING ✓
- ALTERNATE MATERIALS ✓
- STOCKPILING
- CRITICAL MATERIAL INDEX

Figure 10.- Options considered in preparation for strategic materials shortage).

OBJECTIVE:

- PROVIDE TECHNOLOGY OPTIONS WHICH WILL SUPPORT THE AEROSPACE INDUSTRY IN MAKING STRATEGIC ECONOMIC DECISIONS AIMED AT SIGNIFICANTLY REDUCING STRATEGIC METAL CONSUMPTION
  - Co, Cb, Ta, Cr, AND OTHERS AS IDENTIFIED

APPROACH:

- DEVELOP UNDERSTANDING OF ROLES OF Co, Cb, Ta, AND Cr IN CURRENT SUPERALLOYS
- IDENTIFY SUBSTITUTES AND LOW STRATEGIC METAL CONTENT ALLOYS
- DEVELOP PROCESS TECHNOLOGY THAT WILL MINIMIZE STRATEGIC METAL INPUT AND WASTE
- IDENTIFY ALTERNATE MATERIALS AND PROCESSES THAT HAVE HIGH LONG TERM POTENTIAL IN REDUCING STRATEGIC METAL USAGE (HIGHER RISK APPROACH)

Figure 11. - COSAM program objective and approach.

ALLOY	TYPICAL ENGINE APPLICATION	FORM	REMARKS
WASPALOY	TURBINE DISK	FORGED	HIGHEST USE WROUGHT ALLOY IN CURRENT ENGINES
UDIMET-700	TURBINE DISK	FORGED	SIMILAR ALLOYS USED IN VARIOUS FORMS AND APPLICATIONS
(LC) ASTROLOY	TURBINE DISK	AS-HIP-POWDER	
(RENE' 77)	LP BLADES	CAST	
MAR-M247	TURBINE BLADES	CAST	CONVENTIONALLY-CAST, D.S. AND SINGLE CRYSTAL
RENE' 150	TURBINE BLADES	DS-CAST	HIGHLY COMPLEX DIRECTIONALLY-CAST ALLOY
HA-188	COMBUSTORS	WROUGHT	HIGH USE COBALT-BASE SHEET ALLOY

Figure 12. - Superalloys selected for initial COSAM activities.

PARTICIPANTS	ALLOY	Ni	Cr	Co	Mo	W	Ta	Re	Al	Ti	Hf	γ'
COLUMBIA UNIV PURDUE UNIV SPECIAL METALS NASA-LEWIS	WASPALOY	58	20	13	4	--	--	--	1.3	3	--	20%
COLUMBIA UNIV PURDUE UNIV SPECIAL METALS NASA-LEWIS	UDIMET-700	53	15	19	5	--	--	--	4.3	3.5	--	40%
CASE-WESTERN RESERVE UNIV TELEDYNE NASA-LEWIS	MAR-M247	60	8	10	.6	10	3	--	5.5	1	1.4	55%
NASA-LEWIS	RENE' 150	59	5	12	1	5	6	3	5.5	--	1.5	65%
(TBD)	HA-188	22	22	39	--	14	--	--	--	--	--	--

Figure 13. - Elements of initial COSAM activities.

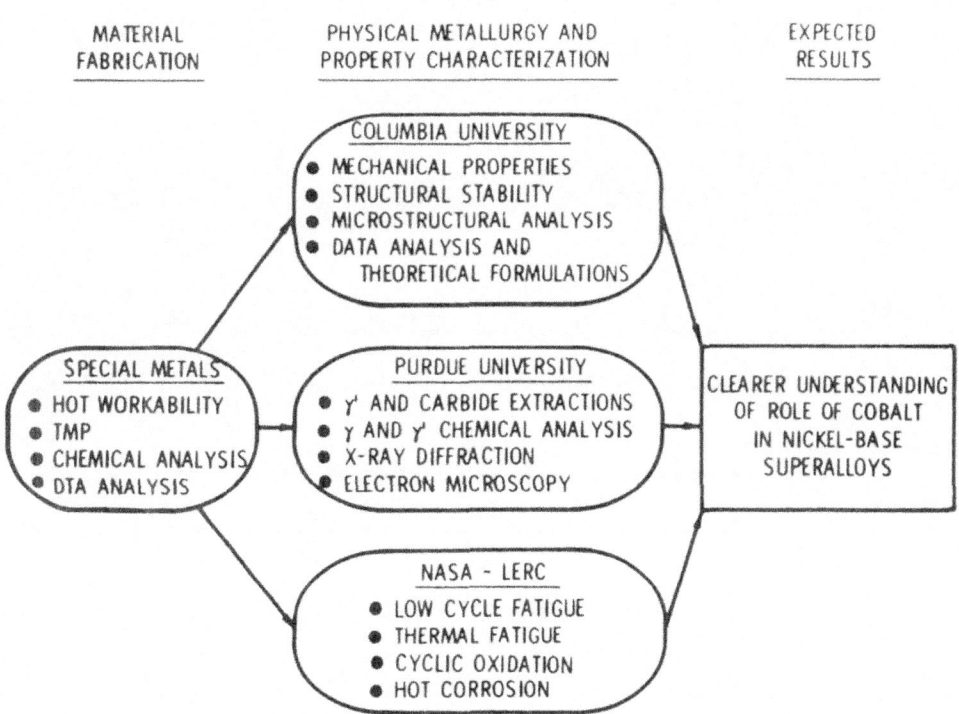

Figure 14. - Cooperative program to determine fundamental role of cobalt in Waspaloy and U-700.

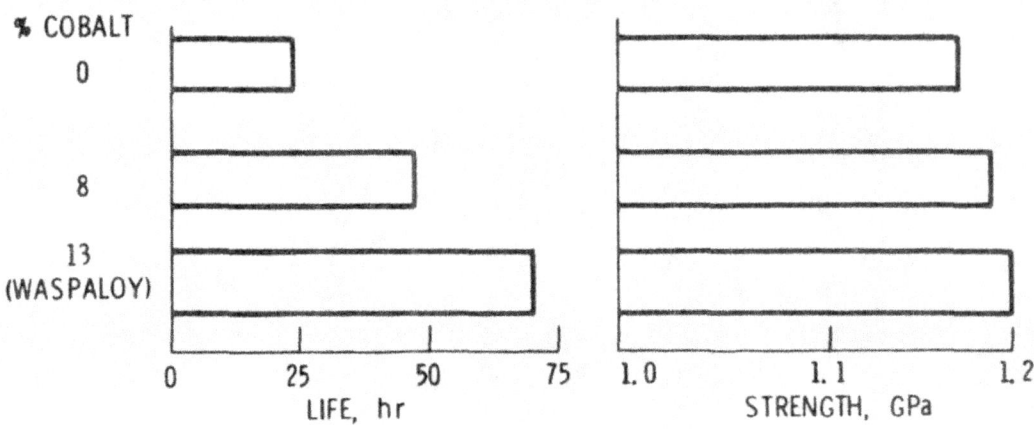

Figure 15. - Preliminary results of reducing cobalt in Waspaloy.

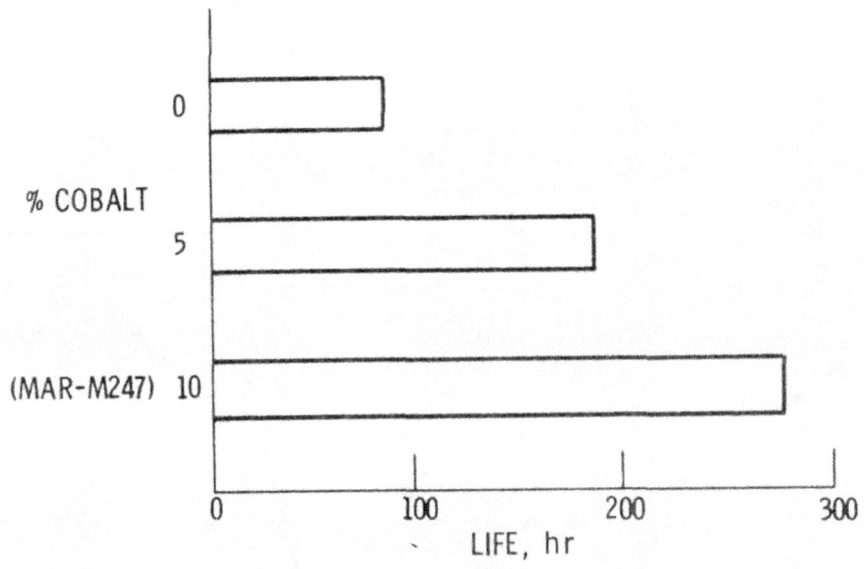

Figure 16. - Preliminary results of reducing cobalt in Mar-M247. (Rupture life, 870°C; 360 MPa.)

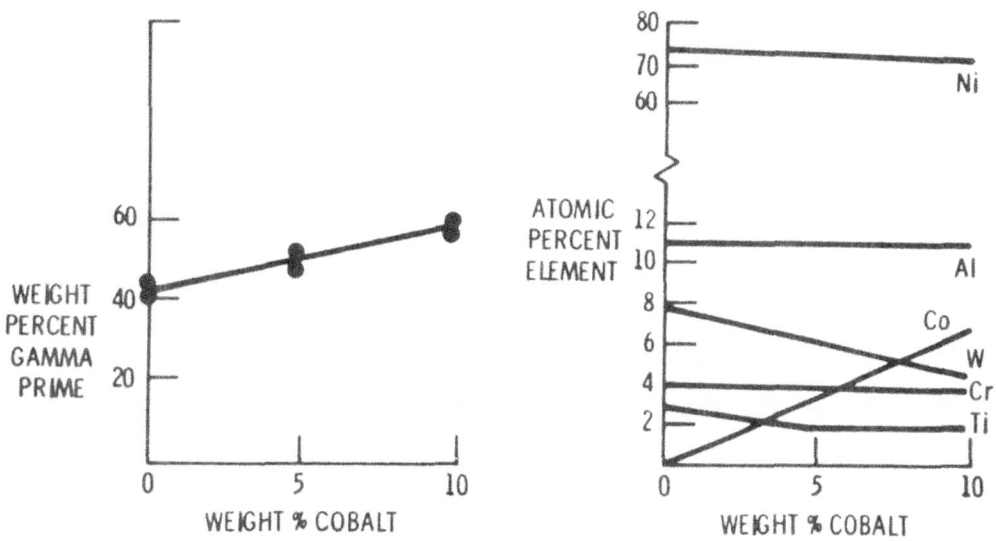

Figure 18. - Planned flow of COSAM strategic element substitution research.

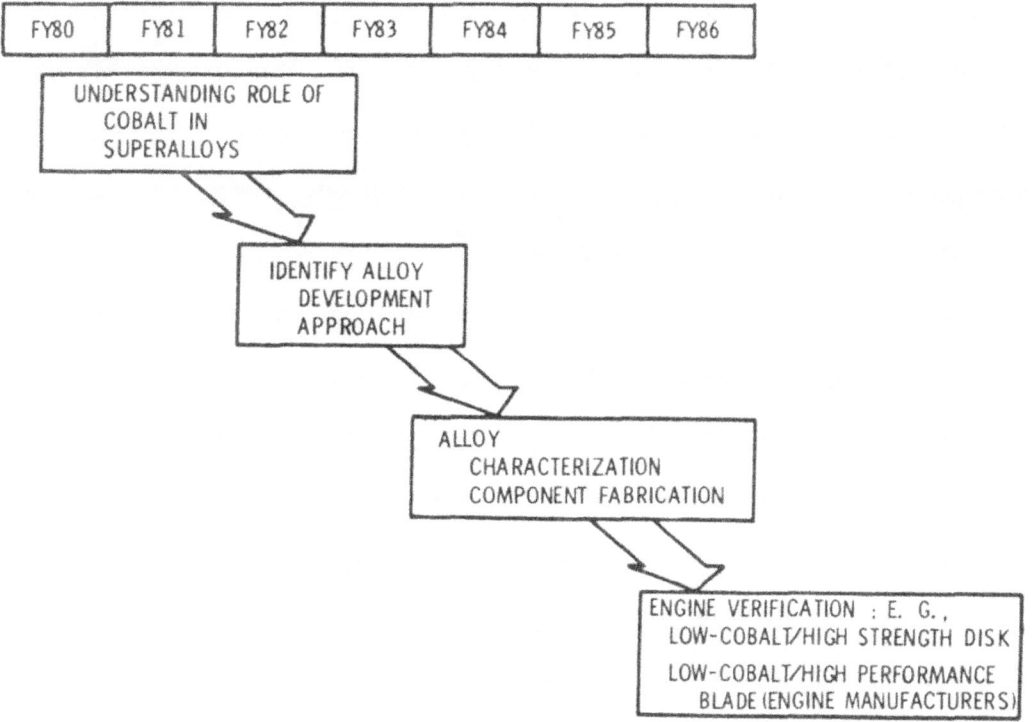

Figure 17. - Amount and composition of gamma prime as a function of cobalt content in Mar-M247.

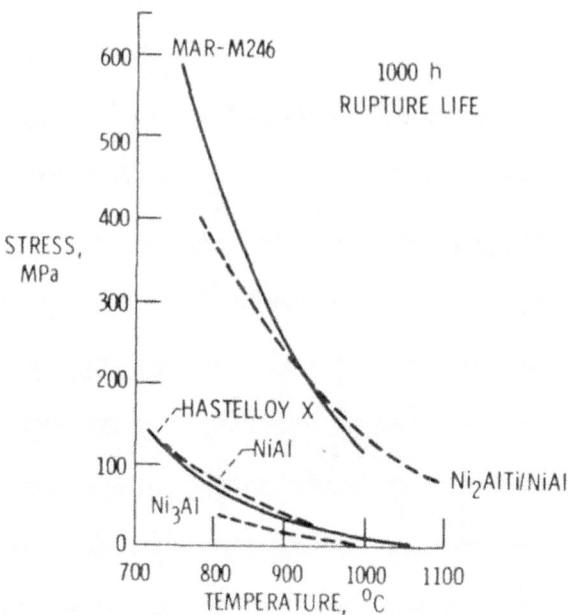

Figure 19. - Typical strengths of aluminides and superalloys.

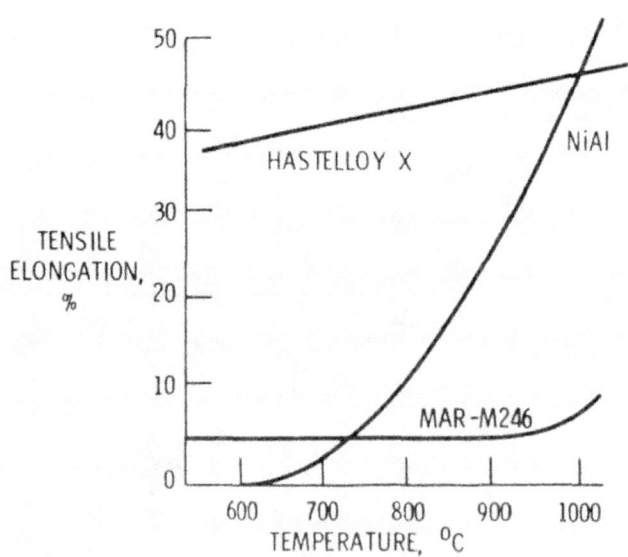

Figure 20. - Typical ductility values for aluminides and superalloys.

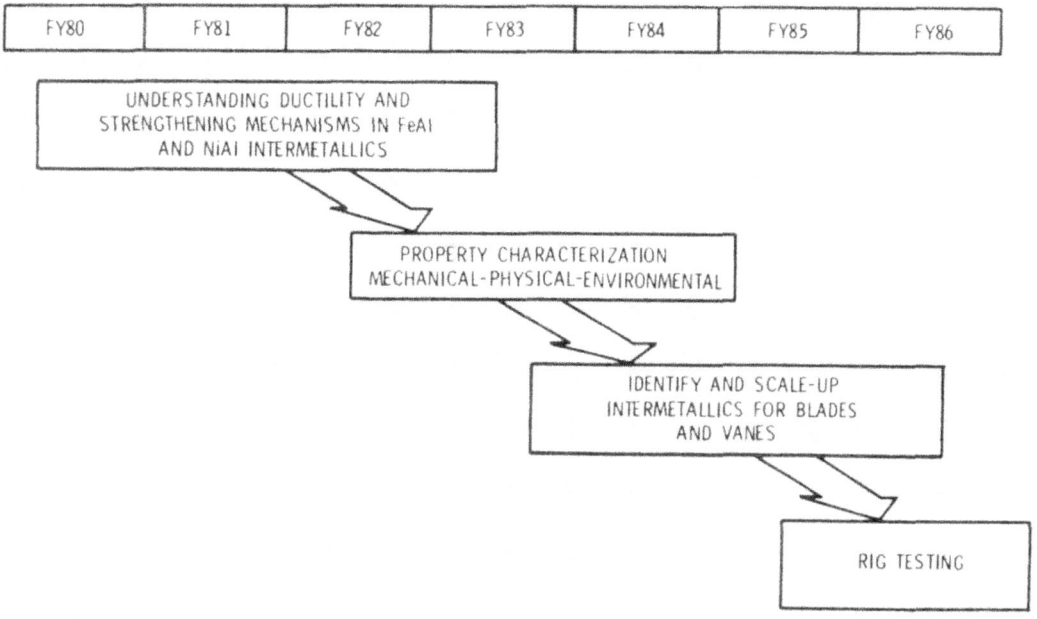

Figure 21. - Planned flow of COSAM alternate materials research.

Page intentionally left blank

WHO NEEDS ENGINE MONITORING?

James L. Pettigrew, P.E., Lt Col, USAF
Wright-Patterson Air Force Base

ABSTRACT

The requirement for Engine Monitoring Systems (EMS) is elusive even for its advocates. Decisions not to invest large sums of up front money in equipment which will be of uncertain value are easily made by conscientious program managers. Even as on-condition maintenance (OCM) is being established as the desired approach in the Air Force, many people in the decision chain doubt the potential value of on-board engine monitoring equipment.

EMS advocates have not provided convincing answers to many hard questions, some of which are: "Should the EMS capability provide on-board GO-NO-GO information? How much engine monitoring is enough? What parameters are required? How will the EMS capability be used to direct maintenance actions? Does the environment require only engine parts tracking, engine usage, or performance trending data?" Answers may not have uncontested technical support but may require judgement based on something like Pareto's 80-20 law applied to operational data.

The true EMS values are certain only in the future operational environments. The EMS advocates' problems are to find for the system managers acceptable up front rationalization for the added EMS cost. Past operational evaluations of a few EMS units for short periods have not all produced convincing results. This presentation will discuss these evaluations and their lessons learned, then review the options for each required EMS phase, and close with a review of the guidance being provided for EMS on new systems.

COST TRENDS

Table I shows the maintenance cost of flying various Air Force engines for thousand engine flight hours. The maintenance cost are in 1980 dollars. In most cases, these costs equal or exceed the acquisition cost for that engine. The acquisition costs shown are first production contract costs in then year dollars. For a true comparison, the earlier engine cost would be corrected for inflation. From these numbers, the throwaway engine might not be such a bad concept, especially when you remember that with increasing engine age performance deteriorates and engine service life between repair shortens. Maintenance manhour per flight hour on the newer engines is also increasing to some very high numbers.

FAILURES

Failures result from wear, leaks, structural damage and human error. There are many things which influence operating time before engine failure. An Engine Monitoring System (EMS) provides a data base from which failures can be predicted, detected, and diagnosed early, before there is a loss in mission capability.

IMPORTANCE OF VIEWPOINT

The viewpoint from which an individual looks on things has a large influence on what he is able to see. This is illustrated by the old saying that "A jackass on a hill can see more than a genius in a valley."

An EMS is more than black boxes full of electronic circuits. The people who look at an engine monitoring system as the black boxes might be considered the genius in the valley. In the total system view, EMS is the executive control system which tells the maintenance supervisor that an individual engine requires diagnostic work to find out why it is abnormal. The EMS data function is similar to the blood pressure check performed by the doctor. If he finds any abnormalities in blood pressure he runs other diagnostic tests to determine what is causing you to be abnormal.

REQUIRED TASKS

Figure 1 illustrates the data flow in an EMS. Data can be obtained in ways ranging from a manual recording of cockpit instrument readings to sophisticated complete electronic systems which automatically records, stores, and transfers the data to ground computers. Airborn engine monitoring system electronics often have decision logic to determine engine status as soon as the aircraft lands. Airplanes with two pilots and mission requirements for a cruise leg are generally able to use manual recording. On single pilot aircraft work load generally prevents the use of manual recording. The ability to get in-flight engine performance is the missing piece for single pilot fighter aircraft. Therefore, current thrust in developing EMS capability is improvement of in-flight data acquisition ability.

Before the in-flight data can be used to predict, detect and diagnose failures, it must be validated, corrected, compressed, displayed, and then interpreted. There are a number of ways of interpreting. The status of engines can be obtained from the data by limit exceedence or by observing trends. The important results from an EMS is the effect of the information on the maintenance system. If we only gather the data, and look at the data, and do not use it to direct maintenance, EMS is of little value to the total system.

WHY OCM

In February 1974, the Department of Defense gave the following logistics and material support guidance:

1. Establish engine maintenance policies to eliminate maximum operating time.

2. Exploit modular designs in new engines.

3. Use on-condition maintenance techniques.

4. Apply to existing engine types wherever practicable.

With the on-condition maintenance you need a methodology to tell you what the existing condition is within the engine so you can schedule it for maintenance. Figure 2 illustrates why on-condition maintenance has an advantage. It can reduce risk and save dollars.

Engine usage varies by the mission being flown. For example, a fighter aircraft on a low-level mission flying at 600 knots, Mach .95, would have its inlet pressure increased by a factor of 1.8. On the low-level mission, the engine with a 20 to 1 compression ratio would have a combustor case pressure of 36 atmospheres. On the other hand, an intercept mission which cruises out at 30,000 feet MSL would only see 12 atmospheres combustor case pressure. The cruise engine obviously is capable of operating more hours before failure because of the less stressful usage. If maintenance is driven by maximum operating time, the additional operating capability of an engine used at the lower rate will not be utilized. If the condition of each engine determines when it must be repaired, then the full engine capability can safely be used.

AIR FORCE EMS PROGRAMS

Air Force EMS programs are divided into three categories: (1) developed with aircraft, (2) contract maintenance, and (3) add-ons to operational aircraft. See Table II for a listing of EMS developed with aircraft.

MRS is a Maintenance Recording System that is applied on the SR71 with a J58 engine. It is an analog recorder that gives a continuous trace of the engine operating parameters throughout that mission. It has an approximately 1100 hours meantime between failure (MTBF), and is considered a successful EMS system. Its data is automatically acquired and formatted with manual interpretation by a technician rolling the strip chart and looking at total trace for each sortie. That strip chart may be 8 to 10 feet long for a sortie. The interpretation of the analog traces is a disadvantage on this system.

The Malfunction Detection Analysis and Recording System (MADARS) was built and developed with the C-5 aircraft. It automatically acquires and formats the data. Interpretation is both manual and automatic. Logis is proved to print out the maintenance action required in many cases. A shortcoming of MADARS is an overall system MTBF. MADARS monitors all aircraft systems. The MTBF for the engine portion of the MADARS which provides engine data is approximately 100 hours.

The Central Integrated Test System (CITS) is another system designed to monitor the total airplane as well as the engine; it has been tested on the four B-1's during their Category I & II flight tests. It is rather complex and there are some differences of opinion on its real potential benefit to the operational weapons system.

The two systems at the bottom of the figure, Events History Recorder (EHR) and Engine Time Temperature Record (ETTR), are different in that they record usage more than they record the traditional performance monitoring parameters. The ETTR infers engine health from counters that pick up the amount of time above a certain temperature and the speed cycles on the engine in terms of core engine speed. This information allows low cycle fatigue tracking. The operational units have some problem of short meantimes between failure. The EHR runs about 600 hours and the ETTR runs about 2500 hours.

Contract maintenance is used on systems with only a few aircraft. Under this approach, the Air Force uses the aircraft and asks the aircraft company to provide all of the support away from the flight line. The maintenance approach used by the Air Force is a threefold approach: flight line, intermediate, and depot maintenance.

Flight line maintenance does remove and replace activities, as well as servicing. The intermediate maintenance shop located at the base does minor overhaul work. The major overhaul facilities does the complete overhaul. In the contracted approach, the contractor provides the intermediate and the depot maintenance.

Contract maintenance systems have a Contractor Operated and Managed Based Supply system (COMBS) at each base operating the type aircraft. Blue suit, flight line maintenance personnel go to the COMBS facility which provides a replacement part over-the-counter. See Table III for a summary of USAF contract maintenance programs.

The T-43 aircraft uses a flight log engine monitoring program with manual data acquisition, automatic computer formatting and both automatic and manual interpretation. The C-9 uses ground trim data from routine ground runs as a basis for determining engine conditions. From the Air Force standpoint, both of these programs are still fairly new. The T-43 is just now reaching the first overhaul on the engines. The KC-10, also contract maintenance, will use flight deck monitoring with manual acquisition, automatic formatting and manual interpreting of the data. The E-4, which is the SAC Command and Control airplane, also uses flight deck monitoring, with manual recording, automatic formatting and manual interpretation.

EMS EXPERIENCE

These applications show the wide range of choices available to accomplish each of the required EMS tasks. Each of the systems discussed currently fulfills the engine monitoring requirements for its weapon system. However, cost benefits from the EMS application are difficult to accurately quantify. The benefits are real, but normal system data has not been defined to break out the results. These systems give insight into how the next monitoring system should be designed and built. A selling point often used to justify an EMS is elimination of all ground support equipment. These programs generally show that ground equipment may even see additional use. Monitoring EMS data does give us additional insight into engine health, and is capable of controlling on-condition maintenance.

ADD ON EMS

Several operational aircraft have added EMS for service test in an attempt to demonstrate the value of the engine monitoring. See Table IV for a summary of EMS add-ons to operating aircraft. The Engine Health Monitoring System (EHMS) was tested on the T-38. It automatically acquired and formatted the data for semiautomatic and manual interpretation. The results of the T-38 test indicates that EMS probably would not be cost effective. The operational use for the airplane is important. The Air Training Command wants to assure highly reliable engines. Therefore, its overhaul interval is shorter. The test was run within the ATC standard operational framework; therefore, there were few failures. If the engines do not fail, the monitoring system cannot prove its capability and benefits.

Engine Condition Monitoring Program (ECMP) employed in SAC is flight deck monitoring. ECMP is being credited with secondary damage savings of $2 million dollars a month and reducing the in-flight shutdown rate on the SAC fleet by better than 50%. The 50% is based on the three year, in-flight shutdown rate average prior to implementation of the program, compared against the three years since the program has been in use.

Again, look at the concept of operation. ECMP is used on a multi-engine aircraft. With multi-engine aircraft, in-flight shutdowns do not have a strong safety indication. Therefore, the overhaul interval is much longer than on a single engine aircraft. Failures do, therefore, occur within the maximum operating time. The ECMP was able to detect these failures before occurrence, allowing repair when the deterioration was in the earlier stages. More than 2000 engines have been repaired solely because of ECMP indications. Only six have been disassembled during this period where no problem could be identified.

The A-10 Turbine Engine Monitoring System (TEMS) has been service evaluated with positive results, and is following on with a squadron level evaluation planned to determine how well that system functions to drive maintenance in the operation scenario.

The electrostatic probe is new technology that came out of the Air Force Institute of Technology (AFIT) about ten years ago. The theory is that rub or errosion in the engine gas path produces an electrostatic charge in the exhaust stream. The quantity measure of electrostatic charge per unit time infers the rate of deterioration within the gas path. The phenomena has been verified but it has not been operationally employed as a monitoring system.

The engine diagnostic system EMS is a monitoring system for the F100 engine in the F-15 aircraft. It is a service test to validate state of the art EMS capability against thirty-two goals. The results proved the system would get the data with accuracy equal to the test stand.

GENERAL RESULTS FROM ADD-ON TESTS

Experience does not show optimistic near term expectations for add-on monitoring systems. EMS generally drives the maintenance cost higher. Start-up problems show that a successful new system takes time to mature. Software problem solutions have taken longer than expected before the EMS successfully records in-flight data. Test plans often are written to conduct the evaluation within normal operating scenario which prevents the test yielding conclusive evidence on EMS value. The test aircraft are used to meet mission requirements in the normal manner. Maintenance is done by the TOs with little flexibility allowed to meet test objectives. Therefore, the test articles may not obtain sufficient flight hours or get appropriate focus.

Many valuable benefits come from a monitoring system. You get design feedback, correlation between the testing and operation usage, and verification of repair effectiveness. Verification of repair is often overlooked in the benefits analysis of the program. Maintenance replaces the wrong part, puts the aircraft back in service and it flies without a squawk, so it is concluded that the repair fixed the original squawk. Data from the monitoring system allows one actually to see the performance trace change providing a powerful quality control capability on maintenance and repair. EMS certainly provides improved knowledge of failure modes.

Technical orders are based on a number of A PRIORI assumptions. These assumptions are presupposed by experience, and are not subject to further examination or analysis. Based on A PRIORI assumptions, technical orders are written as if the A PRIORI knowledge illustrates the true behavior of an engine.

A monitoring system may provide data which causes one to question A PRIORI assumptions. EGT margin is believed to have full capability to effectively identify an engine as good or bad. EMS data shows that the EGT does go through the red line just before the engine is torn asunder. However, experience with a SAC monitoring program showed that severely deteriorated engines with basket case turbines often run cooler with a greater EGT margin. The cooler operating engine can be explained by the facts that the EGT probes are not covering the total exhaust stream and that the turbine nozzle areas change with deterioration. Engines were found by the SAC program with missing first stage nozzle and burner center cones broken off and laying back against the first stage nozzle. These engines passed EGT tests. In fact, two-thirds of them passed complete test runs and were certified for flight. Teardowns later found the bent and broken hardware within the engine. See reference 1 for an example from the SAC ECMP.

The ECMP showed that the beginning failure in the majority of the J57/TF33 engines started with fuel nozzles. Some fuel nozzles in a couple of burner cans would plug with the engine continuing to meet performance specs. The good burner cans got more fuel causing hot spots which resulted in burning and bowing of the vanes. Hours later, a vane would eventually burn through. The piece of broken vane would have about an 80% probability of making it through the turbine without engaging in the stationary and rotating vane rows. That is hard to believe, but under the monitoring system, many engines were missing a half first stage turbine nozzle vane on tear down. The missing piece had marked the turbine stages as it passed through. In other cases, the piece would engage between the rotating turbine wheel and stationary nozzle with sufficient force to break a blade. The engines are amazingly tough.

The SAC ECMP uncovered a change in depot maintenance procedures. Fuel nozzles were designed to be repaired in matched sets. It was decided that overhaul of the fuel nozzles in matched sets was too costly. So, like parts were worked in batches. Nozzles were assembled randomly from the batches. Tolerance control was gone from the batch repaired fuel nozzles. The result was a very short service life on badly mismatched sets.

Within six months after depot changed overhaul procedure, fleetwide ECMP monitoring on SAC engines identified the problem. The fuel nozzle overhaul problem potential will never be known because it was not allowed to exist long enough to have its full impact on the fleet. ECMP identified engines with bad nozzles for repair before other parts were damaged. How do you value something that is responsible for turning a problem around before its impact is documented?

MANAGEMENT LESSONS LEARNED

Responsibilities should be defined at the outset of an EMS program. Keep on board equipment simple which may be aided by limiting the in-flight task to data acquisition. Do the formating and interpretation of the data in the ground system. Remember that every pound of weight on a fighting aircraft costs performance. The mission of the Air Force is to fly and fight. Man should be in the loop so he is able to understand what the output from the monitoring system means. Provide realistic time and training, support equipment, and EMS spares. Organize a realistic, timely base monitoring team to use the in-flight data to drive maintenance actions. Effectiveness is improved if the EMS system is built-in versus retrofit. One should not wish to monitor everything.

TECHNICAL LESSONS LEARNED

If the necessary parameters can be defined, it is possible to minimize sensor requirements. Insure that the output of the in-flight equipment is compatible with the existing test equipment. Provide flexibility so the necessary data can be obtained to track a new failure mode. Provide self-check to isolate the bad data. Trending does allow you to determine deterioration within the engine. Increasing fuel prices are emphasizing the need to obtain the engine data while the engine is in the revenue service, to use the airline term, rather than do a ground run. If the engines don't have problems, you don't need monitoring. Good engines receive no benefit from being monitored. If you know what the engine's performance parameters are doing, you can determine its reliability potential and therefore enhance flight safety.

THE ADVOCATES PROBLEM

Why is it such a problem to get EMS on AF equipment? (See Figure 3)

The figure shows the time line for a weapons system versus accumulative or life cycle costs. Air Force System Command (AFSC) is responsible for the acquisition process until Program Management Responsibility Transfer (PMRT). then, Air Force Logistics Command (AFLC) takes over for logistic support. The process begins with an approved operational requirement for a specific weapons system to do a job. The System Project Office (SPO) director is assigned the responsibility for the acquisition. He is given a certain budget and has to acquire the required capability within that budget. An engine monitoring system adds an immediate cost increase to the system which is apparent. EMS benefits accrue in system operation after PMRT. Several years of operation may pass before the meantime between failure for the major items of the system is reached. During the acquisition phase there is no way of knowing the correct slope on the operations cost curve. Therefore, the SPO director on his watch sees only the impact of EMS cost on his system. EMS potentially available benefits accrue in service after PMRT when AFLC has the watch.

ELEMENTS OF SYSTEMS EFFECTIVENESS

Earlier the importance of viewpoint was discussed with the idea that the "jackass on the hill could see further than the genius in the valley." Analyze that idea from a standpoint of system effectiveness. (See Figure 4) Three people are involved in the Weapon System Effectiveness Problem: The overall field commander decides what weapon will be employed on what target at what time, and the branch on the right of figure 4 represents his interest; The wing commander has to implement the field commander's orders as the center branch represents his interest. He wants X equipment on the line and ready to meet the mission requirement. The Deputy Commander

Maintenance (DCM) is charged with the responsibility of making that equipment available. His interest is in the branch on the left. The EMS system in order to be judged cost effective and worthy of purchase by the SPO director must clearly improve each of these elements for total system effectiveness. That is the heart of current EMS development guidance that is being given to industry for the new weapon system starts.

EMS DEVELOPMENT GUIDANCE

The following general guidance for the development of an Engine Monitoring System (EMS) was provided by the Propulsion Director of Engineering, 30 January 1981, to maximize system effectiveness of our new weapon systems.

The EMS will be dedicated primarily to the performance of "Engine Monitoring," i.e., capture of in-flight engine operating data. The EMS function operating within the planned logistics/maintenance concept will not be compromised by over sophistication of tasks and multiple roles for the EMS hardware. Where airframe monitoring systems are to be used, the EMS must be compatible and compliment that system. However, an option for independent operation of the EMS should be planned in the event that an integrated airframe/engine monitoring system is not included. The current development guidelines for on-board EMS capability are:

1. Simplify on-board equipment by limiting in-flight requirements to data capture with data interpretation on the ground.

2. Limit EMS design goals to evaluation of engine suitability for continued service rather than fault isolation to an individual module/component.

3. Plan use of ground test equipment, e.g., borescope, chip detector, to confirm EMS indications and enhance diagnostics prior to engine removal.

4. Integrate the EMS output from an operational weapons system viewpoint by use of ground station data processing with the man in the loop for interpretation and direction of maintenance.

PLANNED APPROACH

During the early phases of each EMS program the engine contractor will be tasked by the Air Force to prepare a detailed feasibility analyses covering the following areas:

1. A list of aircraft/engine parameters to be monitored/recorded in-flight by the EMS.

2. Feasibility of performing the following engine monitoring/diagnostic functions using the parameters recorded in-flight:

- Engine Documentary Data
- Parts Life Tracking
- Parameter Tracking/Trending
- Engine Suitability for Flight
- Warranty Validation (if required)
- Suitability for Flight

In addressing the feasibility of performing each of the above functions with the EMS, the contractor must direct his analysis to answering the following questions:

a. Does the technical expertise exist currently to adequately perform each function without causing a negative impact on the planned maintenance/logistics concept for the application?

b. How would each function's data product interface with the planned maintenance/logistics concept?

c. Where and by what means would the EMS data product be converted into useful information?

d. Who would eventually use the information?

e. What will an EMS do for system effectiveness?

Once this feasibility analysis is provided, a complete review will be conducted by the Air Force. The direction for the development of the EMS for the engine will then be established.

TOTAL SYSTEM VIEWPOINT

The engine contractor should be tasked with the responsibility for developing all aspects of the EMS system with Air Force assistance. This includes all hardware required on-board, on the flight line and in the ground station, plus all software required for the EMS to function satisfactorily. The EMS must work hand in hand with the planned engine logistics/maintenance concept. As such, both systems or programs must be developed concurrently to insure optimum utility of the EMS. It is essential that a total system perspective (airframe, maintenance, logistics) be the overriding consideration in the development process and that the EMS and the maintenance concept be concurrently developed. The overriding question is: "How much EMS is enough for system effectiveness optimization while remaining affordable?" Pareto's criteria can help zero in on the answers during the acquisition phase.

Using the weapon system approach as an evaluation criteria early in the acquisition phase will hopefully help get the genius out of the valley onto the back of the jackass on the hill so that they can together gallop toward realization of potential EMS capabilities.

REFERENCE

1. McCord, Robert M.; Engine In-Flight Monitoring (Part II). Maintenance Magazine, Air Force Inspection and Safety Center, Norton AFB, California, 1977, pp. 31-38.

USAF TURBINE ENGINES
MAINTENANCE VS ACQUISITION COST TRENDS

INITIAL OPERATING CAPABILITY	ENGINE NOT MISSION CAPABLE %	MMH/1000EFH	REMOVALS 1000EFH	1980 MAINT COST $ DOLLARS/1000EFH	ACQUISITION COST-THEN $
NON AB					
1956 (TJ)	9	650	1.26	$161K	$175K
1960 (TF)	2	700	0.65	$88K	$210K
1969 (TF)	13	1060	0.36	$728K	$888K
1975 (TF)	28	620	1.2	$681K	$570K
AB					
1959 (TJ)	14	144	3.1	$320K	$160K
1962 (TJ)	22	87	4.2	$53K	$85K
1967 (TF)	18	269	7.4	$876K	$730K
1976 (TF)	17	323	8.2	$2,000K	$1,960K

TABLE I - MAINTENANCE COST TRENDS FOR USAF TURBINE ENGINES. AS A BASIS FOR COMPARING ACQUISITION COST, THE COST FROM THE INITIAL ACQUISITION CONTRACT IS PROVIDED IN THEN YEAR DOLLARS. THE COST DATA IS EXTRACTED FROM THE 1980 ASD ENGINE ADVISORY GROUP (EAG) MINUTES DATED 23-24 SEPTEMBER 1980.

ENGINE MONITORING SYSTEMS
DEVELOPED WITH AIRCRAFT

SYSTEM	DATA ACQUIRE	DATA FORMAT	DATA INTERPRET	AIRCRAFT/ENGINE	STATUS
MRS	A	A	M	SR71/J58	OPERATIONAL MTBF 1100 HOURS
MADARS	A	A	M/A	C5A/TF39	OPERATIONAL MTBF 100 HOURS TOTAL MADAR SYSTEM 8 HOURS
CITS	A	A	M/A	B1/F101	OPERATIONAL IN CAT I&II TESTS COMPLEX NOT READY FOR OPERATIONAL DEPLOYMENT
EHR	M	M	A	F15/F16/F100	OPERATIONAL MTBR 600 HOURS
ETTR	M	M	A	A10/TF34	OPERATIONAL MTBR 2500 HOURS

TABLE II - SUMMARY OF ENGINE MONITORING SYSTEMS DEVELOPED WITH AN AIRCRAFT.

ENGINE MONITORING SYSTEMS
ADD-ONS TO OPERATIONAL AIRCRAFT

SYSTEM	DATA ACQUIRE	FORMAT	INTERPRET	AIRCRAFT/ENGINE	STATUS
EHMS	A	A	M/A	T38/J85	JUDGED NOT COST EFFECTIVE
ECMP	M	M	M	KC135/TF33 B52D G/57 B52/H/TF33 C 141	OPERATIONAL SAVES S2 MILLION IN SECONDARY DAMAGE EACH MONTH REDUCED IFSD RATE BY NEAR 50%
TEMS	A	A	M/A	A10/TF34	SERVICE EVALUATION ON 5 AIRCRAFT WARRANTS FOLLOW ON SQUADRON EVALUATION
ELECTROSTATIC PROBES	A	A	M	NUMEROUS ENGINES	PHENOMENA VERIFIED
EDS	A	A	M/A	F15/F100	SERVICE TEST ON 5 AIRCRAFT MEETS MANY OF THE 32 DESIGN GOALS ACCURACY EQUALS TEST STANDS

TABLE III - SUMMARY OF ENGINE MONITORING SYSTEMS TESTED AS ADD-ONS TO OPERATIONAL AIRCRAFT.

ENGINE MONITORING SYSTEMS
CONTRACT MAINTENANCE

SYSTEM NAME	DATA ACQUIRE	FORMAT	INTERPRET	AIRCRAFT/ENGINE	STATUS
FLIGHT LOG CIEMAS*	M	A	A/M	T43/JT8D 9	FLT LINE BLUE SUIT COMBS***UNITED/SFO OCM CYCLE LIMITS
GROUND TRIM DATA TRENDED	M	M	M	C 9/JT8D 9	FLT LINE BLUE SUIT COMBS HARD TIME CYCLE LIMITS
EPM**	M	A	M	KC 10/CF 6 50C2	FLT LINE BLUE SUIT COMBS OCM CYCLE LIMITS
EPM	M	A	M	E4A/CF6 50	FLT LINE BLUE SUIT COMBS OCM CYCLE LIMITS

* CENTRAL INFORMATION ENGINE MONITORING AND AIRCRAFT SYSTEMS
** ENGINE PEFORMANCE MONITORING GE
*** CONTRACTOR OPERATED AND MANAGED BASE SUPPLY INCLUDES INTERMEDIATE AND DEPOT ACTIVITIES

TABLE IV - SUMMARY OF ENGINE MONITORING SYSTEMS USED WITH USAF AIRCRAFT OPERATED THE CONTRACT MAINTENANCE CONCEPT.

ENGINE MONITORING SYSTEMS
REQUIRED TASKS

```
                    [engine diagram]
                                              SYSTEM IMPACT
                                              • IMPROVED
                                                MISSION
                                                CAPABILITY?
                                              • LOWER LCC?
         ⇓                                    ⇑
  ┌─────────────────────┐        ┌──────────────────┐
  │    ACQUIRE DATA     │        │   TAKE ACTION    │
  ├──────────┬──────────┤        │ • DIRECT SPECIFIC│
  │ • GAS PATH│• MECHANICAL│     │   REPAIRS WITH   │
  │  PARAMETERS│ PARAMETERS│     │   WORK ORDERS    │
  └──────────┴──────────┘        └──────────────────┘
         ⇓                                    ⇑
  ┌─────────────────┐    ⇒     ┌──────────────────┐
  │  FORMAT DATA    │          │  INTERPRET DATA  │
  │  • VALIDATE     │          │  • CHECK LIMITS  │
  │  • CORRECT      │          │  • CHECK TRENDS  │
  │  • COMPRESS     │          │  • DIAGNOSE      │
  │  • DISPLAY      │          │  • PROGNOSICATE  │
  └─────────────────┘          └──────────────────┘
```

FIGURE 1. THE TASKS WHICH MUST BE ACCOMPLISHED FOR DURING THE OPERATION OF AN ENGINE MONITORING SYSTEM. THERE ARE OPTIONS AT EACH LEVEL WHICH RANGE FROM PENCIL AND PAPER TO CAPABLE ELECTRONICS. THE MOST IMPORTANT LINK IS INTERACTION WITH THE OPERATIONAL SYSTEM.

WHY?
ON - CONDITION MAINTENANCE (OCM) REDUCES
RISK .. SAVES $$

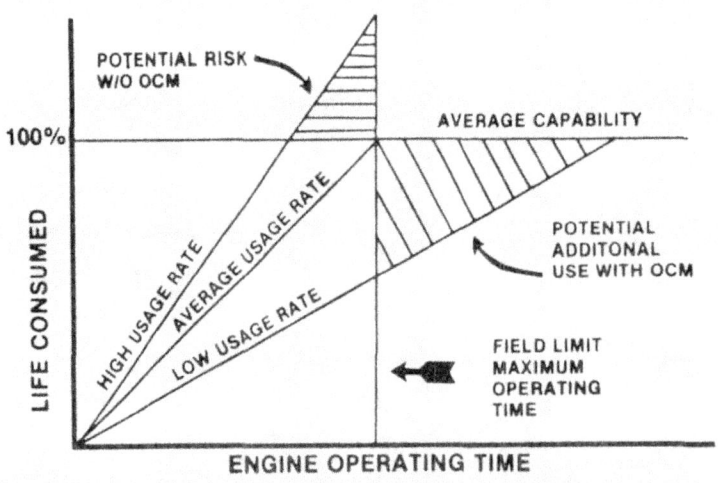

FIGURE 2 - WHY USAF WANTS ON-CONDITION MAINTENANCE (OCM). IF IT IS NOT BROKEN, WHY FIX IT? MAXIMUM OPERATING TIME MAY FIX ONE GOOD ONE WHILE ANOTHER FLIES TO FAILURE. OCM IDENTIFIES THE EXTREMES AND MAKES THE REQUIRED REPAIRS AT THE APPROPRIATE TIME.

ELEMENTS OF SYSTEM EFFECTIVENESS

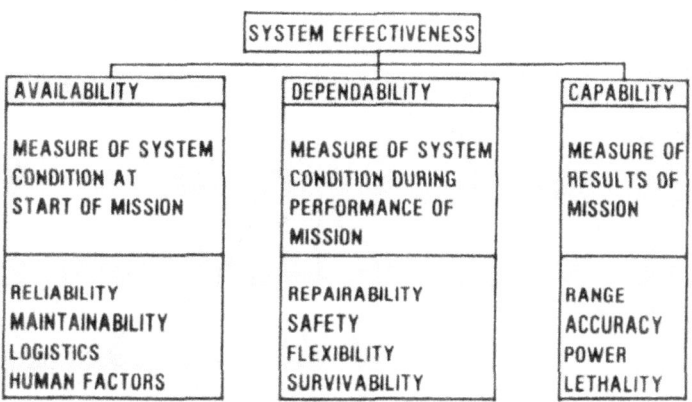

FIGURE 3 - THE ELEMENTS OF SYSTEM EFFECTIVENESS. POSITION IN STACK EXERTS GREAT INFLUENCE ON INDIVIDUAL VALUE PLACED ON EACH ELEMENT OF SYSTEM EFFECTIVENESS. THE ULTIMATE VIEWPOINT IS SYSTEM EFFECTIVENESS.

THE ADVOCATE'S PROBLEM
PROVE EMS VALUE FROM SYSTEM VIEWPOINT

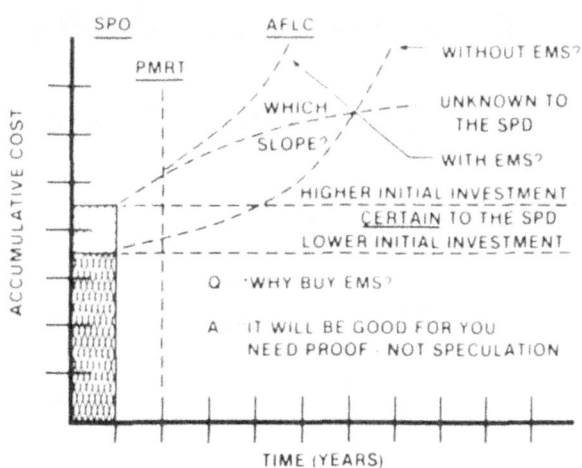

FIGURE 4 - THE ADVOCATE'S PROBLEM. CONVINCE THE SPO DIRECTOR TO SPEND THE UP-FRONT MONEY REQUIRED TO BUY AN ENGINE MONITORING SYSTEM.

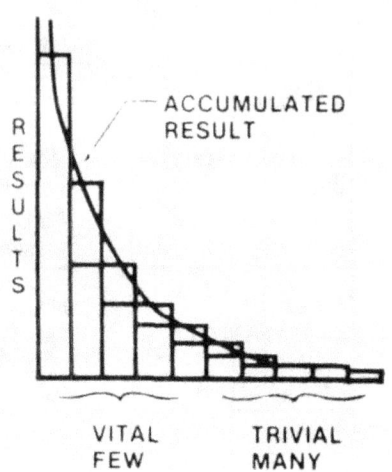

FIGURE 5 - THE TOTAL SYSTEMS VIEWPOINT AND PARETO'S LAW MAY IN COMBINATION BE THE KEY TO ANSWERING CRITICAL QUESTIONS NEEDED TO DEFINE THE REQUIRED ENGINE MONITORING SYSTEM.

F100 ENGINE DIAGNOSTIC SYSTEM STATUS TO DATE

James A. Boyless
Wright-Patterson Air Force Base

SUMMARY

An engine diagnostic system, proposed for the F100 engine, is being tested in five specially modified Tactical Air Command F-15 aircraft during a 16-month flight evaluation at Langley AFB, Virginia. After more than 3300 engine operating hours encompassing almost 900 flights during the flight evaluation, these aircraft provided a data base, still being analyzed, that has shown successful demonstration of the original functional characteristics. Table IA presents the general system evaluation in six areas while Table IB presents a more detailed look at these functional characteristics through March 81. Those areas listed as partially demonstrated are now being further tested at Langley AFB. Four general design requirements; recording engine operating time/low cycle fatigue event detection, engine trim and trend and performance data collection were demonstrated. It also successfully demonstrated validation of maintenance actions taken and indicated needed maintenance.

INTRODUCTION

The U.S. Air Force On-Condition Maintenance (OCM) concept, defined in AF Regulation 66-14, directs maintenance on the basis that the condition of the equipment dictates the need for maintenance. To adequately perform OCM, inputs from engine maintenance management tools such as oil analysis, borescope inspection, parts tracking, periodic and phase inspection, monitoring, and diagnostics are required. Of these, monitoring and diagnostics are, by far, the most difficult to achieve. Monitoring and diagnostics development activities have encompassed aircraft/engine systems from the F100/J57 to the recent F-15/F100. On each system, various parameters, both airframe and engine, have been used to provide a summary of information for maintenance personnel, logistic support and provide a feedback loop for future engine development. A review of the F-15/F100 Engine Diagnostic System (EDS) through a system description and status to date is presented.

BACKGROUND

HISTORY

1. As gas turbine engine technology increased in complexity, so too, did the need to assist maintenance personnel to perform and diagnose problems for maintenance. In addition to assisting on-base maintenance through increased emphasis on On-Condition Maintenance, logistic requirements for improved engine life usage data dictated a means of acquiring that data be developed.

2. Preliminary studies by the Air Force Propulsion Laboratory indicated that an Advanced Fighter Diagnostic System (AFDS) could prove feasible in an

application for an advanced design jet engine. The AFDS results led to definition of both hardware and software requirements as well as researching existing capabilities. Additional studies were then conducted to evaluate the significant areas of on-board processor, engine sensors, and use of existing equipment for system development. This system (renamed F100 Engine Diagnostic System (EDS)) was targeted for the F100 engine in both the F-15 and F-16 aircraft. These two aircraft powered by the same basic Pratt & Whitney F100, were chosen for the complexity and the operational environment envisioned for the engine.

SYSTEM DESCRIPTION

The F100 engine is well suited for the complex task of engine diagnostics. It is a modular engine designed for operational readiness and maintainability. It is also complex and requires knowledgeable maintenance personnel for repair. To ably assist the maintenance personnel, the F100 EDS went through an extensive review of Failure Modes and Effects Analysis (FEMA) and cost effective analysis. Thirty eight engine and airframe parameters were included in the EDS. Once these parameters were selected, hardware was developed to monitor the required information. The F100 EDS has eight primary elements that visually present cues of engine status, and/or provides a means to collect and diagnose engine anomalies. These elements shown in Figure 1 are:

Onboard: EDS Engine Sensors   Ground: Data Collection
          Engine Multiplexors (EMUX)   Unit (DCU)
          Data Processor Unit (DPU)   Diagnostic Display
          Status Panel   Unit (DDU)
          Cockpit Advisory Lights
          Cockpit Pilot Option Switch

ENGINE MULTIPLEXER (EMUX)

The engine multiplexer unit was developed under Air Force contract to collect, condition and multiplex sensor signals serially to the onboard Data Processor Unit (DPU). The EMUX replaces both the present F100 Event History Recorder (EHR) and the junction box (J-Box) for engine aircraft electrical connections. The unit is fuel cooled using existing EHR cooling lines and is hard mounted in the area vacated by the J-Box. EMUX reliability and durability to perform its functions is achieved through internal vibration isolation.

DATA PROCESSOR UNIT (DPU)

The DPU is an airframe mounted, air cooled unit consisting of a central processor, Intel 8080, core memory, and interface circuits. Both cooling and electrical power requirements are provided by the aircraft. This unit is the nerve center of the inflight monitoring system. It is programmed through software logic to detect a limit exceedance, declare an event, and store that event for later collection/diagnosis.

AIRCRAFT COMPONENTS

1. There are three aircraft components that are integral parts of the EDS; the cockpit warning lights and pilot option switch, status panel, and transfer

receptacle. These components are also depicted in Figure 1. The DPU can, on command, store a data record by means of the pilot option switch located to the left of the pilot. In addition, a cockpit warning lights indicate Fan Turbine Inlet Temperature (FTIT) Overtemperature or excessive temperature occurrences.

2. To aid Flight Line personnel to quickly determine if an aircraft can be turned around, an EDS Status Panel is located in the existing maintenance access door, 48L. The status panel has latching indicators that can be set by either the DPU or EMUX. Either built-in-test for DPU and EMUX will set these latches as well as Hot Start detection for either engine.

3. The transfer receptacle, located in the same access door as the status panel quickly connects the DPU to either the DCU or DDU for extraction of stored data. Average transfer time is six seconds. Either collection or diagnostic operation can be performed under the "wing."

DATA DIAGNOSTIC UNIT (DDU)

Just as the DPU is the nerve center of the onboard system, the DDU serves that function on the ground. It is a portable ground unit with an alphanumeric display screen and keyboard for interfacing the maintenance personnel. For storage of flight data the DDU has the capability to maintain five records. The unit shares common components with the DPU for increased maintainability. The interface of maintenance personnel and the under the "wing" is accomplished by providing power through batteries. The unit can also be used with 115 volts AC in test areas or engine shop.

DATA COLLECTION UNIT (DCU)

The DCU is small, light weight, and portable unit that uses internal battery power. It is used to collect and transfer data stored by the DPU. There are indicators for successful transfers of data from DPU and the presence of any maintenance advisory information. The DCU is designed to collect data from 10 to 15 aircraft and shares common modules with the DPU and DDU.

SYSTEM CAPABILITIES

The entire Engine Diagnostic System functional characteristics are designed to perform in five specific areas. These areas include Time and Cycle recording, Event Detection, Diagnostic and Troubleshooting, Engine Trim, and Trend & Performance data collection. A capsulized view of these capabilities vs either installed or uninstalled engine is shown in the following table:

FUNCTIONAL CAPABILITIES VS ENGINE INSTALLATION

	INSTALLED	UNINSTALLED
Time and Cycle	DPU	DPU
Event Detection	DPU	DDU
Diagnostic & Troubleshooting	DDU	DDU
Engine Trim	DPU/DDU	DDU
Trend & Performance	DPU	

A schematic view of how EDS data was collected is shown in Figure 2. The components shown detail the units involved in the airborne and ground portions of the system description.

FLIGHT EVALUATION

TEST ENVIRONMENT

1. The test environment was a Tactical Air Command operational base, where the EDS was an adjunct to the existing base level maintenance organization. The base level maintenance organizations were involved throughout the Flight Evaluation Program (FEP) but the impact of EDS on maintenance was to be on a non-interference basis. This basis was justified in the fact that EDS was in a validation phase rather than actually being incorporated into the entire fleet.

2. Both test equipment, and Auxiliary Ground Processor (AGP) were procured and installed in the EDS Laboratory. Nine permanent party individuals were on-site during the FEP.

TEST AIRCRAFT

Five Tactical Air Command (TAC) F-15 (10 F100 Engines + one spare) were specially modified with EDS equipment. A control group of 12 non-EDS F100 engines were identified for comparison.

TEST METHOD

1. The objective of the test can be summed by Figure 3. The functional capabilities of the F100 EDS were to be validated through actual inflight collection, ground transfer, and on-site evaluation of data. As a basis for validation, 3000 engine operating hours was set as a goal. Furthermore, a detailed Flight Evaluation Plan (FEP) was used as a tool in evaluating the inflight data. Every diagnostic find was verified and validated by performance of a resulting maintenance action.

2. Time and cycle recording functional capability was to be accomplished automatically by the EDS. Transfer of the recorded data would take place from the DDU to a teletype in the proper format of the present AFTO form 93.

3. To accomplish the test method the present Maintenance Action Cycle used at Langley was to be integrated with the EDS. Figure 4 shows schematically how this occurred. During the test the crewchief would check the EDS status panel to determine aircraft availability. If any work would be needed the flight dispatcher would send a technician with the DDU or DCU. The DDU provided information would be reviewed by the propulsion maintenance unit with advice and/or assistance provided by the EDS team.

4. To gather the data, there were five F-15 aircraft and eleven engines specially modified for the test (See Figure 5). These aircraft were to average between 45 and 60 engine operating hours per month (See Figure 6).

5. Validation of the data included an indepth critique of the inflight data. There were five categories in which the data were grouped. These groups include Hits, Goods, False (I & II) and Misses. Hits would be scored as follows; An event was declared only by EDS and/or the event was confirmed by the present reporting system, a pilot or maintenance write up. A good is an event not declared by EDS nor reported by the present system. On the other hand, False I is an event declared by EDS and not by the system while False II is false but a known "fix" is in work to remedy the cause. Finally, a Miss is a pilot or maintenance write up not detected by EDS when it should have been.

6. Engine trim both installed and uninstalled would be performed in place of the present trim procedures using the present M-37 test stand. Careful monitoring of the time used to trim, and fuel used were recorded for comparison against non-EDS engines.

7. Diagnostics and troubleshooting was to be evaluated by careful review of actual usage of the equipment by the Maintenance personnel. If maintenance was declared once an event detection occurred, the procedure called for the repairman to use the DDU. Once validation of the event occurred, the DDU was to be used to diagnose or "troubleshoot" the malfunction. Maintenance records would be screened and data kept that expressed the amount of usage the DDU attained, time to troubleshoot, and diagnose malfunctions and compared to the control group.

8. Once the diagnostic and troubleshooting scenario ended and an Engine trim was required, the fourth capability was to be evaluated. In fact, this evaluation included all engines to be trimmed. Records were to be kept for manhours required to trim and fuel used.

9. Finally, the ability of the system to collect trend and performance data was to be tested.

RESULTS AND DISCUSSION

1. The Flight Evaluation Program (FEP) test results are presented in the succeeding figures. These results are based upon the period 1 Apr - 12 Dec 80 and the data gathered at Langley AFB, Virginia. The FEP, because of software complexity, was divided into a debug and actual validation period. Figure 7 gives a comparison of the actual vs projected engine operating hours. There were 2577 hours attained by 12 Dec 80 and an additional 738 hours through 26 Mar 81. Time and Cycle data was automatically printed by a printer to correspond with the actual AFTO Form 93 required by the engine management information system (see Figure 8). From the beginning of the program there were 13 events that were continuously monitored from start-up to shut-down of the engine. As the program progressed, however, there were lessons learned that deleted one event and switched four from No-Go to Maintenance Advisory. These four events are depicted in Table III. The system's ability to detect events accurately was extremely important. Figure 9, Event Detection Accuracy, shows how the accuracy of the system progressed. The check sum on the abcissa is an identity associated with software improvements of the basic event detection logic. As can be seen accuracy increased from a point of 88.7 percent to 99.7 percent at 12 Dec 80. The 99.7 percent assessment is based upon

77 transfers of data (See Figure 10) where 63 Hits were recorded. These Hits inturn were used to recommend maintenance action. There were 1006 goods reports with only 3 False I events detected.

2. These results demonstrate the successful capability of EDS to detect events. By using the last operable checksum, 0119, the entire evaluation period was reviewed and summarized as shown in Table V. Of the twelve events, continuously monitored by EDS there were five events that were detected on numerous flights during the evaluation period. Three events clearly stand out. Stalls were very prevalent during the evaluation period. How extensive they are can only be speculated at the present time. Detailed analysis of this event continues.

3. Fan Turbine Inlet Temperatures (FTIT) Spread events were also numerous during the evaluation. These EDS modified engines have reported numerous FTIT Spread Events. Investigation continues to determine cause and effect on the engine. One attempt to seek information on effects has been to change borescope (Visual) inspection of the three engines to a 50 hour interval rather than the 100 hour interval presently used.

4. Just as Stalls and FTIT events have been prevalent, Rear Compressor Variable Vane (RCVV) events have also been numerous. These events have been tracked throughout the evaluation period. Investigation as to cause continues. Diagnostics and Troubleshooting capability have been demonstrated by EDS personnel in the laboratory and to a limited extent by maintenance personnel. Diagnostics and Troubleshooting by the EDS personnel throughout the evaluation period occurred on a daily basis to confirm detected events and recommend maintenance. Maintenance personnel used EDS in a limited manner as a maintenance tool as well as a training aid.

5. Tables V, VIA and B show the potential engine and maintenance saves credited to EDS during the flight evaluation. Engine saves included a high scavenge pressure event that the pilot was unaware of. Had the discrepancy continued the engine could have reached the catastrophic state of complete loss. The most obvious save for maintenance is preventing mis-directed maintenance. Four of the eight pilot assessments included dual engine anomalies whereas EDS confirmed only one engine had the anomaly not both.

6. EDS engine trim capability was demonstrated. Both uninstalled and installed engine trim was performed using EDS. After five partially successful attempts at uninstalled trim, identified software changes have been made and testing continues. Installed engine trim has been successfully demonstrated after several attempts. The entire trim procedure with exception of Engine Pressure Ratio (EPR) check was performed. EPR check could not be accomplished due to a false sensor reading.

7. Finally, trend and performance data was collected. Accumulated data indicates that 183 data points were captured EDS and the resulting analysis revealed that 74% of the data points lay in the lower PLA range (See Figure 11). This range, $30°-40°$, is the area of idle reset area where the augmentor is wide open. The conditions for data to accurately reflect trending and performance **required** a stabilization time of 180 seconds and was often reached in a landing

approach. But due to the reprogramming ability of EDS a change of constants for PLA was approved and capture of data will be in the desired PLA range, 40-80°.

SYSTEM EVALUATION
SUMMARY

- OVER 650 FLIGHT SETS OF DATA ANALYZED
- DEVISED/IMPLEMENTED SOFTWARE DIAGNOSTIC TECHNIQUES
- EVALUATED 8 SETS OF SOFTWARE - 3 OFPs, 2 ODPs, 2 OCPs
- VALIDATED SYSTEMS CONTRIBUTION IN AREA OF
    - ENGINE ANOMALIES
    - MAINTENANCE ACTIONS
- VALIDATED SYSTEMS ABILITY TO COLLECT DATA
    - TIME/CYCLE
    - PERFORMANCE/TREND
- DEMONSTRATED SYSTEM FLEXIBILITY THRU REPROGRAMMABILITY

TABLE IA

CAPABILITIES
SUMMARY OF RESULTS
(1 APR 80 - 26 MAR 81)

- DATA COLLECTION - 87% OF FLIGHTS RECORDED
- EVENT DETECTION - 99% ACCURATE
- DATA ANALYSIS - DEMONSTRATED
- TRIM CAPABILITY - PARTIALLY DEMONSTRATED
- GROUND DIAGNOSTICS - DEMONSTRATED
- USER EVALUATION - PARTIALLY DEMONSTRATED
- TREND AND PERFORMANCE - DATA COLLECTED

TABLE IB

EDS TOTAL SYSTEM DESIGN
FUNCTIONAL CHARACTERISTICS

- DETECT EVENTS AND LIMIT EXCEEDANCES
- COLLECT IN-FLIGHT TREND DATA
- PROVIDE IN-FLIGHT PERFORMANCE CHECK CAPABILITY
- RECORD OPERATING TIME AND LCF COUNTS
- PROVIDE "NO-GO" INDICATION AT THE FLIGHT LINE
- CONDUCT FAULT ISOLATION AND DEFINE MAINTENANCE ACTIONS
- PROVIDE HARD COPY OF ENGINE RECORDS FOR INPUT INTO CENTRAL DATA SYSTEMS
- PROVIDE CAPABILITY TO PERFORMANCE ENGINE TRIM

TABLE II

EVENT MENU

EVENT TYPES	AT START OF FLIGHT PROGRAM 13 EVENTS (11 NO-GO)		AT END OF FLIGHT PROGRAM 12 EVENTS (6 NO-GO)	
	NO-GO	MAINTENANCE ADVISORY	NO-GO	MAINTENANCE ADVISORY
HOT START	X	—	X	—
N_2 OVERSPEED	X	—	X	—
FTIT OVERTEMP	X	—	X	—
FTIT SPREAD OUT OF LIMITS	X	—	—	/
OIL PRESSURE OUT OF LIMITS	X	—	X	—
SCAVENGE PRESSURE OVER LIMITS	X	—	—	/
VIBRATION OVER LIMITS	X	—	—	/
EEC FAULT	X	X [1]	X	/ [1]
ENGINE STALL	X	X [2]	—	/
AUGMENTOR BLOW-OUT-MISLIGHT	X	X [2]	—	/
RCVV OUT OF LIMITS	—	X	—	/
MAIN FUEL PUMP DETERIORATION	—	X	—	—
MAIN FUEL PUMP FAILURE	X	—	X	—

[1] If cleared by pilot   [2] If out of envelope

TABLE III

EVENTS* (HITS) DETECTED PER ENGINE
10 JULY THRU 18 DECEMBER

		ENGINE S/N											TOTALS
		160	311	330	415	470	528	639	694	722	801	907	
ENGINE NO-GO	HOT START					2							2
	O'SPEED												0
	O'TEMP			1					1				2
	OIL PRESS.	1					2					1	4
	EEC	4		7		4	8			1			24
	MFP FAIL												0
MAINTENANCE ADVISORY	STALL			1		1	11	2	3	3			21
	SPREAD	23	22	38				1	1				85
	SCAV PRESS.					9							9
	AUG BO/ML			1			1						2
	RCVV		5	19		2	3	1				16	46
	VIBS		4		3	13	4	4	7	6		1	42
RECORDS	TREND	16	15	29	9	17	9	24	27	18	6	13	183
	PERF	3	3	17	4	3	2	6	10	12	4	3	67

*Hits — as determined by checksum 0119 in use at the end of the program

TABLE IV

POTENTIAL ENGINE "SAVES" BY EDS

EVENT	ENGINE S/N	PILOT REPORTED	CORRECTIVE ACTION
OIL PRESS, LOW	311	YES	SERVICED OIL TANK
SCAV PRESS., HI	470	NO	VAC CHECK NO. 4 COMPARTMENT. FOUND FOREIGN MATERIAL IN ENGINE OIL SYSTEM.
O'TEMP	330	YES	EDS DETECTED O'TEMP LEVEL HIGHER THAN REPORTED BY THE PILOT. EEC CHANGED.
FTIT SPREAD	160 311 330	YES* YES* YES*	BORESCOPE EVERY 50 FLIGHT HOURS AS A PRECAUTION UNTIL PHENOMENON AND CONSEQUENCES CAN BE QUANTIFIED.
FAILED FTIT PROBE	311 694 694	N.A. N.A. N.A.	REPLACED NO. 4 PROBE, VERIFIED REPLACED NO. 4 PROBE, VERIFIED REPLACED NO. 5 PROBE, VERIFIED
FAILED COCKPIT WARNING LIGHT		NO	NON BILL OF MATERIAL RELAY PANEL BLOCKED AN EEC FAILURE WARNING TO COCKPIT.

*FTIT Spread is not monitored in aircraft other than EDS equipped aircraft.

TABLE V

MAINTENANCE "SAVES" BY EDS

ENGINE S/N	PILOT ASSESSMENT	EDS RECORD
694	HAVE TO MISMATCH THROTTLES TO MATCH RPM	PILOT OPTION DATA RECORD CONFIRMED MISMATCH IN RPM, FTIT. PLA RIGGING.
470	NO COMPLAINTS	REPEATED EEC LEVEL 1 FAULTS. ODU CABLE SHORTED.
639	NOT APPLICABLE	SEVERAL FALSE RCVV EVENTS ON RECENT FLIGHTS, TT2.5 ERROR. MISSING AP2 PLUG.
722	LOW THRUST RPM IN STABILITY	LOW OUT OF TRIM, PILOT OPTION RECORD.
801 (528)	A/B BLOWOUT ON BOTH ENGINES (528/801)	NOTHING WRONG WITH 801. ENGINE S/N 528 HAD A "HARD LIGHT/BLOWOUT" FOLLOWED BY A STALL. EPR HIGH 0.11.

TABLE VIA

MAINTENANCE "SAVES" BY EDS

ENGINE S/N	PILOT ASSESSMENT	EDS RECORD
311	A/B BLOWOUT ON 311, TOOK PILOT OPTION.	STALL FOLLOWING AN AUGMENTOR "HARD LIGHT/BLOWOUT." RCVV's OUT OF BAND, AXIAL ON STALL, RCVV AND PILOT OPTION EVENTS.
907 (528)	A/B BLOWOUT ON BOTH ENGINES (907/528).	NOTHING WRONG WITH 907. ENGINE S/N 528 HAD A "HARD LIGHT/BLOWOUT" FOLLOWED BY A STALL.
722	PILOT REPORTED AUGMENTOR ANOMALIES ON THREE FLIGHTS. ON SECOND FLIGHT DOUBLE HARD LITE ON BURNERS.	EDS DETECTED STALLS IN AUGMENTATION ON EACH OF THE THREE FLIGHTS FOR ENGINE S/N 722 ONLY.

TABLE VIB

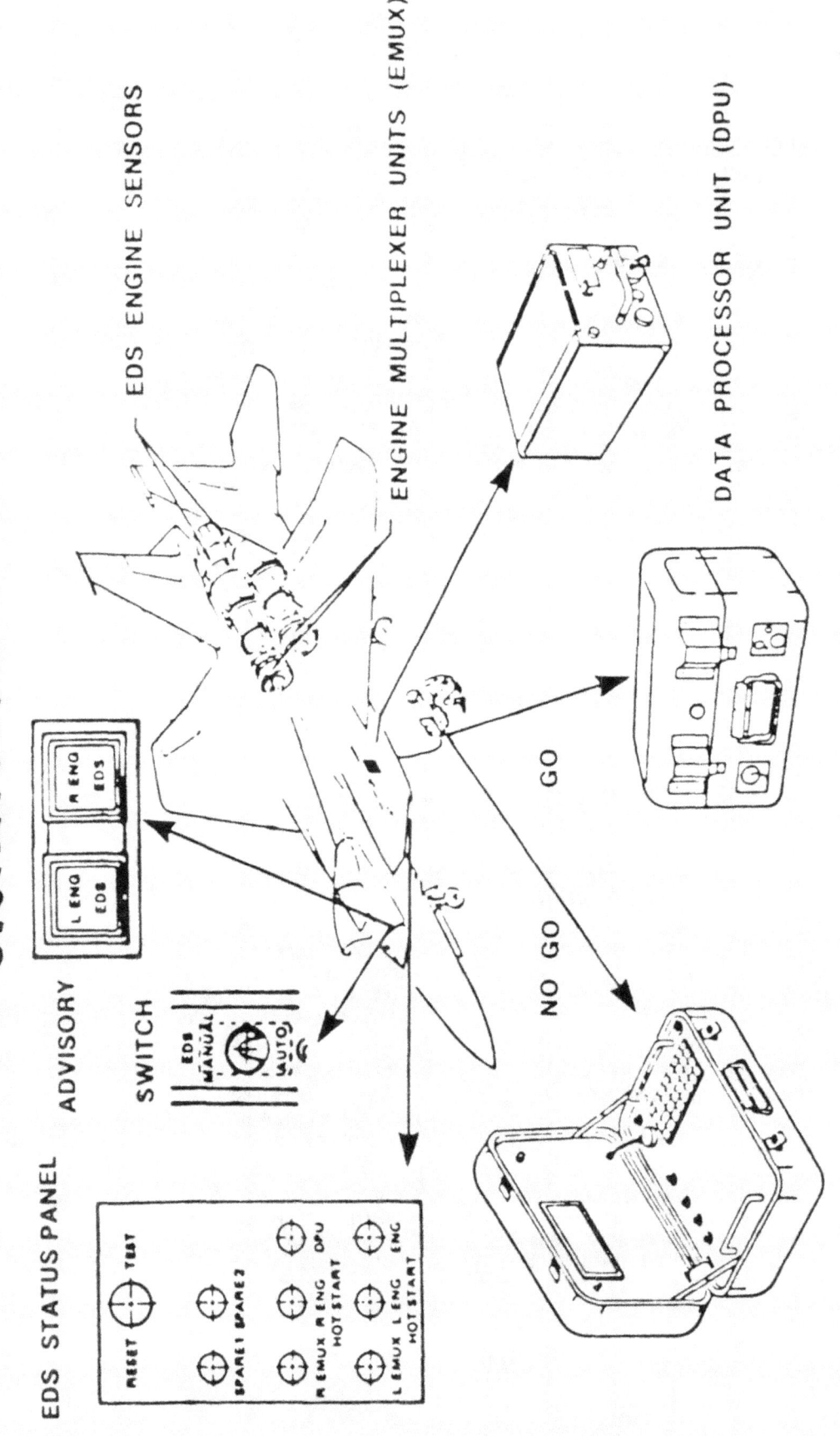

FIGURE 1

EDS DATA COLLECTION

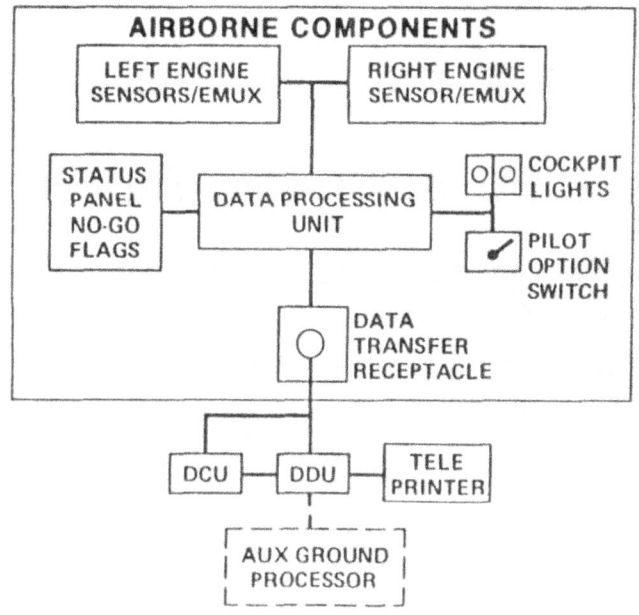

FIGURE 2

OBJECTIVE

- OVERALL
  - VALIDATE SYSTEM CONCEPT, DESIGN, AND CAPABILITIES

FIGURE 3

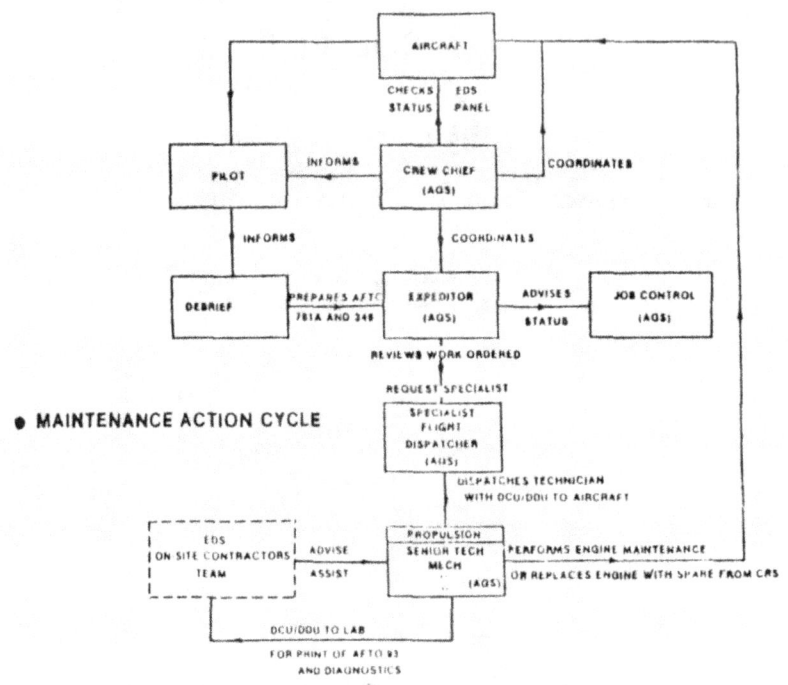

• MAINTENANCE ACTION CYCLE

FIGURE 4

METHOD
AIRCRAFT/ENGINE ASSIGNMENT

AIRCRAFT (5)
74-099
74-103
74-105
74-107
74-108

F100 ENGINES (11)

PW E680160	PW E680639
PW E680311	PW E680694
PW E680330	PW E680722
PW E680415	PW E680801
PW E680470	PW E680907
PW E680528	

FIGURE 5

EDS EXPECTED FLIGHT EVAL
ENGINE OPERATING HOURS

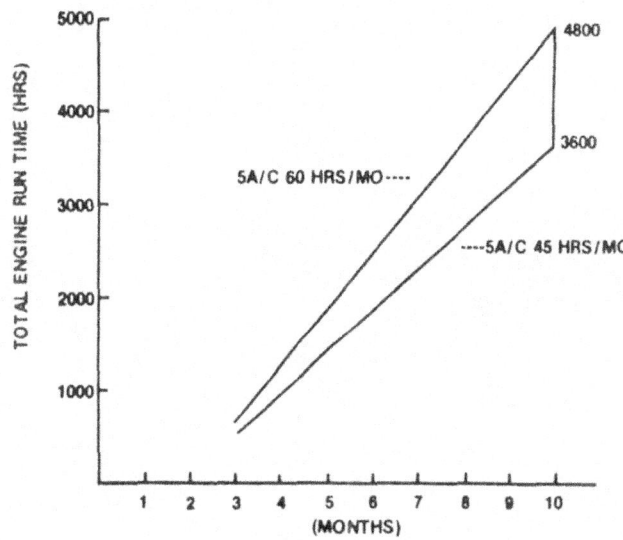

FIGURE 6

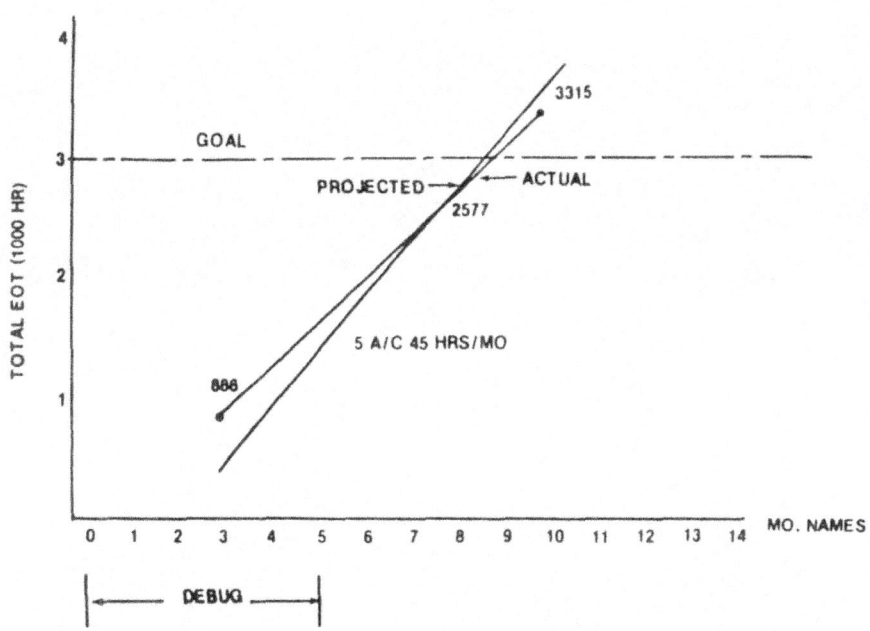

FIGURE 7

ENGINE TIME AND CYCLE DATA RECORDS

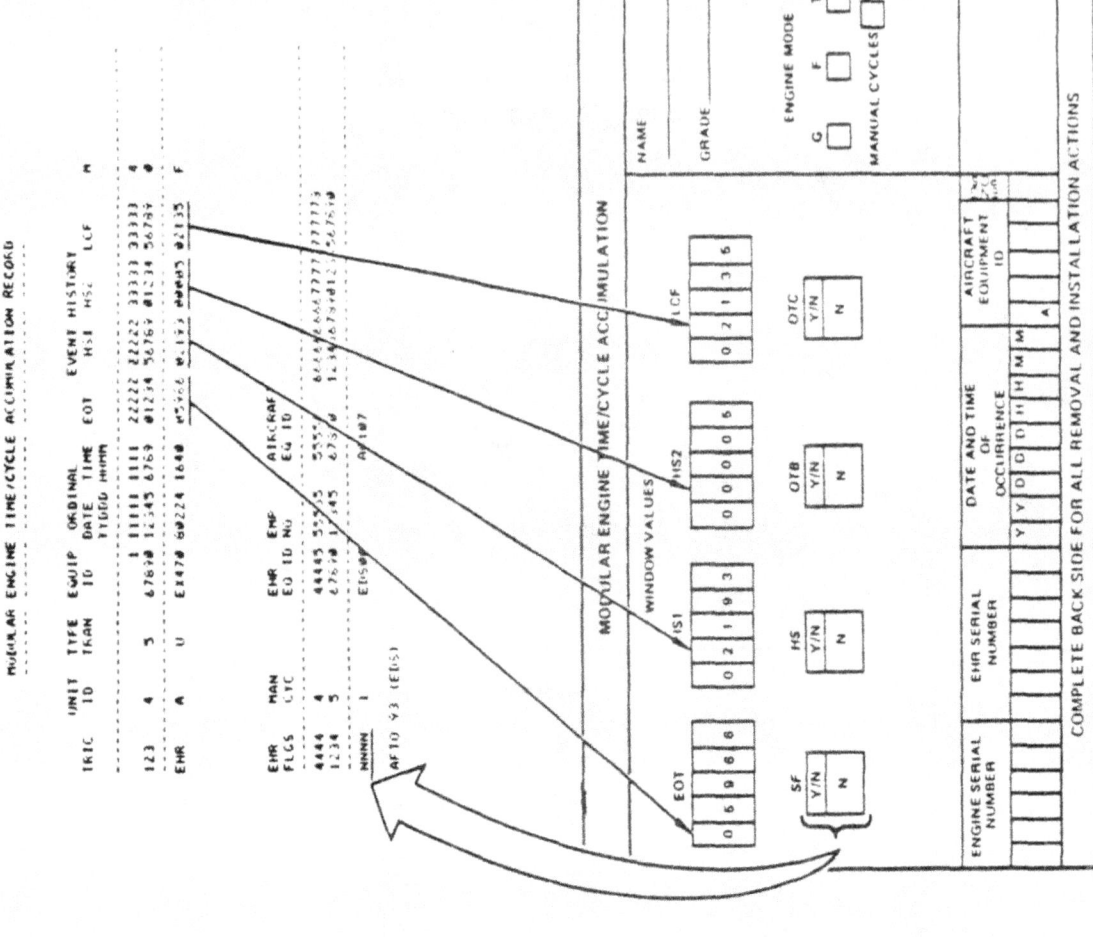

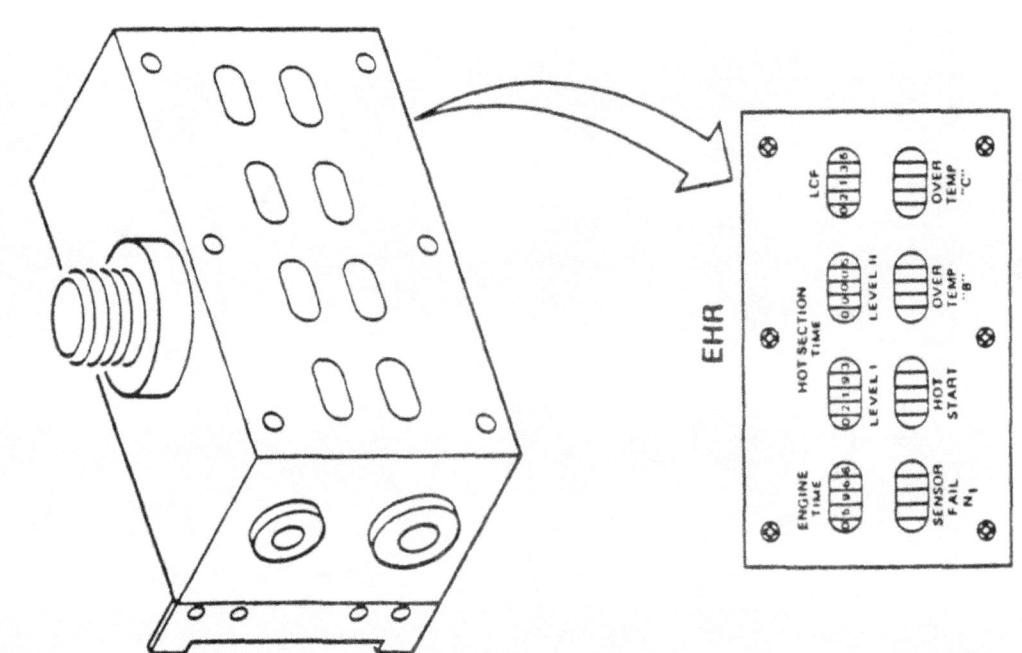

FIGURE 8

EVENT DETECTION ACCURACY

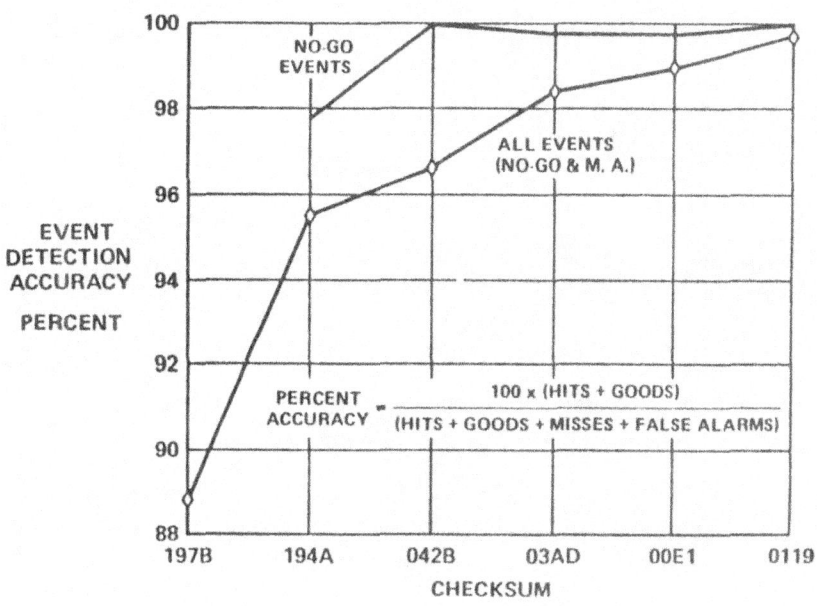

FIGURE 9

SUMMARY OF
EVENTS DETECTED
(THRU DEC 80)

	HITS	GOODS	FALSE II	FALSE I	MISSES	ACCURACY %
TOTAL	63	1006	6	3	0	99.7

FIGURE 10

TREND/PERFORMANCE RECORD SUMMARY

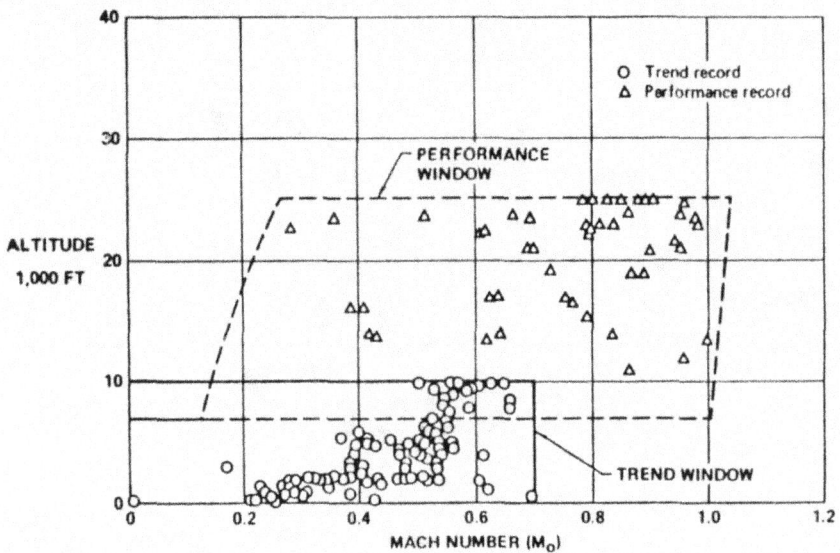

FIGURE 11

TURBINE ENGINE PERFORMANCE ESTIMATION AND ITS ROLE IN FUTURE SYSTEMS

Ronald L. DeHoff
Systems Control, Inc.

Charles A. Skira
Wright-Patterson Air Force Base

(Text was unavailable at the time of printing.)

OBJECTIVES

- EVALUATE DATA ACCURACY, REPEATABILITY AND SENSOR VARIATIONS
- DEVELOP ALGORITHMS TO REDUCE PERFORMANCE DATA TO USABLE PARAMETERS
- ASSESS APPLICABILITY OF RESULTS TO THE ENGINE MAINTENANCE PROCESS

FIGURE 1

MIMS INFORMATION ARCHITECTURE

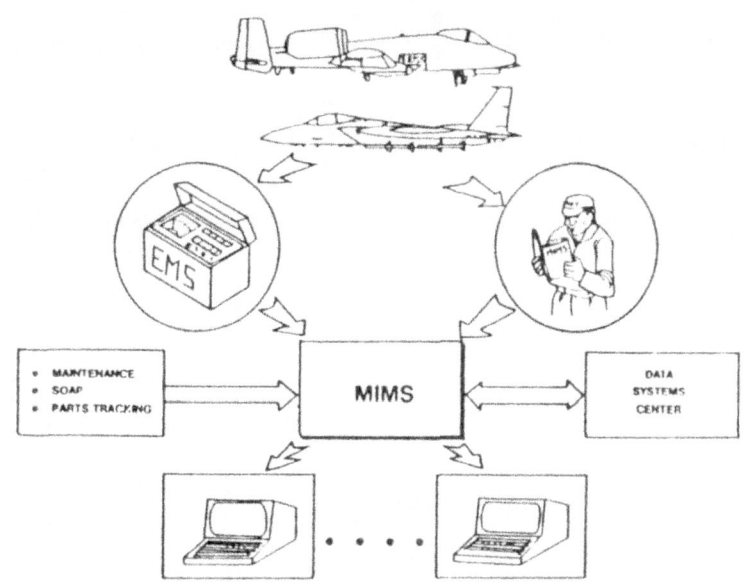

FIGURE 2

PERFORMANCE DATA PROCESSING FLOW PATH

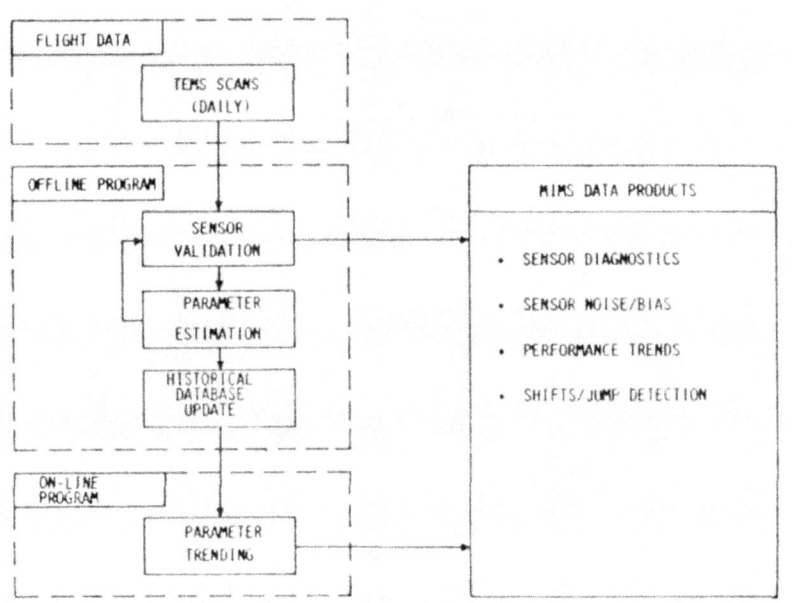

FIGURE 3

ENGINE MODEL

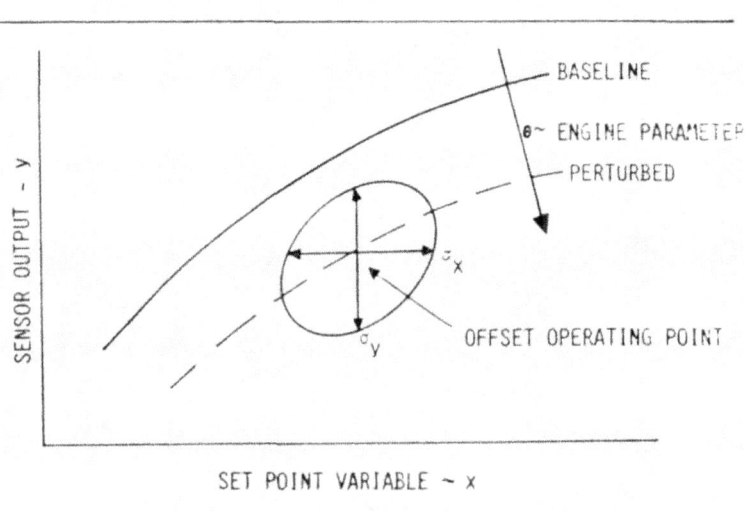

$$y = \underbrace{f_0(x)}_{\text{BASELINE}} + \underbrace{f_\theta(x)\theta}_{\text{PERTURBATION}} + \underbrace{v}_{\text{NOISE}}$$

FIGURE 4

MODEL DEVELOPMENT TECHNIQUES

- BASELINE ($f_0(x)$)

    FIT POPULATION OPERATING DATA WITH A GROUP OF CORRELATED EXPLANATORY VARIABLES

- PERTURBATION ($f_\theta(x)\,\theta$)

    FIT PERTURBATION DATA GENERATED AT REPRESENTATIVE FLIGHTPOINTS BY VARYING ENGINE PARAMETERS SINGLY

FIGURE 5

SUBSET REGRESSION - A MODEL GENERATION ALGORITHM

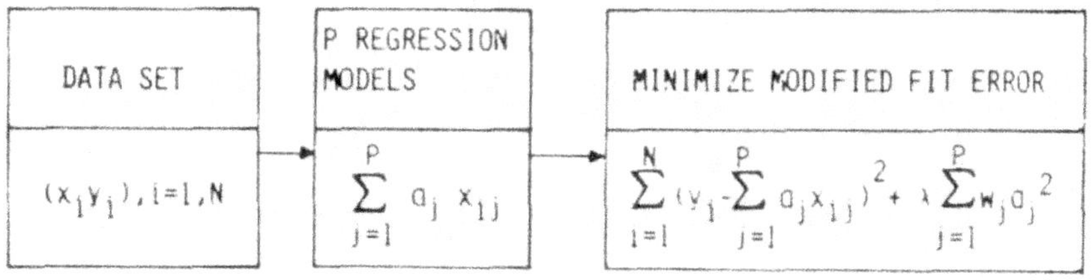

DATA SET	P REGRESSION MODELS	MINIMIZE MODIFIED FIT ERROR
$(x_i, y_i), i=1, N$	$\sum_{j=1}^{P} a_j x_{ij}$	$\sum_{i=1}^{N}\left(y_i - \sum_{j=1}^{P} a_j x_{ij}\right)^2 + \lambda \sum_{j=1}^{P} w_j a_j^2$

MODEL FOR P = 1: $\hat{y} = a_1 x_1 + v_1$

MODEL FOR P = 2: $\hat{y} = b_1 x_1 + b_2 x_2 + v_2$

MODEL FOR P = 3:   ETC.

FIGURE 6

TYPICAL BASELINE MODEL

$$W_F = a_1 (T_{2C})^2 N_F + a_2 P_{S3} N_F + a_3 T_{2C}^2 + a_4 + v$$

FIGURE 7

DATA SCREENING

- ELIMINATE FAILED CHANNELS

- VERIFY DATA STATIONARITY

- NORMALIZE POPULATION

FIGURE 8

FLIGHT SERVICE EVALUATION DATA PROCESSING

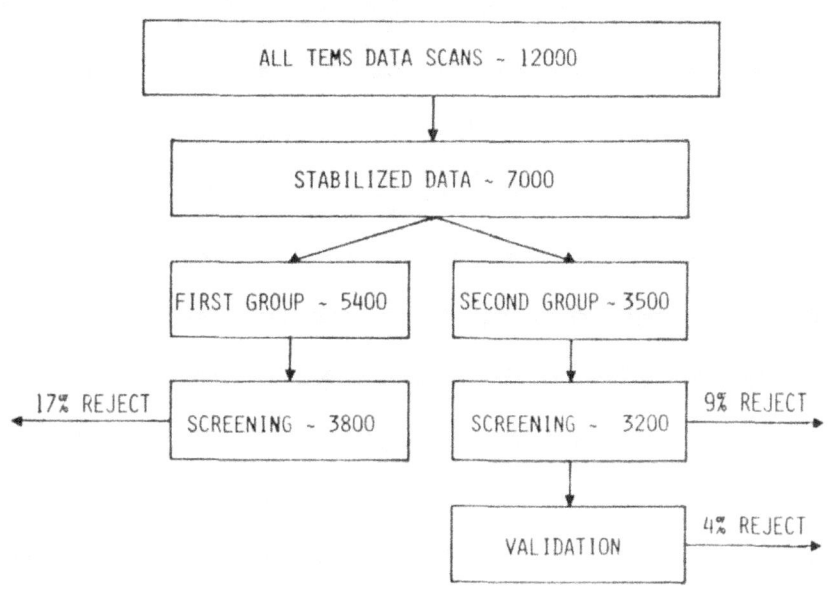

FIGURE 9

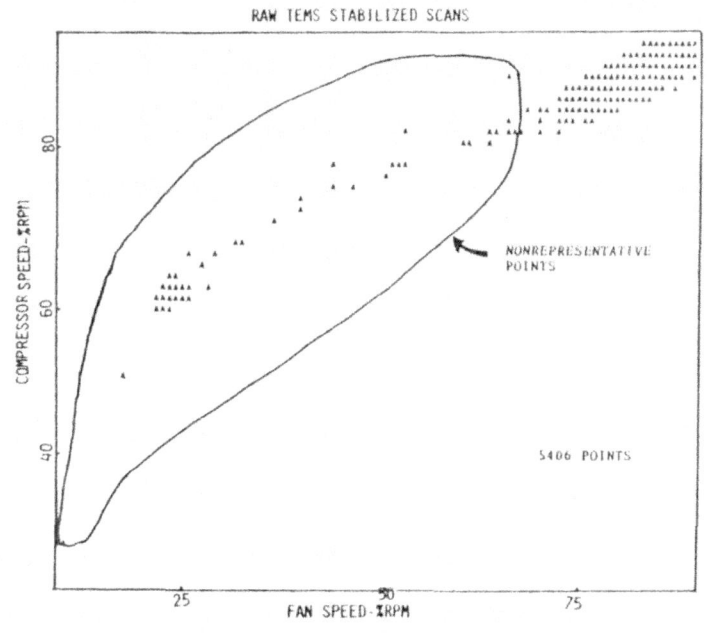

FIGURE 10

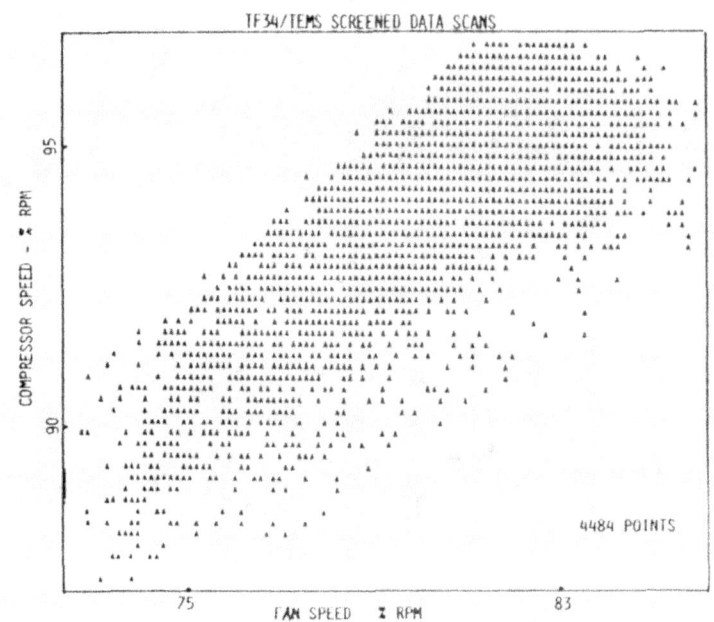

FIGURE 11

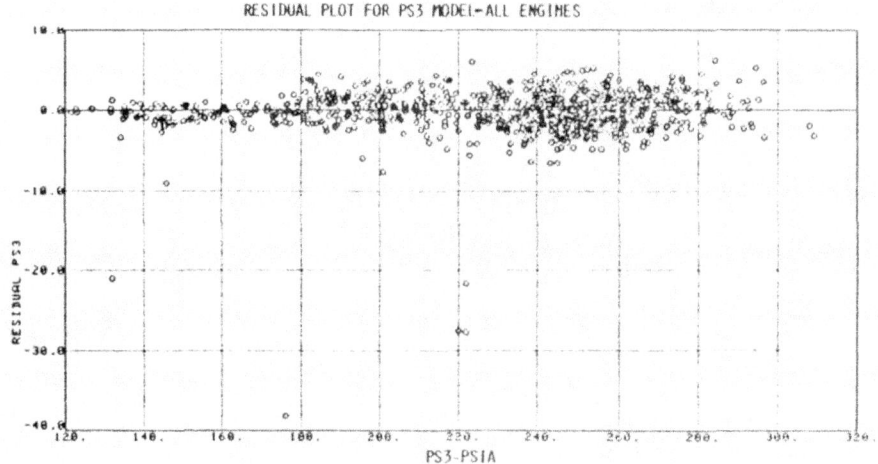

FIGURE 12

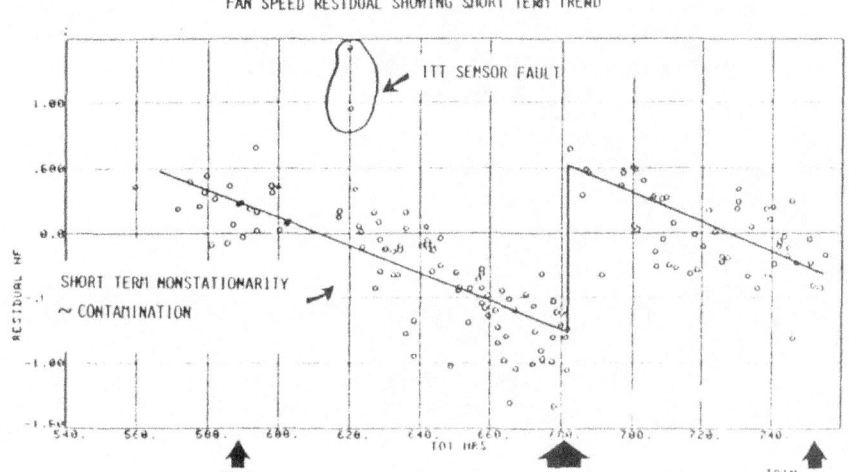

FIGURE 13

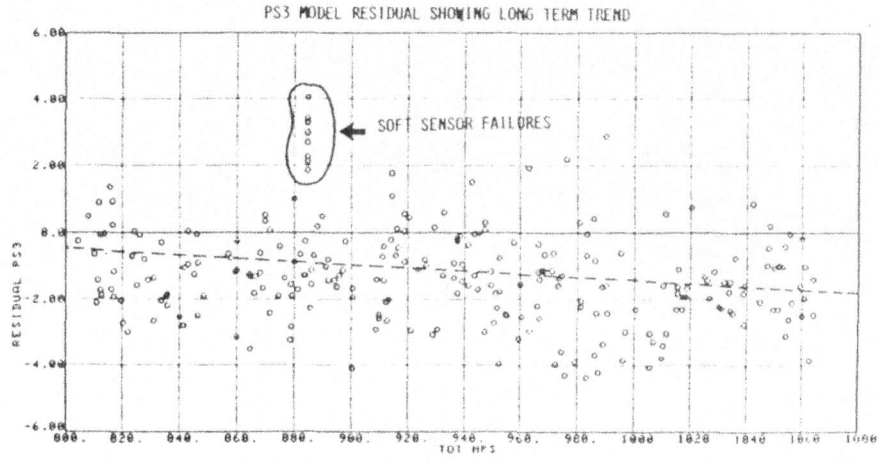

FIGURE 14

SENSOR VALIDATION ALGORITHM

- USES ENGINE MODEL DERIVED FROM OPERATING DATA INCLUDING CHANNEL ERROR VARIANCE

- USES RANGE CHECKS FOR HARD FAILURES

- USES CHANNEL ERROR STATISTICS FOR SOFT FAILURES

- DETECTS MULTIPLE FAILURES WITHOUT FAULT TREE

- ESTIMATES FAILED CHANNEL READING FOR SUBSEQUENT DIAGNOSTIC UTILIZATION

FIGURE 15

SENSOR VALIDATION FLOWPATH

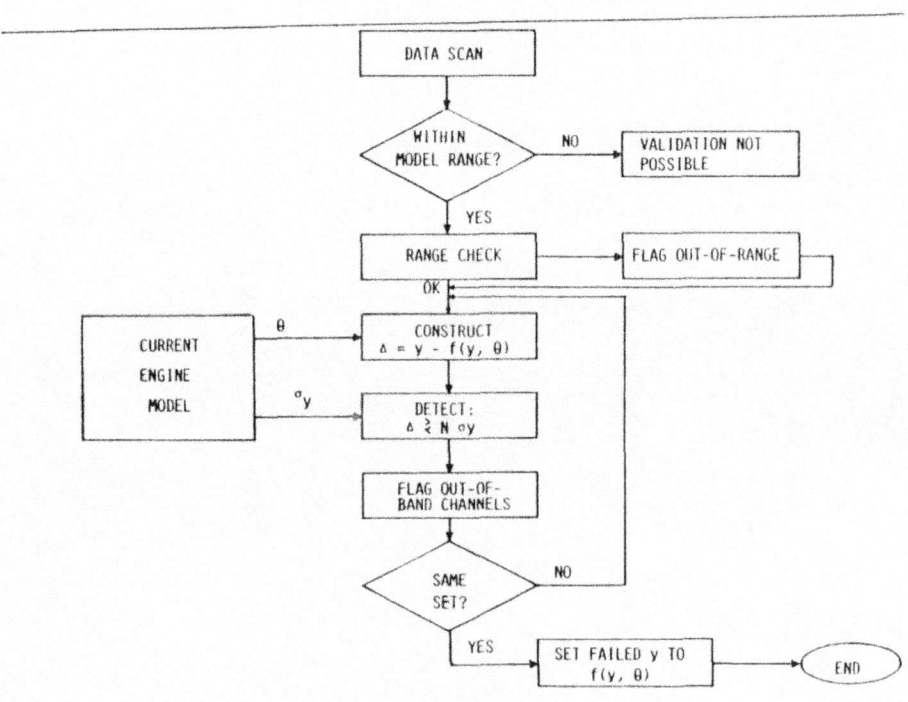

FIGURE 16

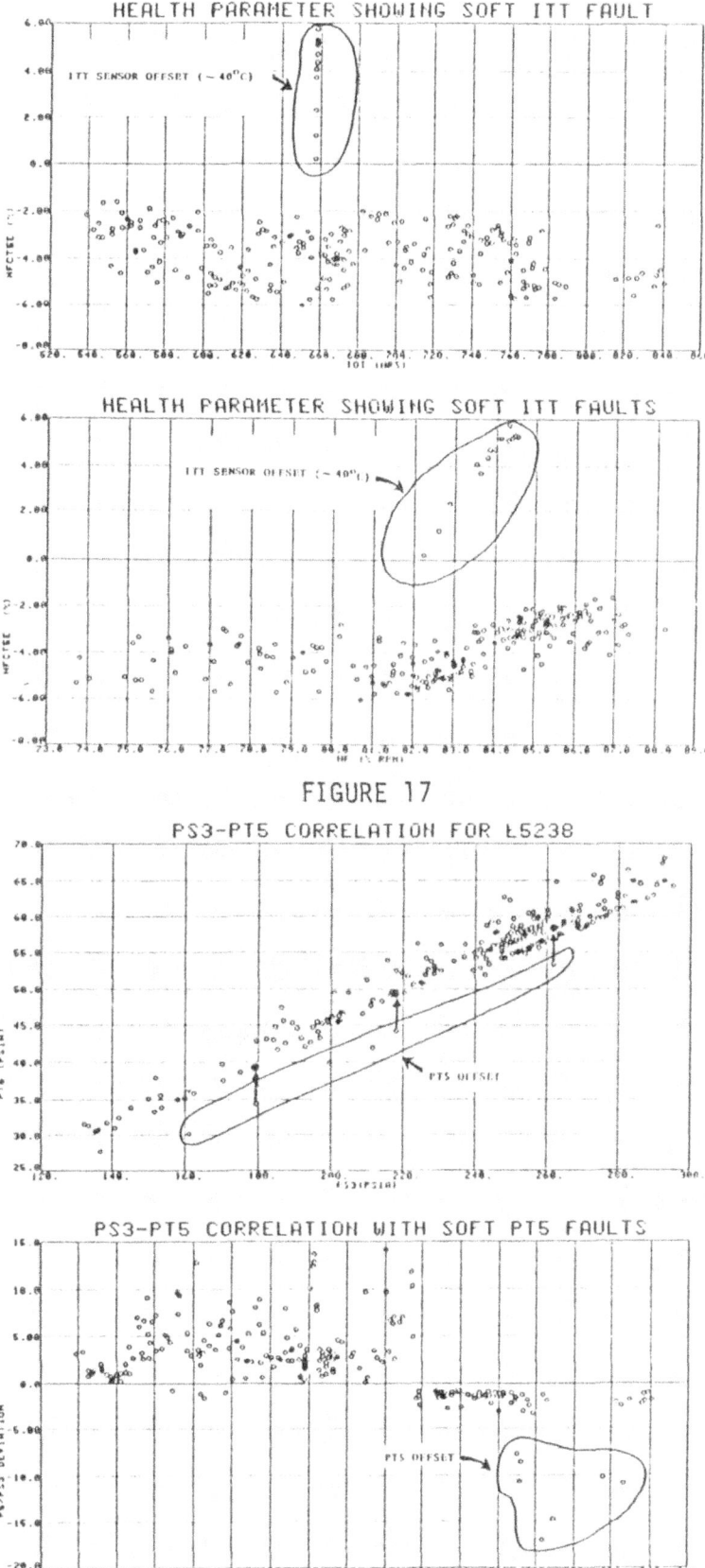

FIGURE 17

FIGURE 18

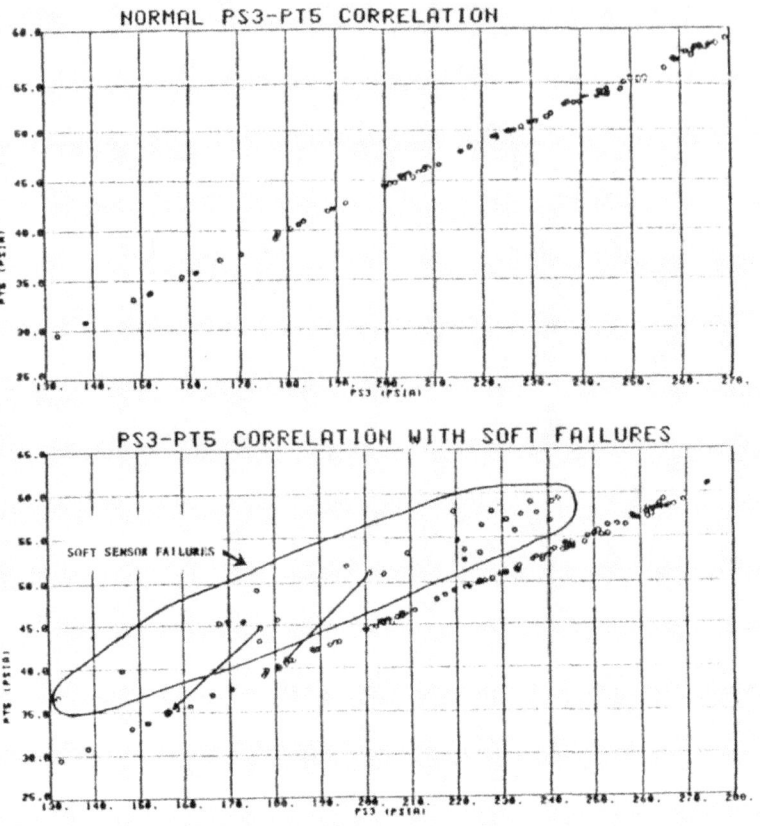

FIGURE 19

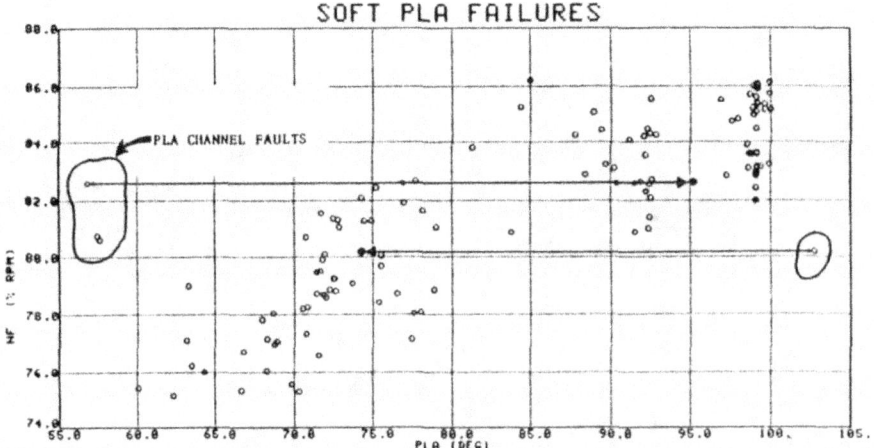

FIGURE 20

PERTURBATION MODEL GENERATION

- SELECT REPRESENTATIVE FLIGHTPOINTS
- RUN SIMULATION IN NOMINAL CONFIGURATION
- PERTURB COMPONENT DESCRIPTORS THROUGH EXPECTED RANGE OF VARIATION
- CALCULATE PERTURBATION FROM BASELINE
- FIT SIGNIFICANT VARIATIONS AS (POSSIBLY NONLINEAR) FUNCTIONS OF SETTING PARAMETERS

FIGURE 21

PERTURBATION ENGINE MODELS

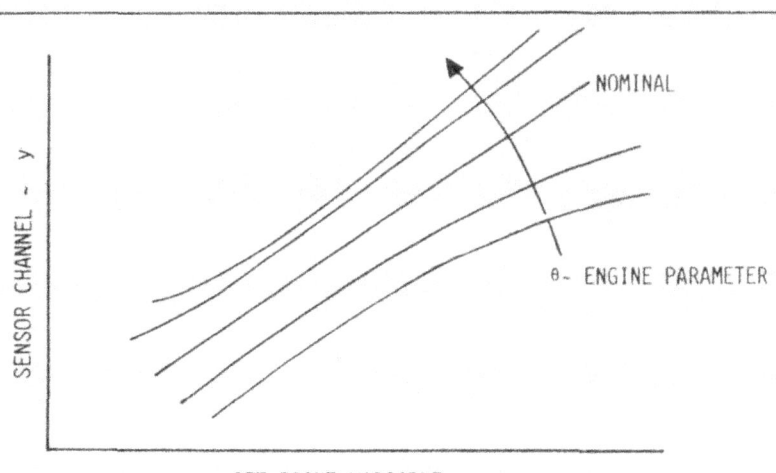

- PERTURBATIONS ARE LINEAR ABOUT NOMINAL
- PERTURBATIONS ARE DEPENDENT ON SET POINT
- PERTURBATIONS REFLECT COMPONENT CONDITION AND ARE ADDITIVE

FIGURE 22

TYPICAL PERTURBATION MODEL

$$W_F = (W_F)_{BL} + b_1 W_F \Delta n_{FAN} + (b_2 + b_3 N_F) \Delta A_{FAN} + b_4 P_{S3} \Delta n_{HC}$$

$$+ b_5 W_F \Delta A_{HC} + \ldots \ldots$$

FIGURE 23

ENGINE PERFORMANCE MONITORING

- DIRECTIONAL VS. DIRECT INVERSION

- MODULE DIRECTED RATING PARAMETERS

- TF34 PERFORMANCE ESTIMATOR

FIGURE 24

APPROACHES TO PERFORMANCE MONITORING

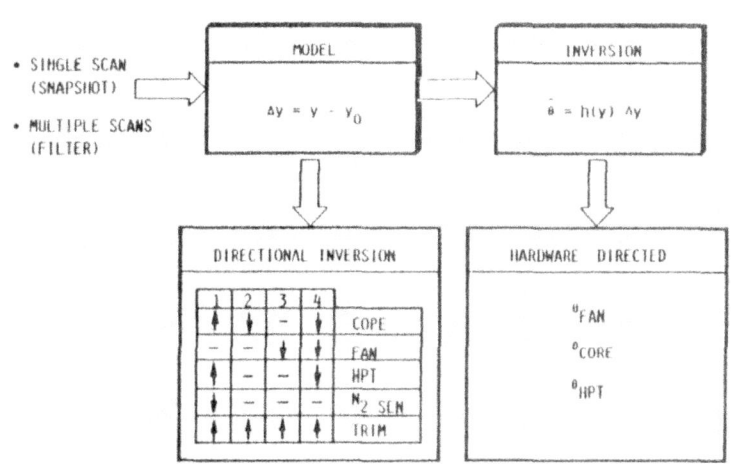

- SINGLE SCAN (SNAPSHOT)
- MULTIPLE SCANS (FILTER)

FIGURE 25

PERFORMANCE ESTIMATOR

- USES VALIDATED DATA SCANS
- AVERAGES OVER FIXED WINDOW
- CALCULATES PARAMETER ESTIMATES AND UNCERTAINTY

FIGURE 26

PERFORMANCE ESTIMATION FLOWPATH

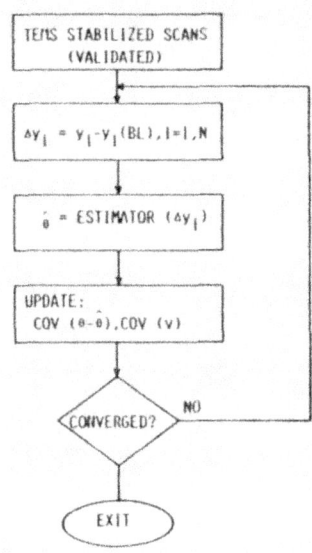

FIGURE 27

PERFORMANCE ESTIMATOR

MODEL: $y = H\theta + v$

NOISE: $v = N(0,R)$

ESTIMATOR: $\hat{\theta}_{i+1} = \hat{\theta}_i + D_i \nu_i$

DISPERSION: $D_i = (D_{i-1}^{-1} + H_i^T R^{-1} H_i)^{-1}$

INNOVATION: $\nu_i = y_i - H\hat{\theta}_i$

RESIDUAL: $\tilde{\nu}_i = y_i - H\hat{\theta}_{i+1}$

COVARIANCE UPDATE: $D_k = D_{k-1} + \infty \left(\text{diag}(\tilde{\nu}_k \tilde{\nu}_k^T) - D_{k-1} \right)$

FIGURE 28

MODULE DIRECTED PARAMETERS

- EFFICIENCY AND FLOW AREA REPRESENT INDEPENDENT CONSTITUTIVE (i.e., MASS AND ENERGY) PARAMETERS

- THEY DO NOT REPRESENT INDEPENDENT FAILURE MODES

- ALL FAILURE/DETERIORATION MODES ARE COMBINATIONS OF INDEPENDENT PARAMETERS

- PARAMETER ESTIMATION CAPABILITY IS LIMITED BY SENSOR SET

FIGURE 29

COMPONENT FAILURE MODES

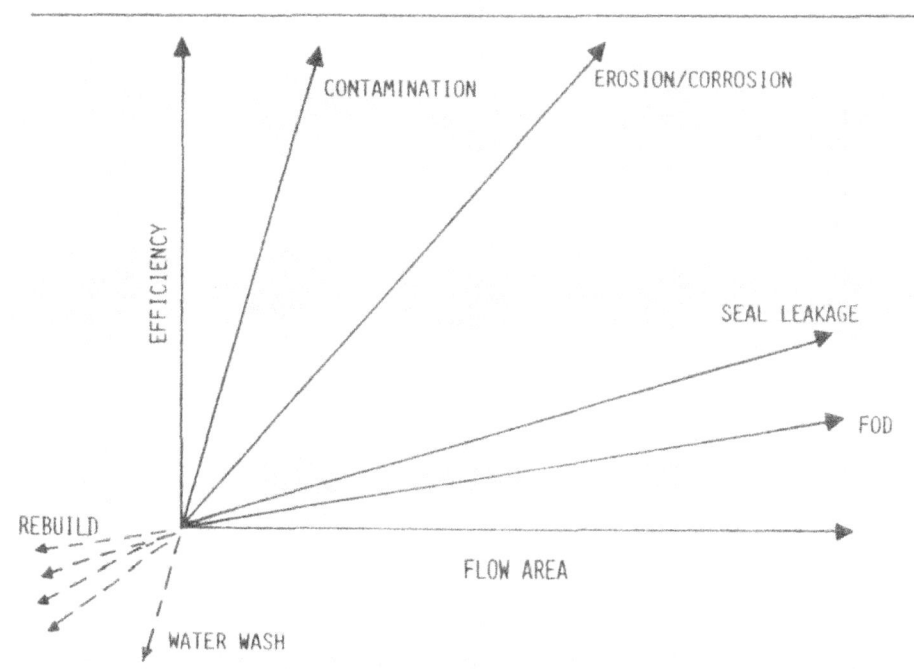

FIGURE 30

MODULE DIRECTED RATING PARAMETERS

- SELECT TYPICAL FAILURE DIRECTIONS
- CALCULATE MOST EASILY ESTIMATED PARAMETER

EXAMPLE:

$$\Delta\theta_{LOW\ SPOOL} = a_1 \Delta n_F + a_2 \Delta A_F + a_3 \Delta n_{LPT} + a_4 \Delta A_{LPT}$$

$$\left.\begin{array}{l} a_1 > 0 \\ a_2 > 0 \\ a_3 > 0 \\ a_4 < 0 \end{array}\right\} \rightarrow \Delta\theta \text{ DECREASES AS LOW SPOOL COMPONENTS DEGRADE}$$

FIGURE 31

SIMULATION RESULTS

- 350 HOURS OF E5186 TEMS OPERATING POINTS

- SIMULATE SENSOR NOISE LEVELS

- SIMULATE HARDWARE FAULTS AND TRENDS

FIGURE 32

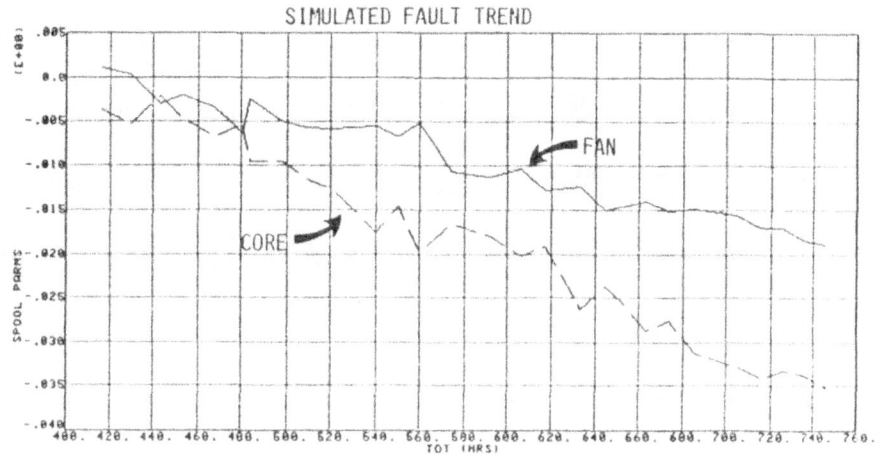

FIGURE 33

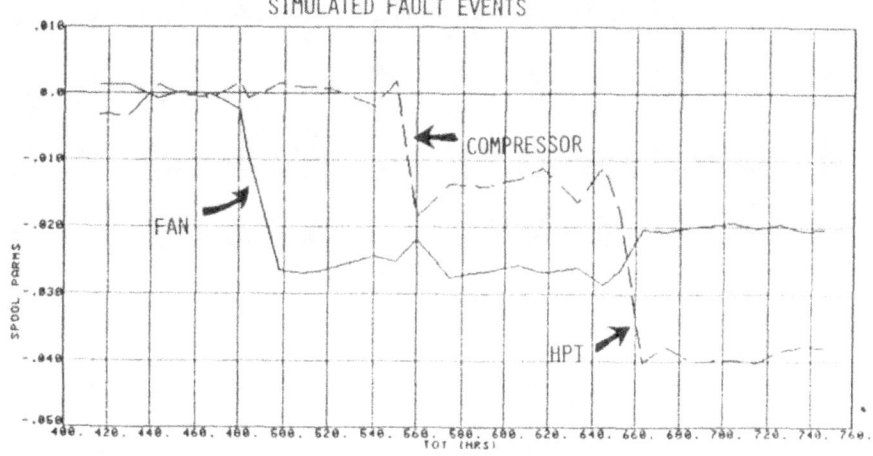

FIGURE 34

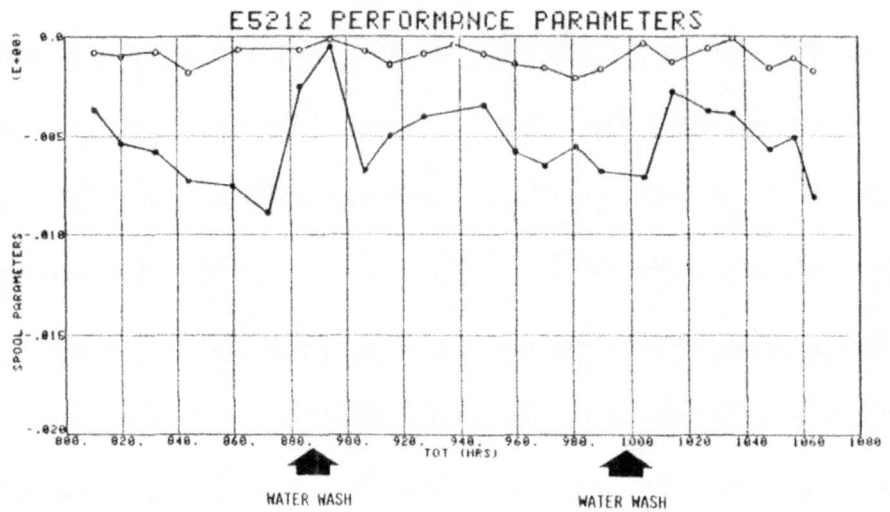

FIGURE 35

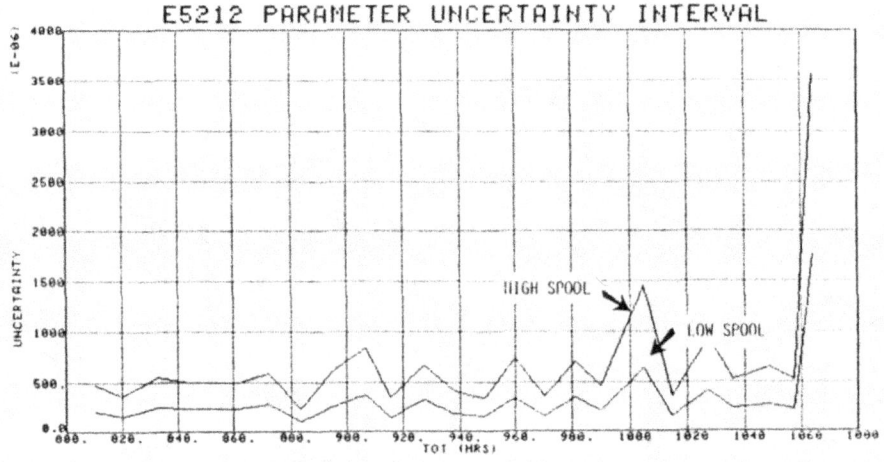

FIGURE 36

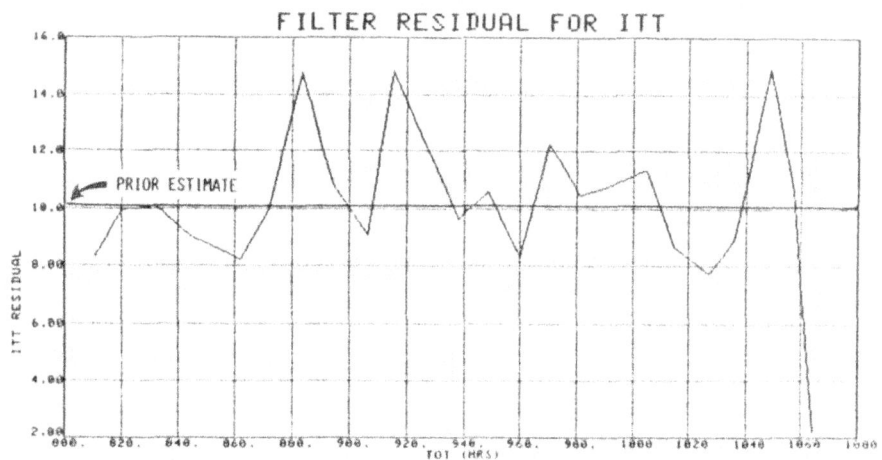

FIGURE 37

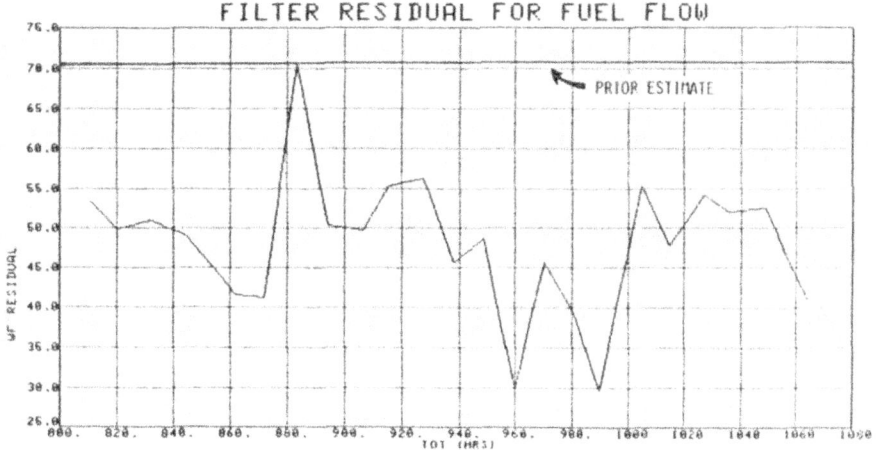

FIGURE 38

SUMMARY

- PERFORMANCE MODEL DERIVED FROM OPERATING DATA AND STATUS DECK
- MODULE DIRECTED PARAMETERS SELECTED FOR LOW AND HIGH SPOOLS
- SENSOR DIAGNOSTICS VALIDATES DATA PRIOR TO PARAMETER ESTIMATION
- ESTIMATION ALGORITHM USES FIXED TIME WINDOW

FIGURE 39

IMPACT OF AUTOMATED ENGINE MONITORING ON RELIABILITY CENTERED MAINTENANCE AND LOGISTICS SUPPORT

Laura E. Baker and W. Earl Hall, Jr.
Systems Control, Inc.

(Text was unavailable at the time of printing.)

DEFINITION OF INTEGRATED SYSTEM

- MAINTENANCE SUPPORT SCENARIOS UNDER OCM
  - TROUBLESHOOTING REQUIREMENTS
  - PREVENTIVE SPECIFICATION REQUIREMENTS
- SYSTEM LEVEL REQUIREMENTS
- INFORMATION INTEGRATION REQUIREMENTS

FIGURE 1

DECISION PROCESSES FOR ENGINE MAINTENANCE

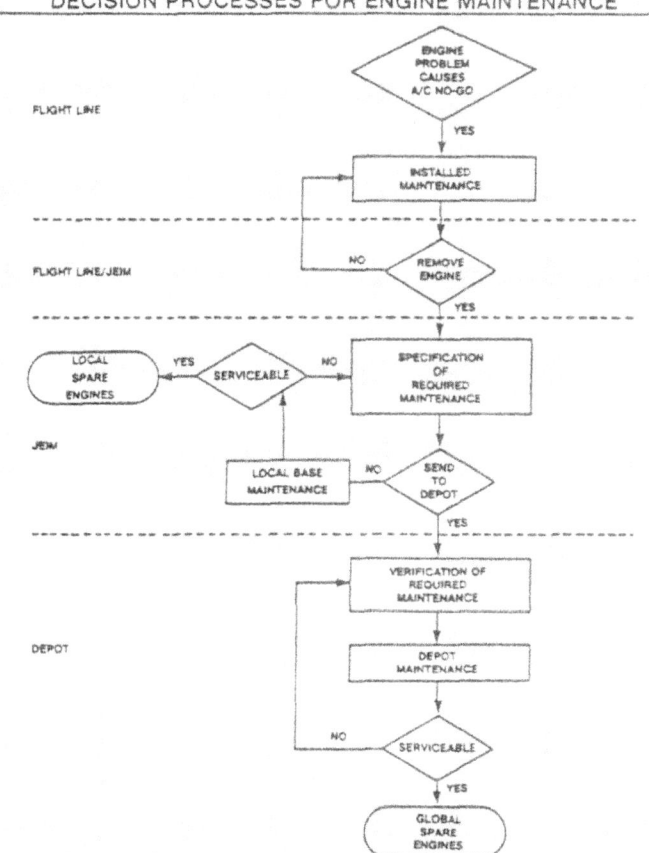

FIGURE 2

ENGINE HEALTH INDICATORS / INTEGRATION REQUIREMENTS

- USAGE FACTORS
  - TIME
  - CYCLES
- EVENT / EXCEEDANCE DETECTION
- VIBRATION
- PERFORMANCE MONITORING / TRENDING
- OIL ANALYSIS
- MAINTENANCE HISTORY
- CONFIGURATION STATUS

FIGURE 3

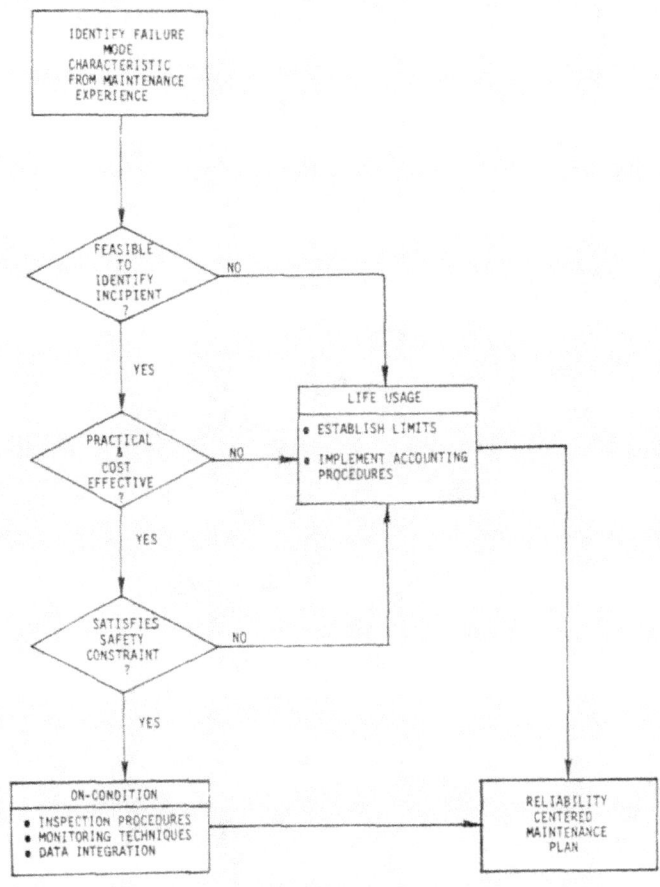

FIGURE 4

MIMS DEVELOPMENT BACKGROUND

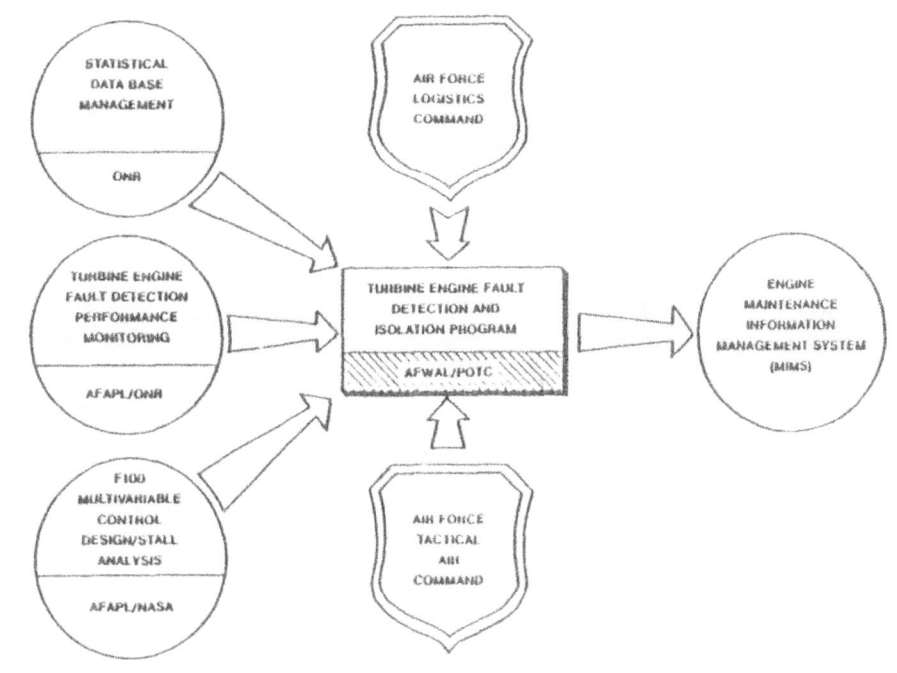

FIGURE 5

TEFDI-A FOUNDATION FOR AN ENGINE MAINTENANCE INFORMATION MANAGEMENT SYSTEM (MIMS)

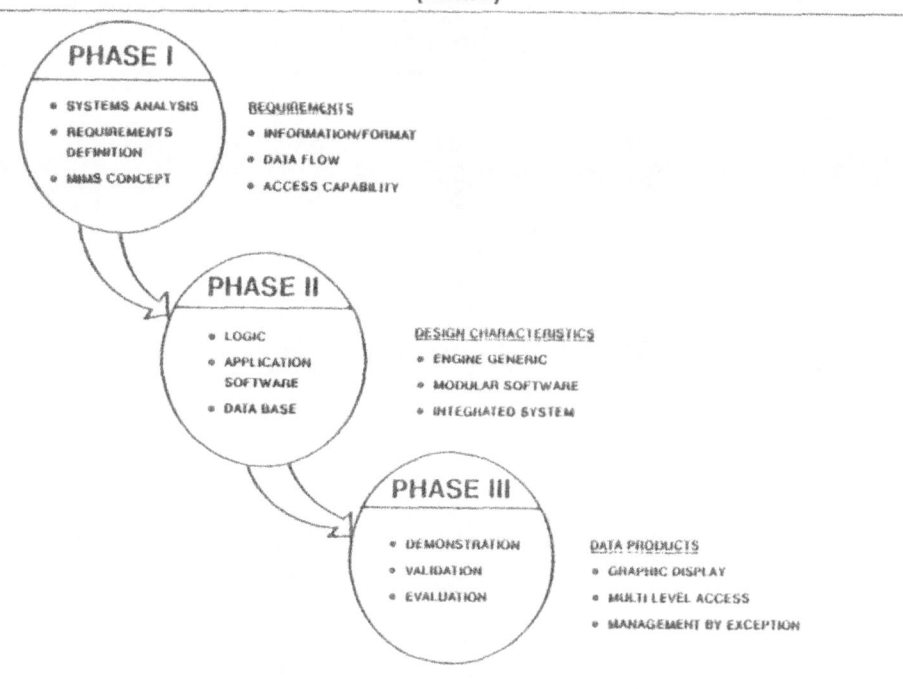

FIGURE 6

MIMS INFORMATION ARCHITECTURE

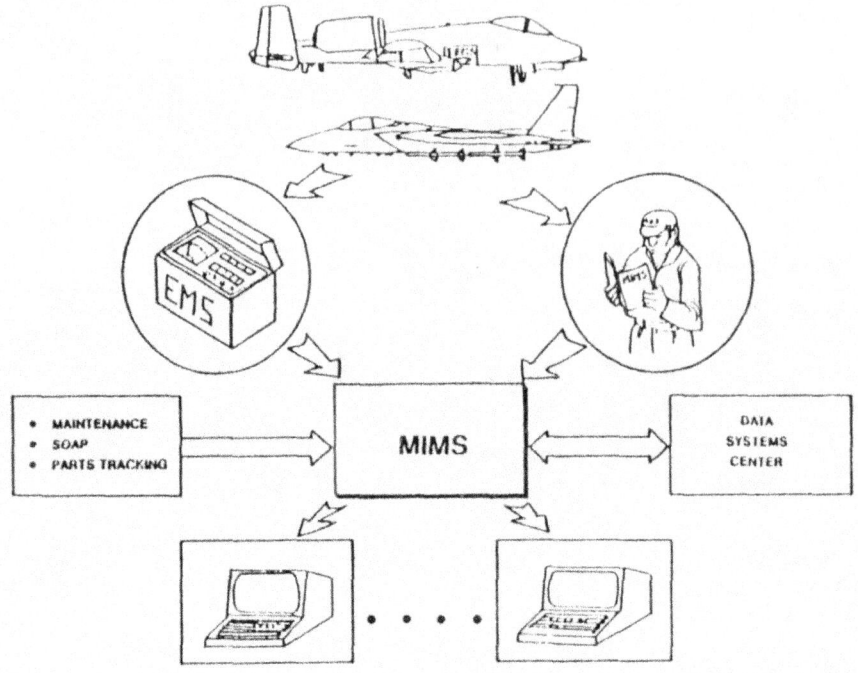

FIGURE 7

REQUIREMENTS

- REDUCTION OF PERFORMANCE DATA TO CONCISE, USABLE PARAMETERS

- SYSTEM OUTPUTS CONSISTENT WITH THE DECISION PROCESSES OF MAINTENANCE PERSONNEL AND LOGISTICS ANALYSTS

- INCORPORATION OF STANDARD ENGINE USAGE FACTORS (E.G., LCF, TIME, ETC.)

- WELL-DEVELOPED PLAN FOR INTEGRATION INTO MAINTENANCE/LOGISTICS OPERATIONS

FIGURE 8

MIMS SOFTWARE DEVELOPMENT

- STRUCTURED AUTOMATIC DATA ACQUISITION, PROCESSING AND TRANSFER TO PERFORM
  - GAS PATH ANALYSIS
  - OIL ANALYSIS
  - VIBRATION ANALYSIS
  - COMPONENT LIFE USAGE

- ADP ARCHITECTURES, DATA BASE INTERACTIONS COMPATIBLE TO CURRENT AIR FORCE LOGISTICS ORGANIZATION

- SOFTWARE MODULES TO PROCESS/ANALYZE DATA AND SUPPORT MANAGEMENT INFORMATION SYSTEM

FIGURE 9

MIMS DATA BASE SOFTWARE

- MODULAR DESIGN

- RANDOM ACCESS FILE STRUCTURE

- CURRENT DISPLAY/SOFTWARE IMPLEMENTATION
  - BASE STATUS SUMMARY
  - USER INTERROGATION – FIND/RANK
  - WATCH STATUS – USER CONTROLLED
  - ENGINE PROFILE
  - HISTORY – LINEAR/POINT PLOTS AND TRENDS
  - TABULAR
  - MAINTENANCE EVENTS
  - MMICS INTERFACE

- SOFTWARE IN CONCEPT DESIGN/DEVELOPMENT
  - ALARM
  - GPA ALGORITHM (SNAPSHOT/TREND)

FIGURE 10

MIMS DATA BASE

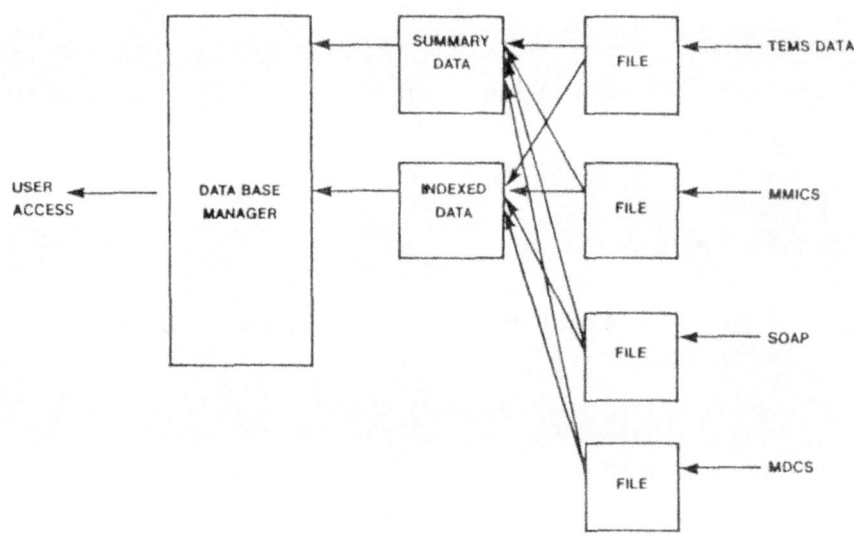

FIGURE 11

MIMS IMPLEMENTATION PROGRAMS

PROGRAM	OBJECTIVES	SCOPE
MIMS/EDS	• EVALUATION OF MIMS HARDWARE AND SOFTWARE INTERFACES • VALIDATION OF TEFDI GAS PATH ANALYSIS ALGORITHM ON EDS GENERATED DATA	• ACQUIRE DATA FROM FIVE EDS EQUIPPED AIRCRAFT AT LANGLEY AIR FORCE BASE — F 100 EDS — SOAP — MAINTENANCE EVENTS — PARTS TRACKING • TRANSMIT DATA TO MIMS AND ANALYZE RESULTS WITHIN TEFDI OBJECTIVES
MIMS/TTP	• DEMONSTRATION OF MIMS DATA PRODUCTS • PRELIMINARY EVALUATION OF "IN-THE-FIELD" INSTALLATION	• BARKSDALE AIR FORCE BASE INSTALLATION OF MIMS DATA STATIONS, DATA AQUISITION PROCEDURES, AND COMPUTER INTERFACE LINKS • TWO A-10/TF-34 TTP AIRCRAFT • DEMONSTRATION OF THREE LEVELS OF DATA PRODUCTS AT BASE AND REMOTE SITES
STEMS/MIMS SQUADRON EVALUATION	• DEMONSTRATION OF SQUADRON LEVEL STEMS/MIMS OPERATIONAL EFFECTIVENESS	• DATA ASSIMILATION FROM 24 A-10 AFRES AIRCRAFT AT BARKSDALE AIR FORCE BASE • AIR FORCE OPERATION OF MIMS • BASE LEVEL COMPUTER UTILIZATION

FIGURE 12

FUTURE IMPACTS / CONCLUSIONS

- PILOT/FLIGHTCREW
- FLIGHT LINE / AMU
- INTERMEDIATE SHOP
- DEPOT
- COMMAND

FIGURE 13

OVERVIEW

- DEFINITION OF INTEGRATED CONCEPT
- EVALUATION CAPABILITIES
- IMPLEMENTATION REQUIREMENTS
- FUTURE IMPACTS / CONCLUSIONS

FIGURE 14

Page intentionally left blank

A-10/TF34 TURBINE ENGINE MONITORING SYSTEM (TEMS)

Robert G. Christophel
San Antonio Air Logistics Center

SUMMARY

The A-10/TF34 Turbine Engine Monitoring System (TEMS) integrates inflight and ground hardware to sense, signal condition, perform computations and analysis, and record various engine and aircraft information and parametric data for the purpose of fault detection, isolation and trending. Basically, the data are collected, processed and stored by the airborne Electronic Processor Unit (EPU) then transferred through the GO/NO-GO indicating Umbilical Disconnect Unit (UDU) to the Diagnostic Display Unit (DDU) for flight line maintenance use before final transfer to the TEMS ground station peripheral equipment for Jet Engine Intermediate Maintenance (JEIM) shop use, actuarial processing and permanent storage. If flight line display and use of the data is not required, transfer to the ground station may be done with the Data Collection Unit (DCU). TEMS data will be used at the flight line to assess engine GO/NO-GO status, aid in troubleshooting and fault isolation and to perform engine trim. Potential JEIM and depot TEMS information uses include engine troubleshooting and fault isolation, test cell trim and data collection, maintenance programming, parts tracking, spare parts forecasting, and actuarial analysis.

INTRODUCTION

In 1974 the Department of Defense (DOD) adopted the Reliability Centered Maintenance (RCM) concept for all military aircraft systems, consequently requiring restructure of existing aircraft scheduled maintenance programs and establishment of RCM programs for all new aircraft. The DOD RCM concept is based on the commercial airline maintenance decision logic called MSG-2 developed by a committee known as Maintenance Steering Group 2 composed of representatives from the commercial airlines, Air Transportation Association, and Federal Aviation Administration (ref. 1). Basically, RCM is a decision logic process which divides scheduled maintenance requirements into the three basic categories of hard limits, on condition, and condition monitoring, followed by a Maintenance Requirements Analysis that translates the maintenance requirements into specific inspections, limits, tasks, and work packages and produces technical data and instructions for the maintenance of a specific system (ref. 2). The United States Air Force (USAF) incorporated the DOD RCM philosophy into an expanded On Condition Maintenance (OCM) concept, defined as maintenance that allows the condition of the equipment to dictate the need for maintenance or the extent of repair/overhaul required (ref. 3). Successful conversion to full OCM for a complex turbine engine requires the use of monitoring systems such as the A-10/TF34 Turbine Engine Monitoring System

(TEMS) (ref. 4).

The fundamental success of OCM is directly dependent on the ability to adequately perform the tasks dictated by the nature of the three RCM categories, continually assessing the OCM data and updating the RCM analysis by transferring items from one category to any other as necessary. The hard limits category requires parts time and cyclic tracking; on condition generates the need for repetitive inspections or tests; and condition monitoring is greatly enhanced by diagnostic and trending capability. The USAF has always practiced OCM to a certain extent with these functions satisfied by a variety of manual and automatic data acquisition systems. However, recent radical developments in microprocessor technology and data processing have made possible completely automated systems capable of acquiring and processing the vast amounts of data needed to support the OCM of a modern, complex turbine engine (ref. 5).

The TEMS being incorporated into the A-10/TF34 system is designed and built by Northrop Electronics Division and was originally flown on the T-38/J85 combination before being upgraded for the A-10/TF34 application. This paper discusses the operation and interfaces of the A-10/TF34 TEMS hardware focusing primarily upon function, capabilities and limitations. The TEMS data types are defined and the various data acquisition modes are explained. Potential data products are also discussed.

SYMBOLS AND ABBREVIATIONS

ITT - inter turbine temperature

N_F - fan speed

N_G - core speed

PLA - power lever angle

P_{S3} - compressor discharge static pressure

RPM - revolutions per minute

T_{2C} - compressor inlet total temperature

VG - variable geometry

W_F - fuel flow rate

HARDWARE AND INTERFACES

System

The hardware used in the A-10/TF34 TEMS, (Fig. 1), is comprised of in-flight and ground equipment to sense, dignal condition, compute and analyze, record and store various aircraft and engine information for fault detection, isolation, diagnostics, trending, and parts tracking (Fig. 2). The basic components are the airborne Electronic Processor Unit (EPU) and the ground used Diagnostic Display Unit (DDU) and Data Collection Unit (DCU). These units are microcomputers that share common components and are based on 8080 microprocessor architecture. The EPU, Umbilical Disconnect Unit (UDU), sensors and signal conditioners, and associated wiring make up the airborne hardware. The ground equipment consists of the DDU, DCU, a printer, Intelligent Disk Unit (IDU), and telephone modem.

In operation, the EPU (Fig. 3) continuously receives and monitors sensor and transducer dignals (Fig. 4) and records and stores a data frame automatically for preselected flight conditions or whenever the pre-established normal limits of a critical parameter are exceeded. Data frames are manually taken and stored upon pilot command through a cockpit data switch or for maintenance record purposes through the DDU (Fig. 5). Data stored in the EPU is retrieved on the ground by either the DDU or DCU through the UDU (Fig. 6), which also provides GO/NO-GO and limit exceedance event indicators. The DDU has real-time display and operation capability and provides maintenance personnel with a display of engine performance parameters, operating conditions, and other information permitting review of routine data and troubleshooting/diagnostic capability at the flight line. Engine trim functions can also be done using the DDU independent of other test equipment. The DCU is essentially the same as the DDU without display capability and both units transfer data to the printer and IDU intthe Jet Engine Intermediate Maintenance (JEIM) shop for permanent storage and further troubleshooting, fault isolation and diagnostic activity as required. Reference 6 contains a complete, detailed description of the hardware and its operation.

Electronic Processor Unit

The EPU provides central administration, execution and regulation of the TEMS. It continuously receives and monitors inputs from aircraft and engine transducers and sensors and performs various functions relating to the signal conditioning, processing and storage of the data. The signal conditioning function converts the sensor signals into scaled direct current values. High impedence isolation between the sensors and conditioners protects on-board instrumentation, allowing the use of existing aircraft instrumentation without affecting the cockpit indicators. After conditioning and multiplexing, the signals are digitized by the Analog to Digital Converter and input to the processor. The processor is the computer portion of the EPU and uses both Random Access Memory (RAM) and Programmable Read-Only Memory (PROM). It constantly monitors and processes the data and, when a maintenance action item has been confirmed, transmits the appropriate information to data storage for ground recovery. The PROM stores the executive routine, equation subroutines, diagnostic logic, signal averaging and instructions. Program constants, calibration data, engine signatures, threshold levels and logic options are stored in the RAM. The RAM also is the working memory and provides temporary data storage for ground retrieval. These memories can be programmed through the DDU to account for engine changes or limit changes without removal of the TEMS hardware. The processor also provides interface control for EPU communications through the UDU to the DDU or DCU.

Umbilical Disconnect Unit

The UDU is mounted in the A-10 nose gear storage compartment for easy access and provides the capability to retrieve data from the EPU, to display aircraft and TEMS status and event mode indicators, and to enter mission configuration information for structural tracking use. EPU data is transferred automatically by connecting the DDU or DCU umbilical to the UDU and depressing the Data Transfer button. Aircraft and TEMS status indicators include red/yellow/green light indicators for NO-GO, Caution, and no limit exceedance events stored in the EPU, respectively. If the NO-GO or Caution indicators are lit, additional information is available in the form of a four digit alphanumeric code, displayed upon command by depressing the status button. This display also indicates TEMS malfunctions.

Diagnostic Display Unit

The DDU is a one-man portable microcomputer unit that communicates with the EPU through the UDU to transfer EPU stored data to the DDU for flightline maintenance use and/or further transfer to the peripheral ground equipment for printout, permanent storage, and processing. The data transfer is simply and expeditiously done and includes automatic data validity checks. The Light Emitting Diode display capability of the DDU provides for flightline review of routine data as well as plane-side troubleshooting and fault isolation when desired, and the performance of engine trim functions independent of other test equipment. The keyboard is used to re-initialize and calibrate the EPU following an engine change or as required by other maintenance action. The DDU microcomputer is the same as that in the EPU and various modules are interchangeable.

Data Collection Unit

The DCU performs the same data transfer function as the DDU but does not have the display capability for flightline data review. The computer and data transfer elements are identical to those in the DDU but the elimination of the display section and part of the power supply results in a much smaller, lighter unit weighing approximately eight pounds that can be easily handcarried whereas the DDU is usually bicycle transported.

Peripheral Ground Equipment

This equipment consists of a Tally T-1612 Printer, a Northrop 094020-301 Intelligent Disk Unit (IDU), and a Vadic VA 3451 Telephone Modem. This equipment provides for permanent hard copy printout for file records and analysis, permanent magnetic floppy disk storage, and transmission of the TEMS data to a central site or more encompassing data system such as the Comprehensive Engine Management System. The IDU has computer logic and programming capability and can provide a variety of printed and plotted data for diagnostics, trending, life usage, and maintenance planning purposes.

DATA ACQUISITION

Automatic Data Collection

During normal flight and ground operation, the various sensor and transducer signals are continuously monitored by the EPU. However, the EPU only records, for ground retrieval, a data frame whenever specific, preprogrammed conditions are satisfied or when commanded by the air or ground crew through the cockpit switch or DDU. There are two classes of preprogrammed, or automatic, data frame recordings: trend data frames and limit exceedance data frames (refs. 7 and 8).

The purpose of the trending data is to obtain operational flight data for comparison with previous records to detect changes in engine performance, collect parts life tracking and usage information, and provide actuarial documentation data. A large supply of data points usually enhances trending accuracy and confidence, but recovery, storage, and analytic capacity limitations restrict the amount of data that can be processed. This has resulted in the present procedure of two trending data categories, each of which may be taken a maximum of once per flight (Fig. 7). The "Liftoff" frame is taken once each flight and consists of the last data scan monitored just before the weight-on-wheels switch indicates liftoff. The "Cruise" frame is taken later in the flight, after satisfying the stability conditions necessary to ensure repeatable data, valid for comparison purposes. Data is taken the first time the stability conditions are met, with no repeats during the sortie. The stability parameters associated with the "Cruise" data frame are elapsed flight time, PLA, N_G, T_{2C}, gun firing, airspeed, altitude, angle of attack and vertical acceleration.

The purpose of the limit exceedance data frames (Fig. 8) is to report abnormal engine operation and to provide supporting data for troubleshooting and fault isolation. The parameters triggering a limit exceedance data frame are ITT overtemp, N_G/N_F overspeed, oil pressure, vibrations, variable geometry schedule, N_G RPM, compressor stall, slow starting, fuel filter by-pass indication, over-g, and fluctuations in oil pressure, N_G, N_F, W_F and P_{S3}.

These parameters were selected from studies of historical failure records, maintenance impact, and detection reliability. When possible, existing USAF Technical Order (T.O.) limits are used for the detection criteria but, in those cases where no T.O. limits exist, reasonable values were determined and assigned through consultation with General Electric Company, the designer and manufacturer of the TF34 engine. In operation, a limit exceedance data frame is recorded upon the initial detection of an out-of-limit parametric value. Data frames are not recorded for a succeeding limit exceedance of that particular parameter but the number of occurrences and total duration of the limit exceedance for that parameter are accumulated and stored for retrieval. Of course, an out-of-limits event by any other parameter will produce a recorded limit exceedance data frame.

Manual Data Collection

Data frames can be manually taken for record purposes by depressing the cockpit switch or upon command through the DDU. The primary purpose of the cockpit switch is to allow the pilot to record data at his discretion to document abnormal or unusual circumstances. A one second depression of the switch produces a data frame. Continuous activation of the switch results in a new data frame every two seconds. Data will be taken by maintenance personnel using the DDU for record purposes during engine trim, engine maintenance or EPU calibration.

DATA

Diagnostic Display Unit

The DDU displayed data is categorized as documentary, measured or computed. This data provides maintenance personnel at the flightline and JEIM shop with engine trim data and troubleshooting, fault isolation, and trending information for performing engine maintenance.

The documentary data includes aircraft and engine serial numbers, flight and record numbers, Julian date, record time, elapsed flight time and flight condition information. The data is primarily for actuarial, record and classification uses.

The measured data consists of the output from each engine sensor. This includes the detected event limit exceedance data and special diagnostic indicators which provide spool differentiation for vibration data, aircraft modes such as slat deployment and out of envelope conditions, and instability information.

The computed data is composed of the results of calculations concerning trim and performance verification. The trim relationships verify airborne and ground fan speed trim, trim margin, variable geometry schedule, and idle trim. The relationships are corrected for bleed air, power extraction, Mach number and droop and, although the airborne checks are valid at part or full power, there are engine minimum speed, maximum pressure, and altitude limitations. The performance relationships have been identified as being effective in measuring specific characteristics through sensitivity analysis pertaining to engine degradation and performance changes.

JEIM Printed Data

All data displayed on the DDU is available by printout in addition to backup data including corrected parameters, calculations, cumulative times, aircraft parameters and fluctuations. Also, information including ITT time above 790°C, ITT time above 810°C, and temperature, fan speed, core speed and compressor static pressure cycles are presented for special parts tracking, life usage, and actuarial functions.

DATA PRODUCTS

Over the period of the past few years, the USAF Aero-Propulsion Laboratory and Systems Control, Inc. (Vt) have been investigating the integration of various data sources, including TEMS, into the USAF maintenance/logistic process (ref. 9) with the objective of developing procedures for reducing and processing raw data elements to provide maintenance decision information to the flight line, JEIM shop, depot, and major command level. The raw data includes maintenance action records, oil analysis results, configuration tracking, and TEMS data. These data are processed into a data file, ranked and sorted, and stored for subsequent access.

A preliminary set of data products have been identified for various user levels. These include summary reports of the operational status of the engine population by base location including such pertinent information as Time Compliance Technical Order completion, spare engine availability, engine and component life data and usage trends. Also, reports for individual engines could be generated with the same type information in addition to maintenance history, oil analysis data and trim and performance trends. Documents prepared specifically for depot and command level use could include a wide variety of actuarial information, parts tracking and forecast usage, fleetwide distribution of maintenance manhours expended for specific failure modes and general fleetwide engine health trends (Fig. 9).

CONCLUDING REMARKS

The A-10/TF34 TEMS hardware and software development is virtually complete and, from the viewpoint of a qualified system, the TEMS is now ready for incorporation to the A-10 force. However, before total retrofit is done, it is necessary to fully develop, validate and establish engine maintenance and management procedures based on TEMS data and to integrate the TEMS data into the mainstream of the USAF maintenance and logistics process. A pilot program is now being initiated with that objective. The program will consist of one full squadron of A-10 aircraft equipped with TEMS and will be done in conjunction with the Comprehensive Engine Management System Increment IV prototype. This will provide for both the development and evaluation of new or modified A-10/TF34 maintenance procedures, capitalizing on TEMS technology, and the engine management data products necessary to provide the basis for a composite, total On Condition Maintenance system for a modern, complex turbine engine.

REFERENCES

1. Reliability Centered Maintenance Analysis Course, User's Guide. Air Force Institute of Technology, Wright-Patterson Air Force Base, Ohio, Undated.

2. Reliability Centered Maintenance, On Condition Maintenance - What Do They Mean. Staff Briefing by HQ AFLC/LOP, Wright-Patterson Air Force Base, Ohio, Undated.

3. Equipment Maintenance Policies, Objectives, and Responsibilities. AFR 66-14, United States Air Force, Washington D.C., 15 November 1978.

4. Report. USAF Scientific Advisory Board Ad Hoc Committee on Turbine Engine Monitoring Systems, 28 November 1980.

5. DeHoff, R.L.; Baker, L.E.; and Hall, W.E., Jr.: Impact of Automated Monitoring on Engine Operations and Support. AIAA/SAE/ASME 15th Joint Propulsion Conference, June 18-20, 1979, Las Vegas, Nevada. Paper No. 79-1276.

6. A-10/TF34 Turbine Engine Monitor System (TEMS) Phase I Final Report. NORT 80-244. Northrop Corporation, Electronics Division. May 1980.

7. Software Definitions Turbine Engine Monitor System (A-10 Aircraft). NORT 79-14B. Northrop Corporation, Electronics Division. October 1979.

8. Software Definitions Turbine Engine Monitor System (A-10 Aircraft). Update of NORT 79-14B. Northrop Corporation, Electronics Division. To be published.

9. Baker, L.E.; DeHoff, R.L.; and Hall, W.E., Jr.: Turbine Engine Fault Detection and Isolation Program - Phase I Requirements Definition for An Integrated Engine Monitoring System, AFWAL-TR-80-2053, Volume I. Air Force Wright Aeronautical Laboratories, Wright-Patterson Air Force Base, Ohio. April 1980.

USAF TEMS
HARDWARE APPROACH

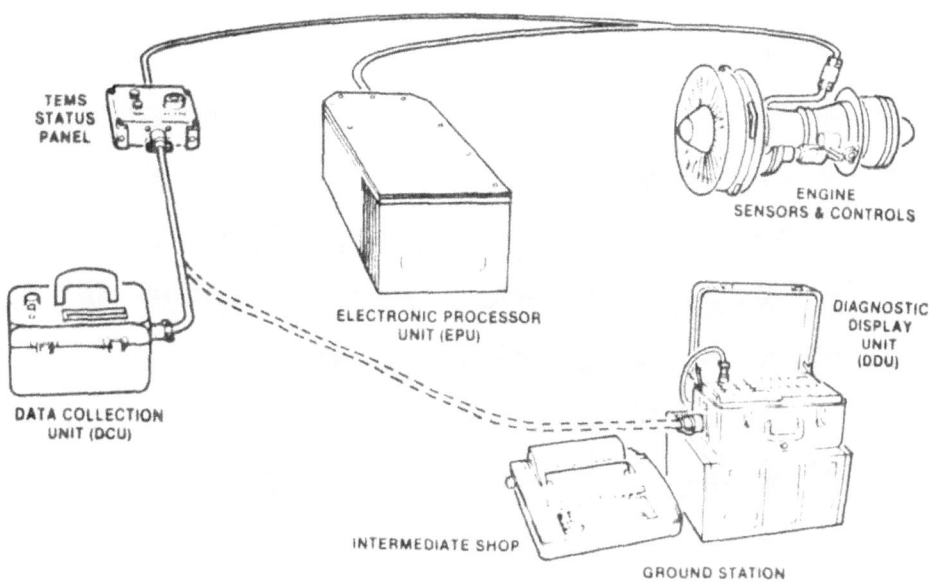

FIGURE 1

A-10/TF34 TEMS

OVERVIEW OF ENGINE MONITORING

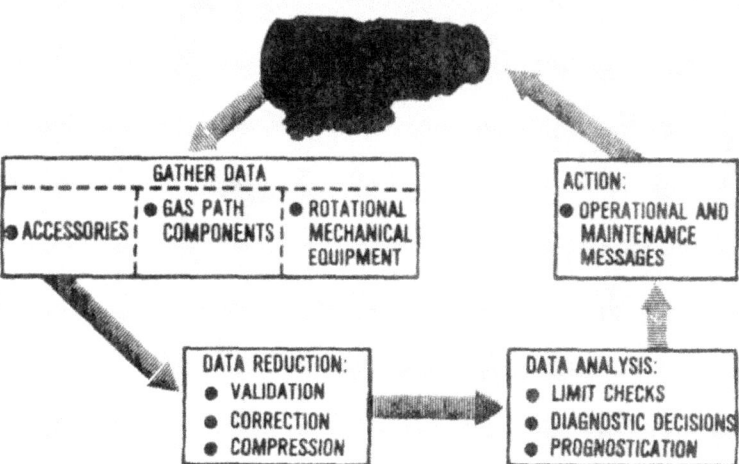

FIGURE 2

FIGURE 3

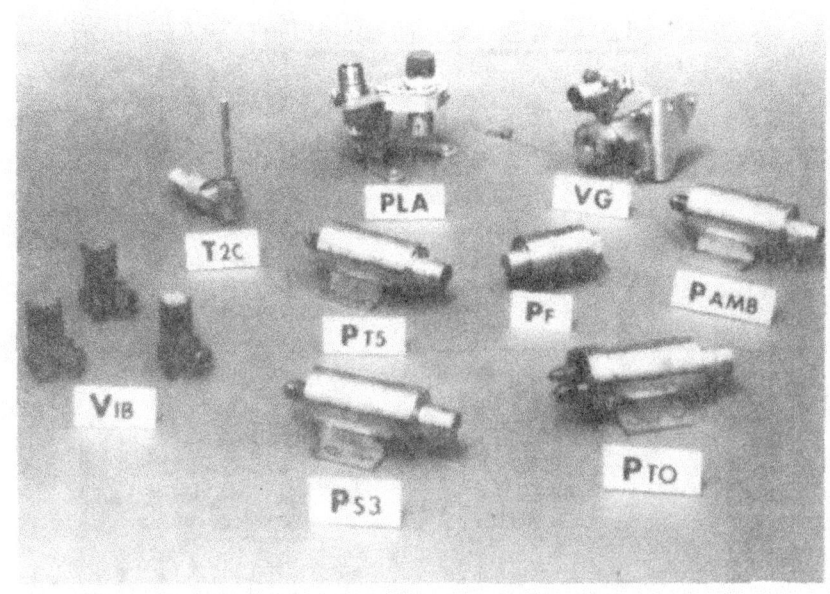

FIGURE 4

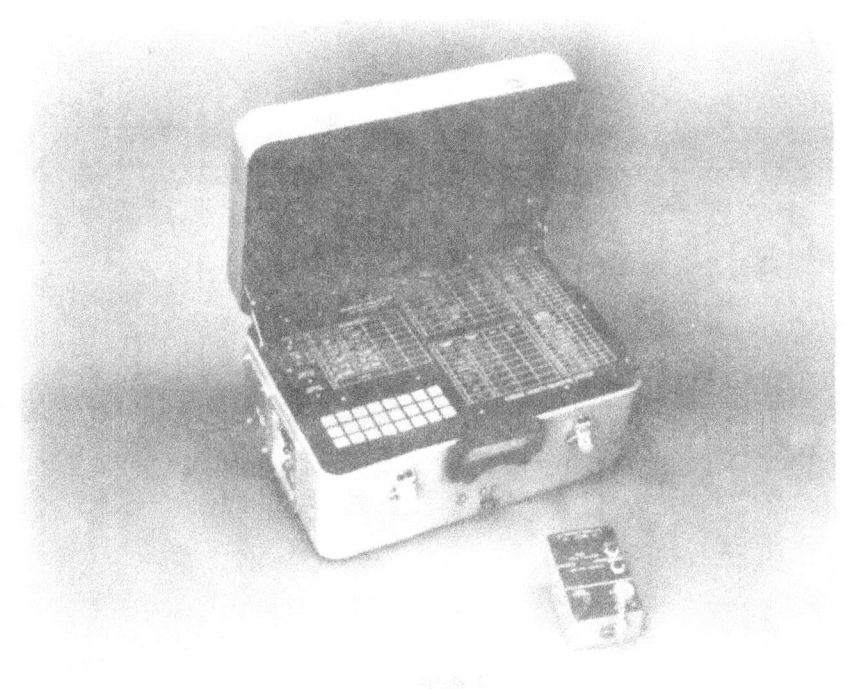

FIGURE 5

FIGURE 6

AUTOMATIC DATA FRAMES

LIFT OFF	CRUISE
WEIGHT OFF WHEELS	TAKEN NOT LESS THAN 15 MINUTES AFTER LIFTOFF
AIRSPEED > 100 KCAS	NG CORR > 85.4% FOR BOTH ENGINES
NG > 56% FOR ONE OR TWO ENGINES	PLA STABLE ± 1°/2 SEC > 16 SEC
	T_{2C} STABLE ± 1°/2 SEC > 16 SEC
	NO GUNFIRE PRECEEDING 16 SEC
	AIRSPEED 200-300 KCAS
	ALTITUDE < 10,000 FT
	ANGLE OF ATTACK < 15 DEG
	VERTICAL G's 1.5 ± 1.0g > 16 SEC

FIGURE 7

DETECTED EVENT FRAMES

ITT OVERTEMP	ENGINE STALL
NG OVERSPEED	SLOW START
NF OVERSPEED	FUEL FILTER
OIL PRESSURE	OVER G
VIBRATIONS	MAXIMUM ITT SHIFT
FLUCTUATIONS	NF VS ITT ERROR
VG SCHEDULE	NG SPEED ERROR

FIGURE 8

A-10/TF34 TEMS

ANTICIPATED BENEFITS
WHAT CAN ENGINE DIAGNOSTICS SYSTEMS DO FOR USAF?

- REDUCE UNWARRANTED MAINTENANCE
- REDUCE PARTS AND FUEL CONSUMPTION
- INCREASE AIRCRAFT AVAILABILITY
- PROVIDE AUTOMATED ENGINE DATA
- FEED BACK REAL OPERATIONAL DATA FOR FUTURE DEVELOPMENTS

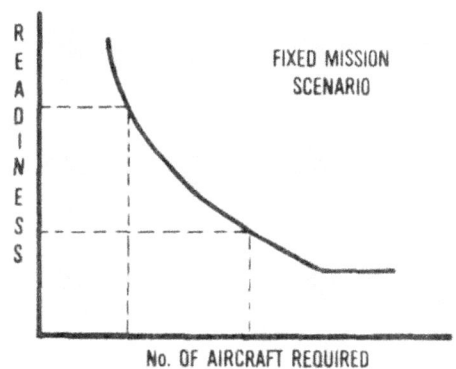

FIGURE 9

Page intentionally left blank

REVIEW OF AIDS DEVELOPMENT

Henk C. Vermeulen
KLM Royal Dutch Airlines

Sven G. Danielsson
SAS Scandinavian Airlines System

SUMMARY

Since the introduction of the wide-body aircraft KLM, SAS and Swissair have been able to collect a mass of experience, meager as will as excellent and in total profitable. All three are very determined to continue with AIDS on A300/310 Airbus (Swissair also on the DC-9-80).
The A300/310 AIDS as selected by KSS (KLM, SAS and Swissair) and Lufthansa has been developed into a very powerful Engine Monitoring System (EMS) and engineering tool capable to enhance aircraft regularity, reliability and economic operation.

INTRODUCTION

KLM, SAS and Swissair started with AIDS at the introduction of the wide-body aircraft in 1970. The AIDS hardware specification for the Boeing 747 and McDonnel Douglas DC-10 aircraft was based on the experience obtained from:

. A digital recording experiment by KLM on a Douglas DC-8 in 1963 and 1964 (3) and DC-9 trials with prototype equipment in 1969.
. The development of the ARINC 573 specification for a Flight Data Acquisition Unit (FDAU) to satisfy the new FAA requirements.

The initial objectives of KSSU with respect to the AIDS were primarily directed to the monitoring of parameters related to:

. the safety of the flight
. the performance of the aircraft
. the performance of the flight guidance system
. the performance and condition of the engines.

The AIDS-EMS function was and still is considered supplemental to the existing monitoring tools.
For AIDS-EMS practically the same parameters were selected as already provided for display on the cockpit instrument panels with a few exceptions. Table 1 provides a list of EMS parameters monitored on KSSU aircraft. The total number of parameters monitored on KSSU 747 and DC-10 aircraft amounts to 380 and 280 respectively of which more than 50% are discretes (on/off signals).

In order to accomodate all these parameters and enable sampling at reasonable rates the system was configured around 3 Data Acquisition Units and a Data Management Unit (DMU) with limited data acquisition capability. Figure 1 depicts the system block diagram.

AIDS OPERATION

In KSS the AIDS is primarily applied as an engineering tool with a strong emphasis on analysis of recorded information. A printer was added to the airborne system because it was recognized rather early that hard copies of exceedance reporting could provide a very effective aid in trouble reporting. The application of on-board processing for limit exceedance monitoring and recording control allowed to add a printer which could provide hard copy reports on request by the crew or automatically.

The tape-cassette is removed every landing made at the home-base and is subsequently transcribed to IBM compatible tape and processed. The routine programs applied comprise o.a.: flights logging, AIDS status reporting, EMS programs, autoland verification, etc.
For a good understanding of the function of AIDS in an EMS, it is essential to give some details on the functions of two specific software programs: the AIDS flight logs and the plot/list program. The first program provides a listing of all recorded flights per aircraft registration and the second program allows users to request a time history of a set of 8 analog plus 8 discrete type parameters either in table format or plotted. The user can call the AIDS flight log, select the airplane, the flightleg and the parameterset of interest and request a listing either for a specified flight mode or a GMT time span using the VDU terminals of KLM's data handling system. The very successful use of this program proved that AIDS is an invaluable engineering tool and fully met the set objectives.

CURRENT AIDS EMS APPLICATIONS

ON BOARD

For short-term trend analysis KLM relies on a trendchart that is updated by the flight engineer on every flight that lasts more than 4 flight hours. The flight engineer then selects a stabilized flight condition to request and engine data print (table 2). The flight engineer uses this print to calculate the trend delta's with the aid of an engine performance calculator provided by the engine manufacturer and enters the delta values in his trend chart. Engine bleed and engine indicating problems, serious compressor/turbine problems and EGT-margin losses can be detected by the flight engineer using these short term trends. In addition these trends are checked by powerplant engineers on a regular basis and in case of crew complaints.

The engine data prints are also automatically presented during take-off and in case of limit exceedance e.g. the print of table 2 shows an automatically reported impending hotstart on engine nr. 1.

The take-off prints are used to monitor the hot-day EGT margin, thrust settings etc. Crew complaints are supported by prints selected by the flight engineer and/or limit exceedance prints. In case of critical engine problems the flight engineer will contact the main-base via a single-sideband company channel for expert advise. With aid of the print he is able to provide exact information on the characteristics of the trouble or exceedance, the exact durations and the peak values of exceedances.

Strict adherence to the manufacturers engine operating limits could increase the number of engine removals because of the capability of EMS to very accurately report exceedances of operating limits that are based on experience, which include the human factor. It is obvious that these limits need to be adapted when an advanced EMS is used to prevent increased removals or inspection rates. KLM was able to obtain the approval to extend the limit on the allowable EGT exceedance time-limit on a particular engine when using the AIDS printer. This printer function has proven an invaluable tool for short-term engine monitoring, incident reporting and trouble-shooting to the extent that powerplant engineers consider this feature alone was worth the investment in AIDS.

GROUND-BASED LONG TERM ANALYSIS

For long term trend analysis the AIDS provides weekly trend reports on KLM's JT9D and CF6 engines. These trend reports consist of 3 parts viz. an engine start trend (fig. 2a), an engine take-off trend (fig. 2b) and an engine cruise trend (fig. 2c). The trends shown apply to the Pratt&Whitney JT9D engine as installed in KSS 747 airplanes.

Engine start trend

Of the engine start trend one important feature should be adressed. After careful analysis using the AIDS plot/list program KLM engineering decided to trend the initial fuel flow (IFF) at the moment of "fuel-on". This analysis showed that by monitoring this value it was possible to relate a positive or negative deviation from the required value to a fuel control adjustment. Although the fuel control adjustment screw was by design meant for shop use only, it is now used for on-wing adjustment. This meant that by monitoring the IFF trend, hot and hung starts and consequent fuel control removals are reduced considerably and related unnecessary engine test runs avoided. Since its introduction more airlines became interested and have requested the engine manufacturer to provide proper means for on-wing adjustment.

Take-off and cruise trends

The take-off and cruise trends are both used for monitoring of engine deterioriation, primarily by checking EGT rise and EGT margin, blade failures, compressor and turbine problems and indicating system errors. The take-off trends is also used to monitor powerlever-alignment which avoids valid crew complaints and allows to neglect invalid crew complaints and save on otherwise necessary follow-on actions.

Gas path analysis

SAS first started using GPA based on testcell data to analyze modular deficiencies on the JT9D engine. After showing positive results KLM installed the same program but to analyze the CF6 engine. After an extensive evaluation and calibration program to determine sensitivity, performance levels and ability to find degraded modules, the program also at KLM is deemed useable to identify problems on preshop tested engines.

The function of the test-cell however has always been primarily to verify that an overhauled or repaired engine meets performance requirements. The identification of modules degraded below limits should therefore preferably occur on-wing such that an engine's work scope can be predicted prior to shop entry.

SAS and KLM has therefore decided to develop the necessary procedures and know how with GPA based on AIDS recorded data.

SAS has started with JT9D-7 engines installed in Boeing 747 while KLM will analyze CF6 engines installed in DC-10. The JT9D program will use only partly instrumented engines while the CF6-DC10 will be fully instrumented. Future aircrafts within KLM, SAS and SWR will always be fully instrumented and in particular the A310 with also a PMUX, presently the only one specified with PMUX, will already from the beginning be monitored by programs capable of modular performance analysis.

Table 5 shows the parameters used in the presently ongoing GPA programs. As can be seen the SAS JT9D-7 GPA uses the least number of parameters and there fore also has the least capability. This program however has advanced the most and a discussion showing some results follows. Because of the small number of parameters that are available to describe the engine operating characteristics, some assumptions have been necessary to do on the modular deterioration that is analyzed. The assumptions are a fixed ratio between change in efficiency and air pumping capacity on the compressor modules and also both the FAN and the LPC are treated as one module. The main disadvantage with the hard coupling between efficiency and air pumping capacity is expected to be seen in the HPC where the front stator vane stages are variable and changes in pumping capacity might be induced this way.

Figure 12 shows the variation in ambient conditions under which data is collected for GPA. Each datapoint being used is a stable frame that has been recorded by the airborne system. The measurements are reduced to sea level static and corrected for effects of Reynolds number, engine service bleed and the offset initially found in the actual installation position. This the corrected value is compared to a fleet average baseline and the percentage difference is calculated. This difference known as "gross delta" is first used to look for apparent sensor errors. If the datapoint is deemed erroneous an appropriate message will be issued and no further analysis is done. The accepted datapoints are used for further analysis.

Figure 13 shows one month worth of data in terms of gross deltas. The result when using each individual gross delta point for analysis is shown in figure 14. As can be seen the scatter is significant and therefore with this combination of sensors and the this way obtained accuracy of gross deltas the result is not accurate enough to correctly analyze module performance based on the data point. The way to get around the problem that is presently used, is that gross-deltas from individual flights are calculated, checked for apparent sensor errors and if found within sensor check limits the gross deltas are passed on to a ten flight average calculation. The gross deltas again in the average calculation will be checked for outliers in a simple correlation analysis. The result from using averaged gross deltas as input to GPA is seen on figure 15.

This particular engine had been installed for several months already at the beginning of this trending, why very limited deterioration is to be expected over the trend period. The 13 trend points now corresponds to 130 flights with stable data, or on this particular route-net approx. 200 cycles.

GPA based on the very limited parameter set used in this trial can not replace pre-shop tests of engines. It gives however additional valuable information on top of the normal trending and will be further studied for use in our preventive removal concept.

LAP, Life Accounting Program

This program is used to bookkeep the amount of damage on critical parts based on actual engine performance and routes flown.

The High Pressure Turbine airfoils on the JT9D-7 are amongst the most critical parts in that engine and are deemed possible to be modelled accurate enough for an analysis. The basic program, written bij PWS, uses precalculated severity factors for each mission and actual AIDS data from each flight to define engine performance levels and routes flown.

The actual life consumption depends upon the mission and the performance status of the engine. The mission is described by a sverity factor that has been precalculated by a Mission Analysis Program using statistical data. Each citypair has several severity factors that varies with respect to season of the year and actual failure mode accounted for, see table 3.

The reason for useage of precalculated statistical severity factors is, that the KSSU airborne 747 AIDS program at the time of specification was not defined to collect mission analysis/life accounting data.

The following failure modes are referred to in the program.

Nozzle Guide Vane crack	- NGV
1st blade creep-fatigue	- 1 BCF
1st blade oxidation corrosion	- 1 BOC
2nd vane deflection	- 2V
2nd blade creep	- 2B

The 1BOC is the only life useage that can be reset by repair e g recoating. All other failure modes refer to life useage that is incremented at each flight throughout the service life of the part.

A typical output generated on a monthly basis is shown on table 4 where the percentage life used is shown for each failure mode. 100 means that expected service life is completely used up.

Experience

The most critical part is 1st blade and therefore, the experience on LAP-life compared to actual failures non-failures is shown in figure 11. A curve ending with F means failure and FO means failure due to overtemperature or where also overtemperature has been confirmed. As shown by the graph there is a good correlation between actual failures and 100% life used for 1BCF. The weak point with this program is that statistical missions are used instead of actual. Missions significantly deviating from nominal as well as engine exceedances for example, higher temperatures than normal during take-off, startup or reverse if encountered is not accounted for.

Autoland verification program

This program produces per aircraft a two-monthly review of Autoland performance. Per line the Autopilot disconnect heights, ILS tracking quality, wind at 100 ft, touch-down dispersion and touch-down maximum g-loads are presented. For the total fleet a two-monthly statistical performance review is presented. The program is used to demonstrate an acceptable Autoland success rate to the authorities and keep control over the maintenance of the Autoland system.

747 APU monitoring

The AIDS system allows to apply a very effective means of health monitoring to the Auxiliary Power Unit. The groundbased computer monitors acceleration time, EGT peak and rotor speed at peak EGT, airduct pressure during air-conditioning system operation with 3 packs and EGT at no-load condition. This program provides indication of and/or clues to mechanical problems, starting problems, airleaks, compressor and turbine inefficiencies.

747/DC-10 aircraft structure lifecycle programs

Since many years KLM collects AIDS recorded data for assessment of service load experience. Results allow comparison to Boeing's fatigue integrity program with the objective to compare the severity and schedule structural inspections on this basis.

Studies on the recorded data also revealed that changes in pressurisation procedures would extend the life of the pressurised structure.

TROUBLE AND INCIDENT ANALYSIS

In reference (2) the analysis and monitoring of JT9D starting problems and the analysis of JT9D auto-accelerations was presented with the cures.

Since that presentation the CF6 compressor stall problem has been solved. The total story is as follows:

CF6 compressor stalls

Although the KSSU CF6 compressor stall-rate reached a low level of 0.06 per 1000 engine hours, the nature of this problem urged to aim for elimination. The major reason is that stalls can occur in critical phases of the flight and thus might endanger the safety of the flight or comfort of the passengers. Several of the experienced CF6 compressor stalls in the winter period of 1979/1980 occurred at the moment the airplane entered a rainshower (fig. 3). Apparently the compressor inlet temperature (cit), measured at the high pressure compressor inlet, drops at constant N2 rotor speed and constant total air temperature, causing the variable stator vane (vsv) to move with the ultimate result that the N1 rotor speed increases at constant N2, causing some individual engines to stall.

First action was to monitor N1 vs N2 to avoid an N1/N2 matching critical to stall. At the same time the study and analysis of all AIDS recorded compressor stalls by a group of specialists was performed with the objective to ultimately eliminate in-flight stalls.

The studies resulted in two actions:

First : Protective rainshields around the CIT sensors were installed figure 4.
Second : In the testcell the Variable Stator Vane (VSV) schedule was adjusted to the more closed position figure 5.

By these actions the engine stall-rate started to drop significantly as can be seen in figure 6.

Today CF6-50 aero stalls are practically eliminated. The cost of aero stalls to the airline were negligable for CF6's with steel compressor casing but amounted to over $ 200 000 each for CF6's with titanium casing. The savings resulting from elimination of aero stalls alone paid for a very substantial part of the AIDS investments if not complete.

It should be noted that all CF6 operators benefit through the AIDS EMS system as used by KLM and the engineering efforts that KLM and other AIDS users put into the task of solving problems like the CF6 compressor stalls.

Flight Technical

Special Flight Technical analysis programs provide means to analyze crew complaints more thoroughly. Programs developed for this purpose are:
- an ILS beam quality check program
- a runway surface analysis program
and - a windshear analysis program

Fuel Consumption Management

In order to ensure accurate flightplan fuel determination KSS uses a computerprogram to monitor the consumption levels per airplane type and per individual airplane. Accounting for consumption levels of individual airplanes allow tighter fuel reserves. In order to more effectively analyze individual high consumers KLM developed a program based on AIDS data that enables engineering to verify effects of maintenance actions on consumption on a short notice.

Figure 8 shows CF6 gas generator curves developed from AIDS recorded data on one flight using a least squares approximation.

It demonstrates the high degree of repeatability achieveable with AIDS recorded data. The maximum deviations of individual data points from the curves are 0,4% corrected fuel flow, 0.13% corrected N2 RPM and 0.05 corrected EPR.

GENERAL ASPECTS

From the examples of the previous paragraphs AIDS appears as a reliable and useful source. Because of its accurate observations, lessons can be learned fast, proper actions can be subsequently applied and cockpit procedures optimized. AIDS also demonstrated on various occasions the ability to observe and report problems outside the observation capability of the crew, it does not conceal human imperfections nor human excellence in performance. Careful treatment of problems where the human factor is involved is a must when disturbance of human relations is to be avoided.

A300/310 AIDS

Development

Ten years of experience with expanded AIDS systems have demonstrated to KLM, SAS and Swissair that the AIDS has matured into an effective engineering tool as predicted. With increasing positive experience it became evident that more effective airborne software was desirable but prohibited by the capacity of the system and the extreme costs of fleet modifications. With the advance of digitalisation of aircraft systems ARINC started to develop new characteristics for these systems known as their 700 series characteristics. In the Boeing 757 and 767 and the Airbus 310 these 700 series digital systems are extensively used. The impact of this development can best be illustrated by comparing the types of KSS AIDS inputs for the 747 and A310:

747 : 210 discretes, 170 analog and 4 digital data busses
A310 : 39 discretes, 49 analog and 38 digital data busses.

Taking advantage of this progress in the application of digital technology, KSSU and the ATLAS European group of airlines, with full cooperation of Airbus Industrie, started to develop an Expanded AIDS for the A310 (ref. 8) based on ARINC 717 (ref. 9). Early 1980 this effort was successfully completed. A block diagram of this system is shown in fig. Subsequently KSSU developed a specification which in more detail specified the desired software functions and specific features derived from KSSU AIDS experience over the years and contracted the system to a major U.S. supplier. In general the increased capacity of the A310, both in terms of parameter inputs and installed software, is used to enhance the AIDS as an engineering tool and expand its troubleshooting capabilities. The relative expansion on parameter inputs partially comes by itself on a digital airplane where most parameters can be sampled from ARINC 429 data busses.

Justification

As previously stated AIDS proved to be an effective engineering tool and provided to KSS an ample return on investments.

Of course the 747 AIDS was not an optimum system compared to the possibilities and the available experience of today. Therefore applying the lessons learned to the A310 AIDS application will result in a still more cost effective system. The conditions have also changed, c.q. the amount of sensor wiring in the airplane has been decreased compared to the 747 and DC-10 AIDS and the capacity of the electronics increased drastically with the result that the costs of a complete installed system is less than of its predecessors.

Translating the investments to costs per flighthour and assuming a reasonable aircraft utilisation these costs will rougly amount to $ 10.- per flighthour. The total costs will double when the AIDS operating costs are added.

A KSS return on investment study for the A310 AIDS did not produce a homogeneous result between KSS partners because there were several differences in estimated savings per individual program or different emphasis on values of benefits.

The ultimate conclusion as derived from past experience can therefore best be presented schematically. Figure 8 shows that the level of quantifyable savings will more than balance the AIDS operating costs.

Characteristics of the KSS A310 AIDS

The control of the recording, printing and display functions is performed by the Data Management Unit (fig. 9). In this unit an Intel 8086 16 bit microprocessor is installed to perform the required functions. The memory comprises 58 K bytes PROM, 18 K bytes RAM, 29 K bytes protected RAM and 39 K bytes EAROM.

Depending on the flight mode the DMU commands continuous recording or selective recording.

All data that is to be recorded passes a 20 second delay buffer such that at detection of specific events always 20 seconds pre-event data is available on the tape-cassette. 75% of the cassette-tape capacity is used for routine recording and the remaining 25% is programmable via the Control Display Unit (CDU). This feature provides to engineering a tool to analyse and solve persistent problems.

The on-board printer can be used by the crew for hard copy engine data but its primary purpose is to provide maintenance with all necessary information. For this reason all maintenance prints are stored during flight and a light on the CDU will inform the maintenance crew that exceedance reports are stored in the DMU memory. In case of specific events as defined in the program, 20 seconds of data prior and 20 seconds of data after the event will be stored and can be recalled by the maintenance engineer using the printer. The memory section used for this feature is called the replay buffer.

From the replay buffer the maintenance engineer can select "canned" sets of parameters prividing a second by second listing of the event occurrance.

For the objectives set for the EMS part of the program it became necessary to provide additional sensors on the engines. KSS and ATLAS in close cooperation with Airbus Industrie and the engine manufacturers succeeded to specify a Powerplant Multiplexer (PMUX) as standard part of an A310 Expanded AIDS.

This PMUX multiplexes temperature and pressure signals, combines these with the output of the Electronic Engine Control (EEC) or Power Management Control (PMC) and sends all this information with the engine serial number via an ARINC 429 dataline to the DMU. This improvement ensures a tighter maintenance control on the quality level of those inputs not monitored by the flight-crew.

A310 ENGINE MONITORING

KSSU formed a team of engine monitoring experts to participate in the specification work on the A310 AIDS. This group being able to take advantage of experience with already existing programs has defined a system incorporating several new features that have never before been available.

Program functions

New functions

Two of the new program functions namely the "history buffer" and the "replay memory" are briefly described already above. One other is known as "stored prints" and works such that the printer does not print in real time, with some very few exceptions, but data that shall be printed is stored in a "print buffer" until a special request is made. There is capacity for storeage of 11 prints with the distribution as defined in table 6

Exceedance Control

The software of course is capable of defining flight modes and to compare selected engine parameters against flight mode related limits throughout the mission.

In case of an exceedance, different actions will be taken. An exceedance can thus cause:

1. Update of exceedance print buffer
2. Recording without 20 sec. pre-event data
3. Recording with 20 sec pre-event data
4. Update of replay memory.

Stable Condition search

The search for stable conditions is, done with a new logic. The method can be described as a window sliding in time over actual and previous data, that are remembered in the computer, looking for stability. Old data is compressed in such a way that it is represented by its average over a 16 sec. period. A maximum of 8 periods will be kept in memory this way. Previously, normally was used a method that stored a reference sample to which actual data was compared over a predefined stable time period or until out of stability occurred leading to a new sample being set.

If a stable period occurs that is not considerably longer, than the predefined stable time period, it is very likely so, that the old logic should not be able to find it. The A300/310 is a short haul aircraft and the time during cruise is so short that an improved search method for stable data was needed.

Divergence monitoring

Another new feature will be the "EGT divergence monitoring" (EDM). The purpose of EDM is to have a method that immediately can recognize a sudden gas path damage through its effect on monitored engine parameters. The influence coefficients for a typical twinspool engine show that regardles of performance deterioration, except for fan flow capacity any gas path damage will affect EGT in an increased direction. Therefore during specified conditions EGT of the two engines are compared.

Both engines are operating in the same ambient conditions so that no correction for that has to be done. One engine is corrected for its offset in thrust setting to the reference engine and the resulting difference in EGT between the two engines is compared to a reference difference established first flight every day. If the difference between actual EGT delta and reference EGT delta exceeds a certain value, the one engine with the increased EGT is automatically pointed out as the unhealthy one by the AIDS system.

Recording Control General

Continuous recording takes place in the following flight modes, engine start, take-off and approach landing. Selective recording is performed in the other flight modes, which means one frame every 100 sec. will be recorded.

Exceedance recording

20 sec. of pre exceedance data and additional data min 20 sec, max 120 sec. or until the exceedance is passed is recorded.

Corner points

Corner points are recorded to better define the mission. The corner points are used in groundbased programs for mission analysis with respect to life accounting on critical parts and refined cycle counting on life limited parts. These points are defined as end of climb or start of descent.

Recording of selected frames for oil consumption monitoring

Either during taxi before and after the flight or prior to engine start and after shut down, data can be recorded for the purpose of oil consumption monitoring.

Recording of stable data

Above a certain altitude, mach number and in cruise, stable conditions are continuously searched for. When stable conditions are found data is automatically recorded.

Recording of APU

Every time the APU is started on external or engine power, 16 frames worth of data will be recorded.

Other planned A310 AIDS applications

As for the current KSSU wide-body aircraft the A310 AIDS will also be used for:

. verification of satisfactory autoland system operation
. trouble and incident analysis
. assessment of service load experience

and the monitoring of:

. the condition of the auxiliary power unit
. the braking and anti-skid system
. aircraft performance deterioration
. safety limit exceedances

A310 AIDS ground system

The system proposed for KLM is depicted in figure 10. It shows again the emphasis KSS lays on the function of AIDS as an engineering tool. The AIDS data as recorded on a cassette continuously will update a data-base in the main EDP center where in paralell also data is stored providing information on crew complaints, maintenance actions, etc. This EMS data base provides periodic trend reports, status reports and automatic exceedance reports and is via terminals accessible by line maintenance and engineering

REFERENCES

1. SAE ARP-1587
"Aircraft Gas Turbine Engine Monitoring System Guide".

2. Vermeulen Henk C. KLM Royal Dutch Airlines
"Current and Future Use of an AIDS integrated EMS"
SAE paper 801219

3. Driessen Ed. A., Vermeulen H.C. and Ledeboer K.H.
KLM Royal Dutch Airlines
"Use of recorders in future aircraft operations"
AIAA paper no. 64-352 June 1964 and Journal of Aircraft Vol II no. 3 1965 pp. 176-184.

4. Urban L.A. Hamilton Standard "Parameter Selection for Multiple Fault Diagnostics of Gas Turbine Engines".
ASME paper 74-GT-62 March 1974.

5. Danielsson Sven G., "Gas Path Analysis applied to pre and post-overhaul testing of JT9D turbofan engine"
SAE paper 77-0093

6. Danielsson S.G. and Dienger Dr. G., A. European view on Gas Turbine Engine Monitoring of Present and Future Civil Aircraft".
AIAA/SAE/ASME paper 79-1200 June 1979.

7. De Hoff R.L., Baker L.E. and Hall Jr. W.E., Systems Control Inc. (Vt) Palo Alto, Ca "Impact of Automated Monitoring on Engine Operations and Support.
AIAA/SEA/ASME paper 79-1276 June 1979.

8. Kalbe H. Messerschmitt-Bölkow-Blohm GMBH "New Aircraft Integrated Data Systems for Airbus A310" Paper presented at the 10th AIDS symposium of the "Deutsche Studiengruppe Für Flugdatensysteme (DSF) March 1980. Aachen Germany.

9. ARINC characteristic 717 "Flight Data Acquisition and Recording System"
March 1, 1979

EMS PARAMETERS MONITORED ON KSS AIRCRAFT

AIRPLANE	BOEING 747		DC 10	AIRBUS A310
ENGINE	P&W JT9D	GE CF6-50E	GE CF6-50C	GE CF6-80A1
ENGINE PERFORMANCE PARAMETERS:				
FAN ROTOR SPEED	X	X	X	X
CORE ROTOR SPEED	X	X	X	X
EXHAUST GAS TEMPERATURE (EGT)	X	X	X	X
FUEL FLOW	X	X	X	X
LPC DISCHARGE PRESSURE	-	-	X	X
LPC DISCHARGE TEMPERATURE	-	-	X	X
HPC DISCHARGE PRESSURE	X	X	X	X
HPC DISCHARGE TEMPERATURE	-	-	X	X
ENGINE PRESSURE RATIO	X	-	-	-
VARIABLE STATOR VANE POS. (VSV)	-	-	X	X
LPT INLET PRESSURE	-	X	X	X
VIBRATION FAN ROTOR	X	X	X	X
VIBRATION CORE ROTOR	X	X	X	X
ENGINE OIL QUANTITY	X	X	X	X
ENGINE OIL PRESSURE	-	-	-	X
ENGINE OIL TEMPERATURE	-	-	-	X
POWER LEVER ANGLE	X	X	X	X
IGNITION	X	X	X	X
FUEL SHUT-OFF VALVE	X	X	X	X
BLEED RELATED PARAMETERS:				
START VALVE POSITION	X	X	X	X
PNEUMATIC BLEED VALVE POSITION	X	X	X	X
ISOLATION VALVES	-	-	X	X
PACK MODE SELECTOR	X	X	X	X
ENGINE INLET ANTI-ICE	X	X	X	X
WING ANTI-ICE	X	X	X	X
VARIABLE BLEED VALVES	-	-	-	X
APU SHUT-OFF VALVE	-	-	X	X
NACELLE TEMPERATURE	X	X	-	X
BLEED PRESSURE	X	X	X	X
BLEED TEMPERATURE	X	X	X	X
AMBIENT PARAMETERS:				
FLIGHT IDENT	X	X	X	X
DATE	X	X	X	X
GMT	X	X	X	X
FLIGHT MODE	X	X	X	X
PRESSURE ALTITUDE	X	X	X	X
TOTAL AIR TEMPERATURE	X	X	X	X
MACH NUMBER	X	X	X	X
STEADY STATE IDENT	X	X	X	X
AUTO THROTTLE ENGAGED	X	X	X	X

TABLE 1

SAMPLE ENGINE DATA PRINT
FROM AIDS ON-BOARD PRINTER

```
A/C 747B      1  MODE          2
FLT         688  ALT        6990
DATE      11.05  CAS          46
GMT        1734  MACH       *.***
SCAN       1781  TAT          29
FLCT       1251
SCAN       1781
EPR       1.007 1.012 1.017 1.016
N1         13.1  23.9  28.0  28.3
EGT       629.5 451.6 408.8 399.1
N2         39.9  60.3  63.7  65.0
FF          153   572   647   635
VIB I        .0    .0    .1    .0
VIB T        .2    .1    .3    .3
BR.P         .2    .2    .3    .9
WATER         0     0     0     0
PWRL      -   1     0 -   1 -   2
PS4        12.4  28.3  37.4  38.1
APU EGT     460  APU RPM   100.2
DUCT PR L    30  DUCT PR R    29
IGN 1         0     0     0     0
IGN 2         0     0     0     0
PN.VALVE      0     0     0     0
OIL T        45    57    85    87
NACT A      1.6   1.7   1.9   2.0
NACT B      2.0   1.7   1.7   1.7
EAI           0     0     0     0
WAI           0  EPR LIM   1.445
EPR MODE      2  APU EGTL    900
EGT L.    649.9 649.9 649.9 649.9
N1 LIM    104.2 104.2 104.2 104.2
BR.T       43.6  61.0  88.7  85.1
```

NOTE: IMPENDING HOT START NUMBER 1 ENGINE
(AT MEXICO CITY AIRPORT).

TABLE 2

TABLE 3

SEVERITY FACTORS PER CITY PAIR & FAILURE MODE

city-pair	flt. time	sea-son1	sea-son2	sea-son3	sea-son4	failure mode
AMS FRA	7.9	2169	2371	2898	2616	NGV
AMS FRA	7.9	1240	1656	3410	2089	1BCF
AMS FRA	7.9	0936	1418	3836	2107	1BOC
AMS FRA	7.9	0190	0350	1440	0571	2V
AMS FRA	7.9	7465	8643	11800	8673	2B

TABLE 4

OUTPUT FROM MAP-CAP

PROCESSING DATE   27MAR81
REFERENCE STOMK-R FRANZEN

SORTED BY A/C REG NO

THE FOLLOWING IS A SUMMARY OF THE DAMAGE ACCUMULATED ON EACH PART

INST POS	ENGINE NO	NGV	1BCF	% LIFE USED 1BOC	2V	2B
DDL1	663074	5.2	6.1	7.3	76.5	3.2
DDL2	662987	18.0	16.0	19.1	16.2	5.9
DDL3	662841	93.3	17.8	7.1	76.6	50.9
DDL4	663067	95.7	37.8	23.9	31.8	23.2
KHA1	662814	95.9	54.1	8.1	83.8	26.5
KHA2	662999	7.3	7.4	9.4	49.7	3.1
KHA3	662750	6.8	63.7	8.5	47.8	39.7
KHA4	662909	8.5	9.0	11.2	76.5	3.9
IGA1	662762	92.3	63.6	2.2	72.4	49.4
IGA2	663073	50.0	62.4	.5	46.4	0.0
IGA3	663070	110.9	23.8	21.5	95.9	67.7
IGA4	662803	96.5	55.8	31.8	52.5	66.5
IGB1	662927	47.1	14.0	7.9	46.9	65.2
IGB2	685637	35.2	25.6	24.9	27.2	11.2
IGB3	662982	95.3	70.7	7.3	40.3	4.1
IGB4	685636	68.2	44.5	8.1	48.2	57.8
BUA1	685616	36.2	18.3	10.2	43.6	36.7
BUA2	663011	90.2	4.3	4.7	49.7	1.8
BUA3	662818	98.2	85.2	.1	95.3	80.1
BUA4	685635	95.2	19.0	16.1	21.7	11.8
BUB1	662754	94.1	87.3	1.9	80.4	71.0
BUB2	662979	79.1	41.9	24.3	32.1	.8
BUB3	662930	91.9	31.5	18.0	54.6	44.5
BUB4	662846	97.1	62.1	8.1	84.1	92.0

TABLE 5

PARAMETERS USED IN KLM.SAS GPA PROGRAM

KLM CF6-DC10	SAS JT9D-B747
N1	N1
N2	N2
WF	WF
EGT	EGT
PS2C	PS4
TT2C	EPR
PS3	
TT3	
PT5.4	
VSV *	

* VSV ON SAMPLE ENGINE ONLY

TABLE 6

PRINT FORMATS / NAMES

COMPARISON REPORT	– PRINT/EVENT	1
	– STABLE	2
	– EXCEEDANCE	2
	– TAKE OFF	1
	– CDU REQUESTED	
TAKE OFF TREND REPORT		1
CRUISE TREND REPORT		1
FAILED ENGINE START REPORT		1
MAINTENANCE REPORT		1
LOAD/FLIGHT CONTROL REPORT		1

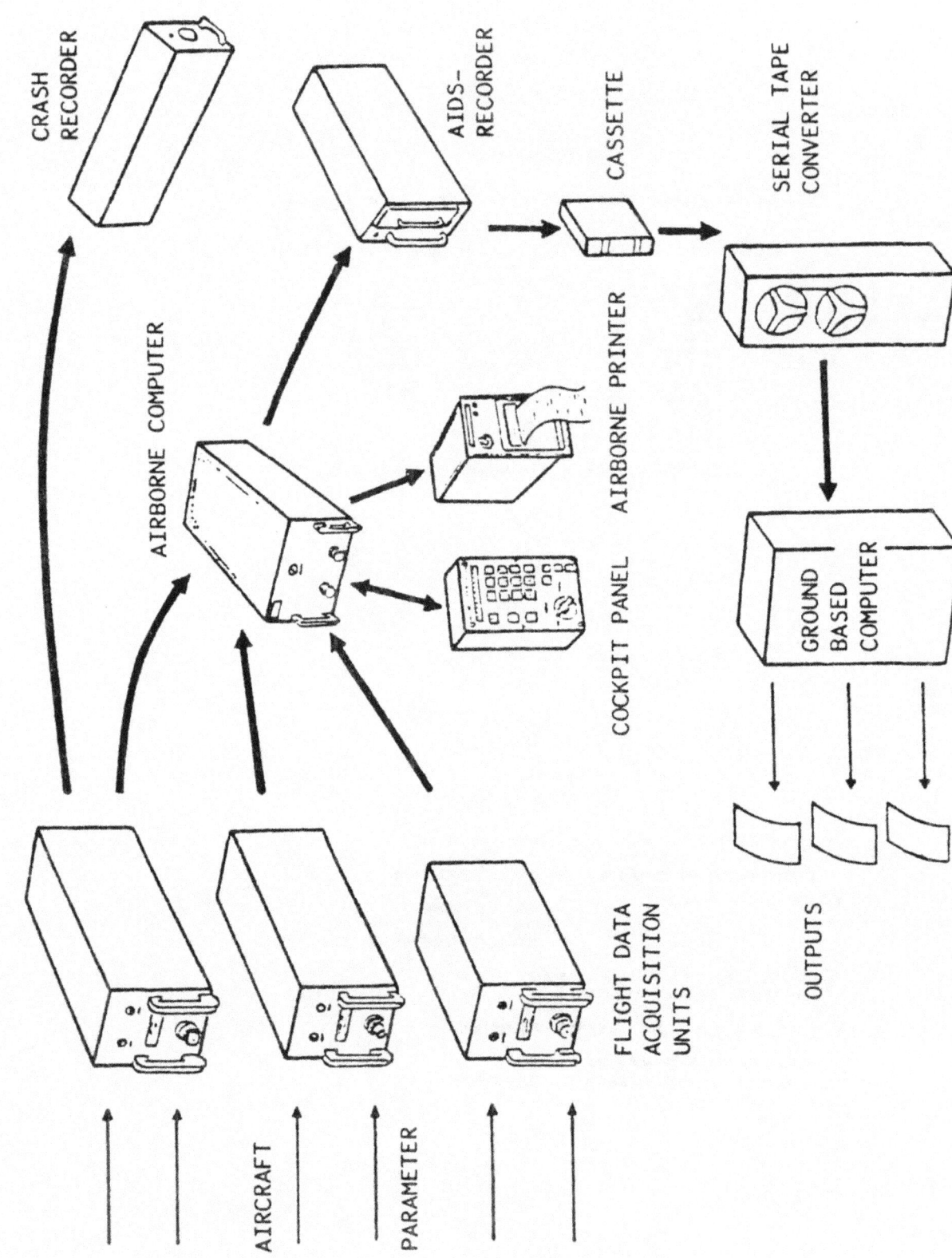

FIG. 1 AIDS SYSTEM AS INSTALLED IN KSS 747 AND DC-10 AIRCRAFT

ENGINE ANALYSIS 747B A/C PH-BUB POS 2 ENGINE 662879 INSTALLED 10JAN. 81 VERSION 30 PRINT DATE 7 APR 81 PAGE 4

OUTPUT TYPE : DAILY

FLIGHT EXCEEDANCES

NO EXCEEDANCES

START TREND

DAY	GMT	FLT	DEP/ARR	T20	T10	T50	BLP	BLT	O/T	PALT	TAT	APUN	EGTP	N1P	N2P	IFF	SFF	BLV 1234
302	723	845	CMB/SIN	17	5	16	29.3	155	91	80	48	101.1	533	14.0	46.2	2460	705	0101
302	1329	845	SIN/MEB	19	6	16	35.8	160	35	90	23	101.0	458	14.0	46.6	376	744	0100
302	2215	845	MEB/SYD	15	5	15	33.7	143	78	483	19	101.0	487	13.9	47.4	345	717	0100
402	455	846	SYD/MEB	14	6	15	35.7	144	56	222	29	100.8	474	14.2	47.2	1555	726	0100
403	707	846	/	***	0	0	35.8	151	91	526	21	100.8	474	14.2	47.2	****	****	0100
202	1859	845	SPL/DXB	18	12	16	36.2	141	12	-333	9	101.2	45V	14.5	47.7	415	825	0100
302	217	845	DXB/	***	10	14	33.6	148	85	-73	17	101.1	460	13.9	46.0	349	737	0100
302	723	845	CMB/SIN	17	5	16	29.3	155	91	80	48	101.1	533	14.0	46.2	2460	705	0101
302	1329	845	SIN/MEB	19	6	16	35.8	160	35	90	23	101.0	458	14.0	46.6	376	744	0100
302	2215	845	MEB/SYD	15	5	15	33.7	143	78	483	19	101.0	487	13.9	47.4	345	717	0100
402	455	846	SYD/MEB	14	6	15	35.7	144	56	222	29	100.8	474	14.2	47.2	1555	726	0100
403	707	846	/	***	0	0	35.8	151	91	526	21	100.8	474	14.2	47.2	****	****	0100
304	1241	641	SPL/JFK	19	6	***	35.7	142	16	-296	13	99.6	469	22.1	58.6	397	762	0101
304	2307	642	JFK/SPL	16	7	15	36.3	143	58	-61	12	100.1	435	13.0	46.0	338	735	0101
504	1210	611	SPL/ORD	16	5	15	37.5	140	42	-285	13	99.6	433	13.9	46.9	401	768	0101
504	2241	612	ORD/SPL	15	5	15	36.6	134	61	488	6	99.6	435	14.4	47.5	329	755	0101
404	1227	677	SPL/YMX	15	6	16	37.7	134	30	-350	12	99.6	424	14.4	47.1	385	793	0101
404	2336	678	YYZ/YMX	16	5	15	35.2	138	72	835	16	99.6	463	13.5	46.4	331	689	0110
304	1241	641	SPL/JFK	19	6	***	35.7	142	16	-296	13	99.6	469	22.1	58.6	397	762	0101
304	2307	642	JFK/SPL	16	7	15	36.3	143	58	-61	12	100.1	435	13.0	46.0	338	735	0101

a

TAKE-OFF TREND

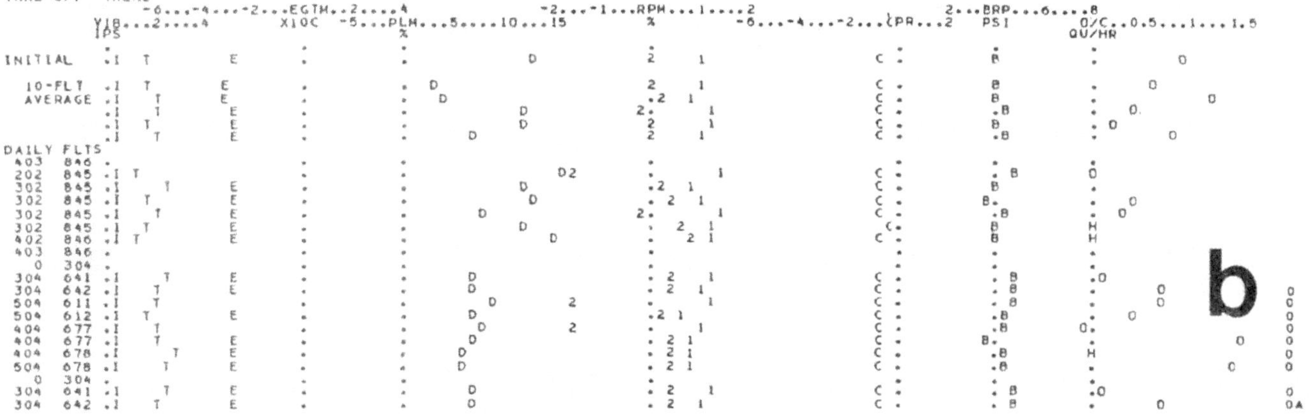

b

TAKE-OFF PERFORMANCE NORMAL NUMBER OF A/C PACKS IS 2

DAY	GMT	SCAN	FLT	DEP/ARR	EPR	EPRL	EGT	F/F	N1	N2	PS4	BPR	O/T	I-VIB-T	THP	PALT	TAT	CAS	TTO	TTA	A-NT-B	ACP 123	BLV	EAI	WI	IGN
304	2321	1060	642	JFK/SPL	1.411	1.42	818	7664	90.4	94.1	291.7	4.9	91	0.6 2.3	69	-56	11	125	0.0	0	1.6 1.8	111	1	0	0	B
504	1223	1897	611	SPL/ORD	1.427	1.44	841	8135	91.0	93.8	298.5	4.8	84	0.6 2.1	72	-275	8	129	0.0	0	1.7 1.9	111	1	1	0	B
504	2251	3599	612	ORD/SPL	1.378	1.39	781	7032	87.7	92.7	276.4	4.7	88	0.5 1.7	63	491	6	127	0.0	0	1.6 1.9	111	1	0	0	B
404	1243	1717	677	SPL/YMX	1.378	1.38	802	7377	88.0	92.4	280.9	4.7	82	0.5 2.3	64	-335	7	121	0.0	0	1.7 1.9	111	1	1	0	B
404	2039	2242	677	YMX/YYZ	1.363	1.39	823	7101	89.6	95.2	270.5	4.0	93	0.5 2.2	71	524	22	134	0.0	0	1.6 1.8	111	1	0	0	B
404	2347	1512	678	YYZ/YMX	1.364	1.38	802	6917	88.6	94.1	266.7	4.5	92	0.6 2.9	66	850	15	124	0.0	0	1.6 1.8	111	1	0	0	B
504	146	657	678	YMX/SPL	1.371	1.38	813	7104	89.3	94.5	272.9	4.5	92	0.4 2.7	67	469	17	125	0.0	0	1.6 1.8	111	1	0	0	B
0			304	XXX/				NO	TO-DATA																	
304	1257	940	641	SPL/JFK	1.410	1.41	818	7710	90.5	94.1	293.8	5.0	84	0.6 2.4	68	-281	10	134	0.0	0	1.6 1.9	111	1	0	0	B
304	2321	1060	642	JFK/SPL	1.411	1.42	818	7664	90.4	94.1	291.7	4.9	91	0.6 2.3	69	-56	11	125	0.0	0	1.6 1.8	111	1	0	0	B

CRUISE TREND

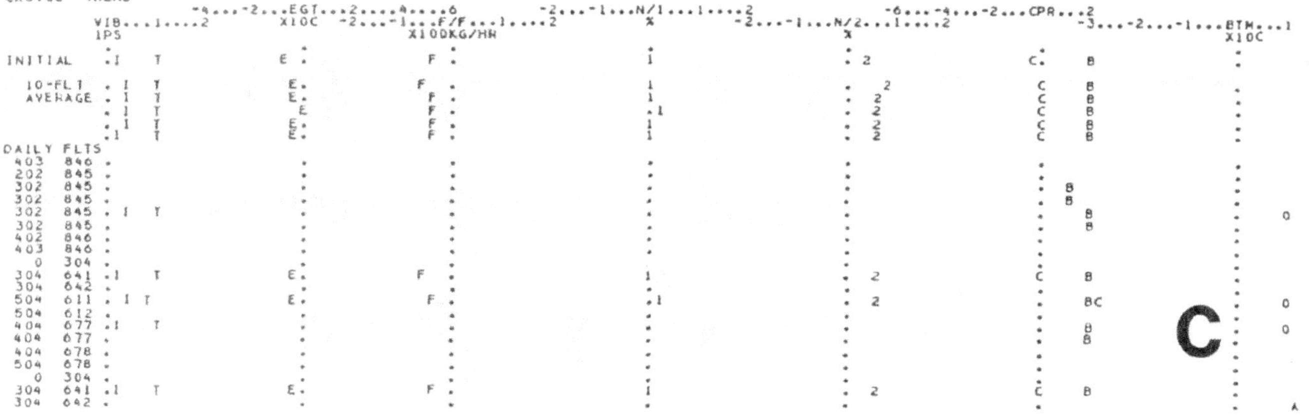

c

CRUISE PERFORMANCE NORMAL NUMBER OF A/C PACKS IS 2

DAY	GMT	SCAN	FLT	DEP/ARR	EPR	EPRL	EGT	FF	N1	N2	PS4	O/T	I-VIB-T	THP	PALT	TAT	CAS	MACH	A-NT-B	ACP 123		AIR-BREATHER-TEMP MARGIN ABT	EPR	TAT	O/T	EBI ALG IVN
304			642	JFK/SPL				NO STABILIZED FRAME														NO BTM DATA				
504	1538	1600	611	SPL/ORD	1.246	1.43	668	2713	86	87	121	93	0.5 1.0	36	309	-19	306	.822	1.6 1.8	101		148	1.361	2	94	011
504			612	ORD/SPL				NO STABILIZED FRAME														NO BTM DATA				
404	1314	3600	677	SPL/YMX	1.309	1.43	692	2931	87	88	127	93	0.3 1.1	41	310	-19	304	.819	1.7 1.8	111	ABN	151	1.340	8	94	010
404			677	YMX/YYZ				NO STABILIZED FRAME														153	1.363	15	94	
404			678	YYZ/YMX				NO STABILIZED FRAME														NO BTM DATA				
504			678	YMX/SPL				NO STABILIZED FRAME														NO BTM DATA				
0			304	XXX/				NO STABILIZED FRAME														NO BTM DATA				
304	1334	3200	641	SPL/JFK	1.340	1.50	711	3074	89	89	131	93	0.3 1.1	42	310	-15	311	.836	1.6 1.8	101		154	1.339	14	94	010
304			642	JFK/SPL				NO STABILIZED FRAME														NO BTM DATA				

FIG. 2 AIDS LONG TERM TREND JT9D ENGINE

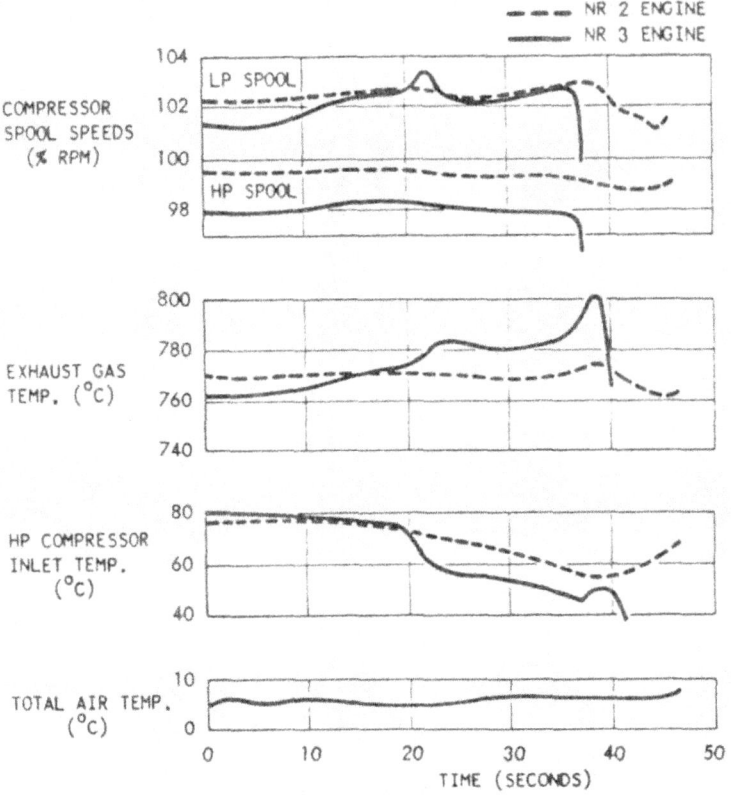

FIG. 3 AIDS TRACE OF COMPRESSOR STALL ON A DC-10 AIRPLANE (CF6-50)

CIT Sensor

With Rainshield   Standard Sensor

FIG. 4 CIT SENSOR

VSV Tracking

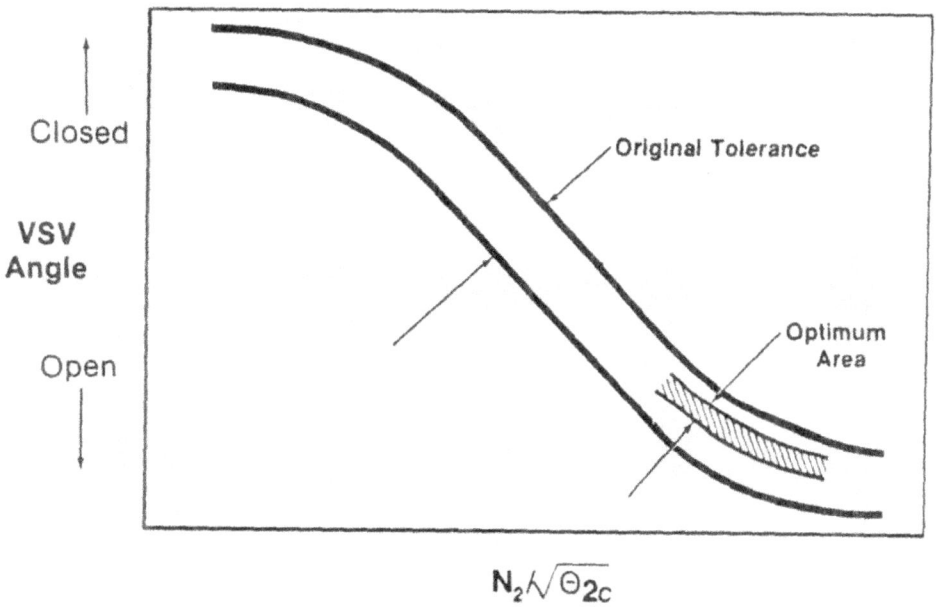

FIG. 5 VSV TRACKING

Aero Stalls
Fleet

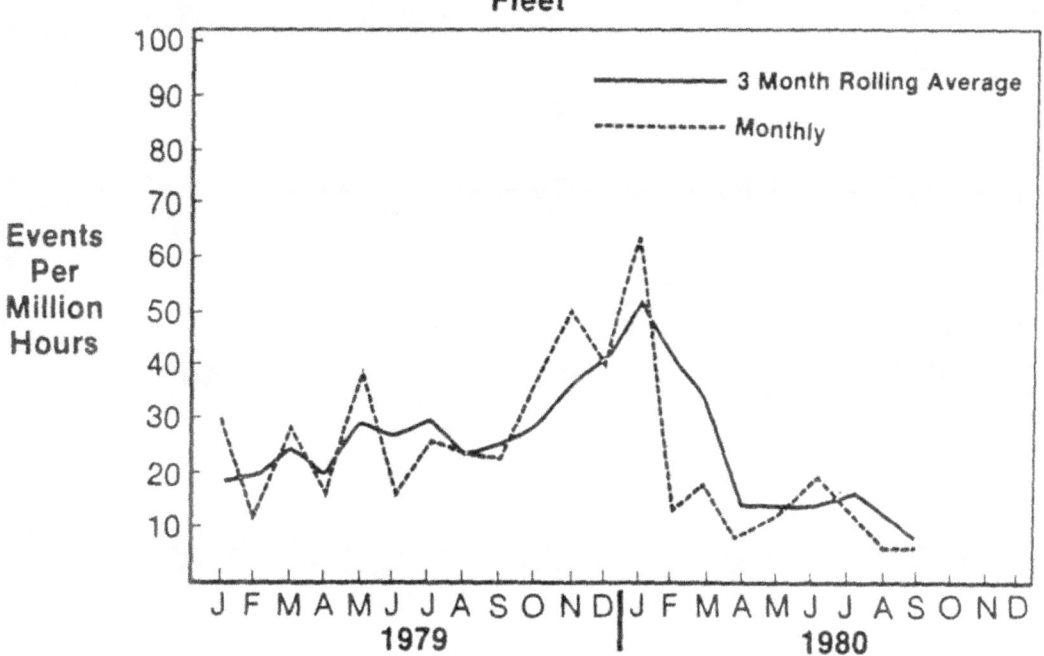

FIG. 6 CF6 AERO STALLS

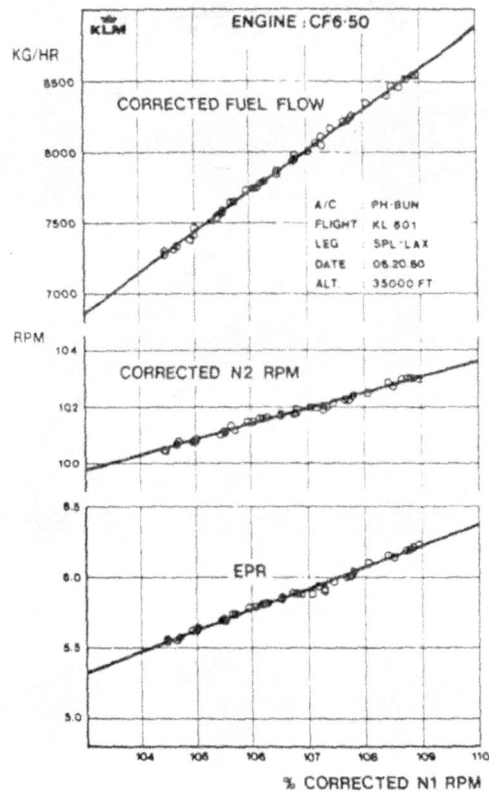

FIG. 7 AIDS CF6 GAS GENERATOR CURVES

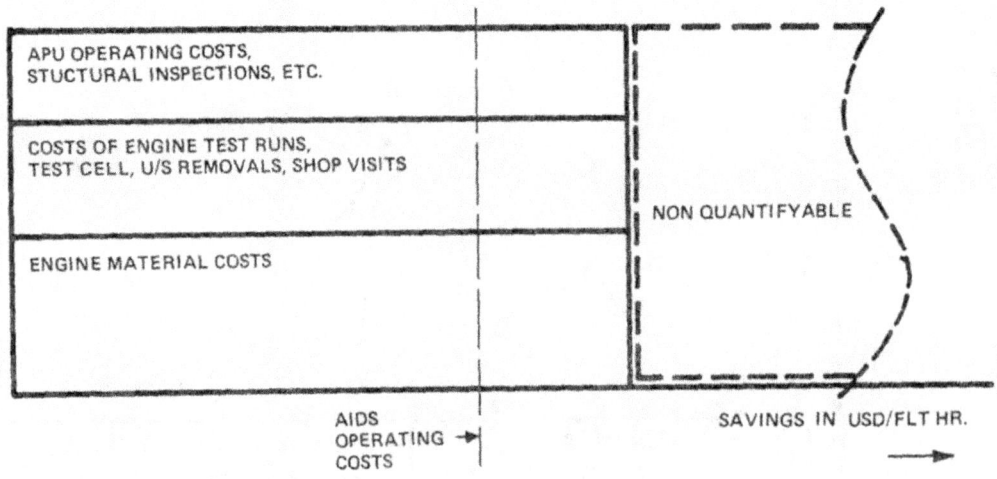

FIG. 8 PROJECTED A310 AIDS RETURN ON INVESTMENT

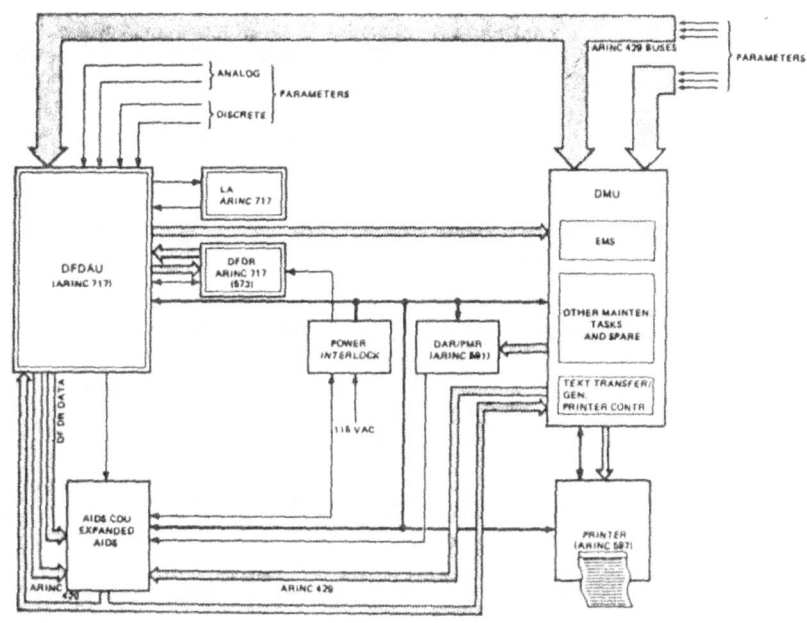

FIG. 9 AIRBUS A310 AIDS SYSTEM DIAGRAM

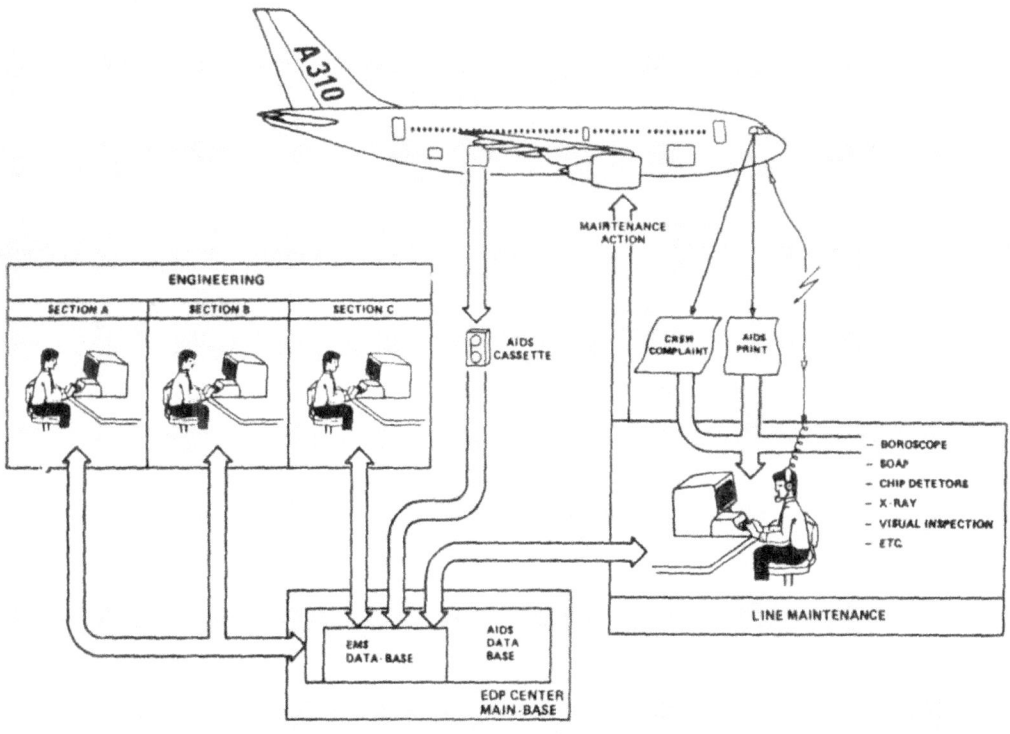

FIG. 10 A310 AIDS GROUND SYSTEM

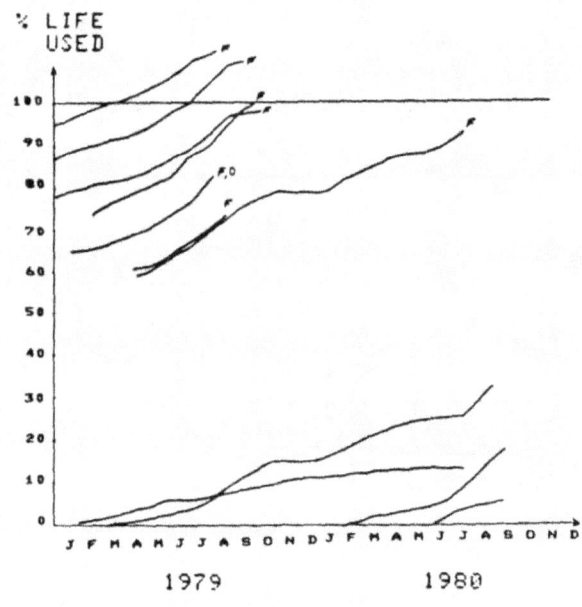

FIG 11

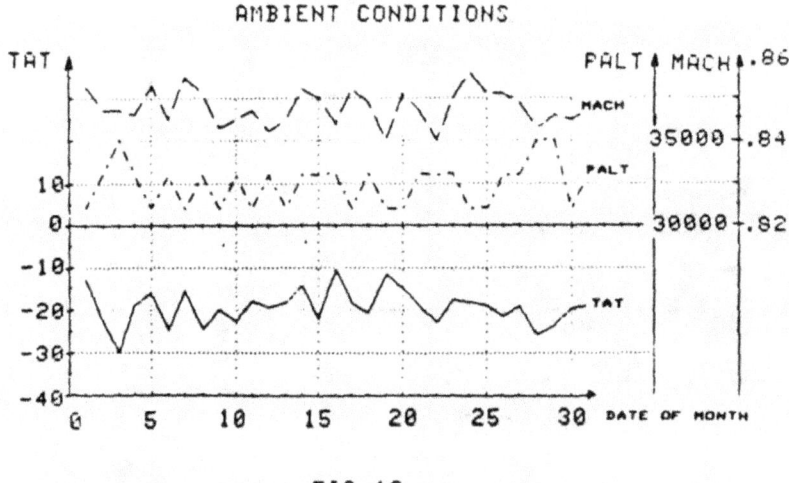

FIG 12

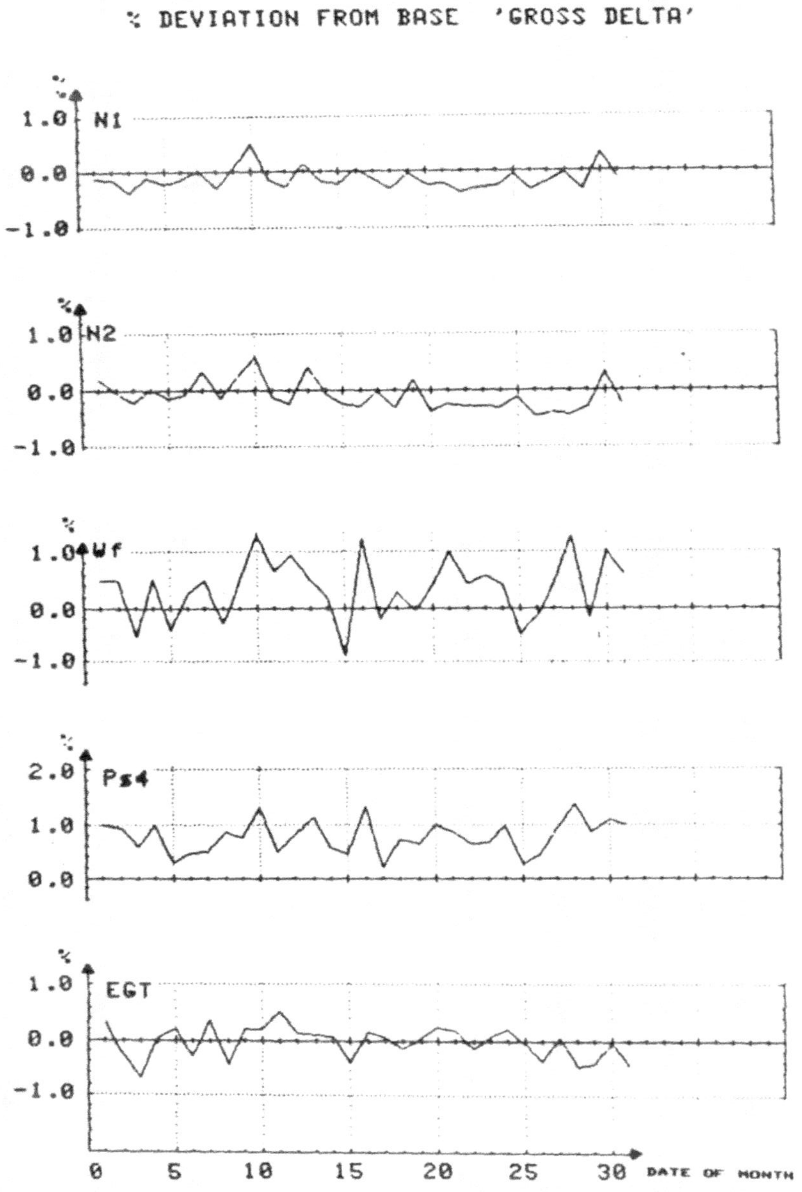
% DEVIATION FROM BASE 'GROSS DELTA'
N1
N2
Wf
Ps4
EGT
%
1.0
0.0
-1.0
2.0
0 5 10 15 20 25 30
DATE OF MONTH

FIG 13

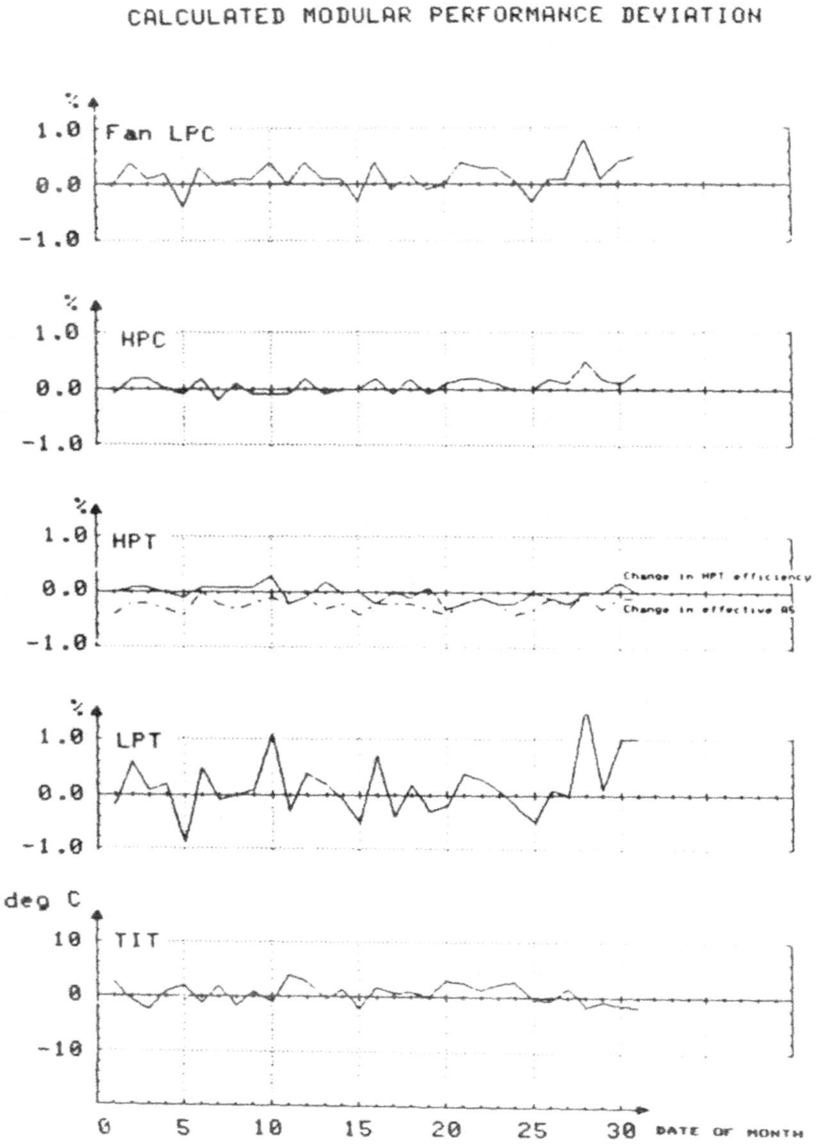
CALCULATED MODULAR PERFORMANCE DEVIATION
Fan LPC
HPC
HPT
LPT
TIT
%
deg C
1.0
0.0
-1.0
10
0
-10
Change in HPT efficiency
Change in effective AS
0 5 10 15 20 25 30
DATE OF MONTH

FIG 14

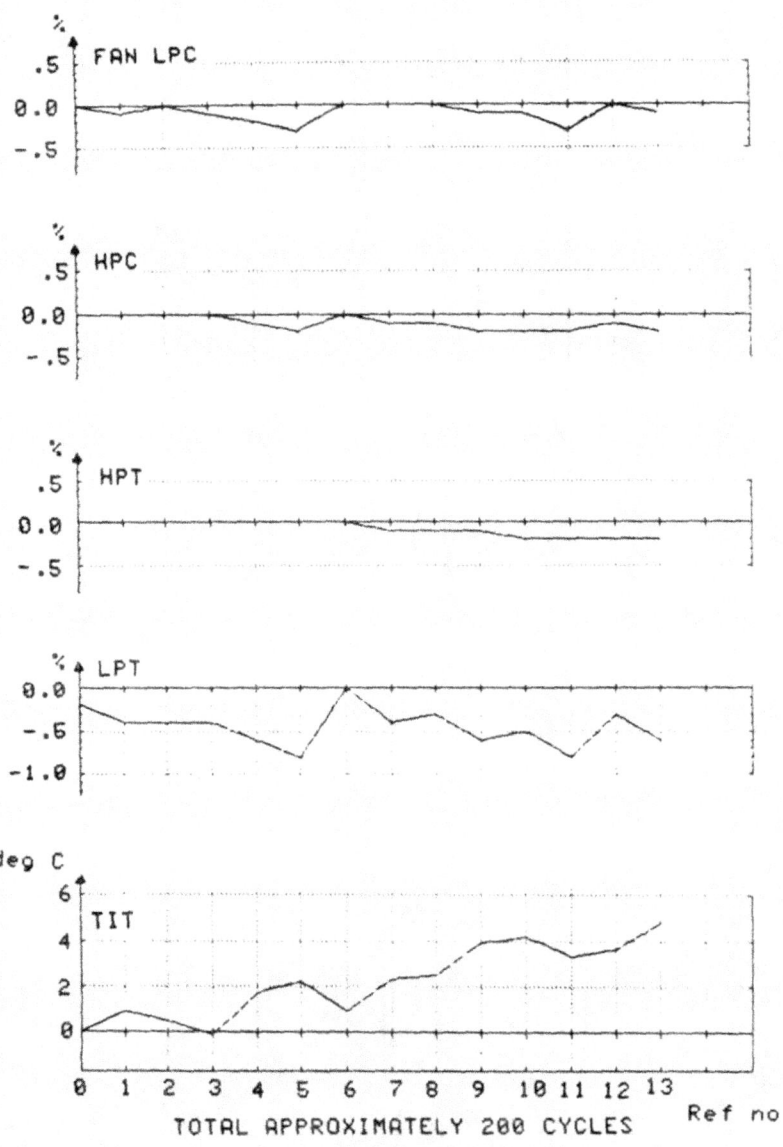

FIG 15

HELICOPTER PROPULSION SYSTEM RELIABILITY AND

ENGINE MONITORING ASSESSMENTS*

John A. Murphy
Bell Helicopter Textron

SUMMARY

Bell Helicopter is conducting a study of helicopter propulsion system reliability problems, specific technology solutions, and engine monitoring implications. The study is approximately 50 percent complete. Engine monitoring implications thus far include consideration of reciprocating engine monitoring, realization of maintenance cost savings due to derated operation, bookkeeping of three-engine installations using two-engine cruise, monitoring of contingency rating usage and integration of drive system monitoring functions. Commercial acceptance will require "up-front" demonstration of cost effectiveness.

INTRODUCTION

Of the various helicopter subsystems, the propulsion subsystem (i.e. engines and drive train) has the greatest impact on reliability, maintainability and safety. A study by Boeing Vertol (Reference 1) of over 1500 FAA Malfunction or Defect Reports (FAA Form 9330) for the U.S. civil turbine-powered helicopter fleet shows the propulsion subsystem accounts for:

- 49.2% of the failures
- 60.1% of the unscheduled maintenance manhours
- 87.3% of the repair parts cost

Similarly, an in-house study by Bell Helicopter Textron (BHT) shows that the propulsion subsystem is involved in 79% of the major, and 49% of the minor, material failure related accidents in a Military light helicopter fleet.

Against this background, BHT has been awarded a contract by NASA - Ames to conduct a study of "Propulsion Systems Reliability and Integrated Engine Monitoring Technology Assessments for Civil Helicopters". The objective of the study is to increase civil helicopter productivity by improving propulsion system life, reliability and maintainability through proper focusing of future research technology programs. Study tasks and schedule are shown by Figure 1. Those tasks which are engine related such as derated engine characteristics, engine design changes, and engine cycle modifications have been subcontracted to Detroit Diesel Allison.

---

* Performed under NASA - Ames Contract NAS 2-10722,
  Dr. John Zuk, Technical Monitor.

PROBLEM IDENTIFICATION

The Statement of Work for this task is to "Identify the major short life, unreliable, and high maintenance engine and power transfer components and subsystems in current civil helicopters. Categories shall include both reciprocating and turbine engines, single and multiple engine configurations, single and tandem rotor vehicles, and light, medium and heavy helicopters".

This task was approached in three ways:

- Accident rate data
- Maintenance rate data
- Direct operator input

Accident Rate Approach

For this approach, the U.S. civil helicopter population was determined, using the latest available data (Reference 2), with results as shown by Table 1. This population was broken out by the various study categories as shown by Table 2. Also shown are the flight hour distributions of the various categories, as determined from Reference 3. Accident data for this population and time period, as reported by the National Transportation Safety Board was then examined. A total of 2302 helicopter accidents occurred, of which 1472 (64%) involved no material failure of the helicopter. However, of the 831 which were material failure related, 586 (70%) had an engine or drive train related cause factor. Table 3 shows a breakdown of accidents by engine type (reciprocating or turbine). Considering the 469 accidents which had a powerplant related cause factor, 354 (75.5%) involved reciprocating engine models, while the remaining 115 (24.5%) involved turbine powered models. These numbers in themselves are meaningless, however when compared to either the population percentages or the flight hour percentages shown in Table 2, it is evident that reciprocating engine powered helicopters are involved in a disproportionately large percentage of powerplant related accidents. Similar results were obtained for drive train related accidents, and for accidents due to all causes. This suggests that perhaps some of the automotive technology which is emerging for monitoring reciprocating engines might well be applied to aircraft.

Similar breakdowns of accident cause factors were made for the other study categories, i.e. single vs twin, light, medium and heavy weight, and single vs tandem rotor. These breakdowns are handicapped by the small twin, heavy, and tandem populations, and did not yield any particularly startling results.

Maintenance Rate Approach

For this approach, an analysis was made of U.S. Navy Maintenance and Material Management (3M) data for the models, systems, and characteristics noted in Table 4. In this group, the twins, heavies, and tandems are better represented than in the civil population, as shown by Table 5. Typical results are shown by Figure 2, which is for engine related problems. The parameters shown across the top are:

MFHBF - Mean Flight Hours Between Failures

MFHBMA - Mean Flight Hours Between (Unscheduled) Maintenance Actions

MMH/FH - Maintenance Man Hours per Flight Hour

MMH/MA - Maintenance Man Hours per Maintenance Action

EMT/MA - Elapsed Maintenance Time per Maintenance Action

The different categories are shown along the left. When categorized by weight, it can be seen that although failure rates are similar, the light helicopter is much easier to repair. When number of engines is considered, it is interesting to note that MMH/FH for the twins is exactly twice that of the singles. Catorizing by type of rotor system has little effect on engine Reliability and Maintainability (R&M) characteristics, as would be expected. Summaries similar to Figure 2 were also prepared for the other systems listed in Table 4.

Direct Operator Input

Direct operator input to the study was obtained by visits to two large gulf coast commercial helicopter operators. The Petroleim Helicopter, Inc. heavy maintenance facility in LaFayette, Lousiana was visited on 1 October 1980. The following day, the Air Logistics facility near New Iberia, Louisiana was visited. Table 6 lists the major R&M concerns mentioned by maintenance management personnel at these facilities. It is significant that no one component, subsystem, or issue was of overriding importance. With regard to engine montoring systems, the concern is that a dedicated piece of electronic equipment, with associated wiring and connectors, will be just another maintenance burden. This feeling is based on a long history of electronic equipment problems in the gulf coast environment. There must be an obvious economic payback from the monitoring system to trade-off, against its possible liabilities.

Problem Identification Conclusion

It is concluded that there is no one overriding R&M issue, but rather a broad spectrum of areas in need of improvement.

Technology Solutions

Several technology solutions to the overall problem of improving R&M characteristics were investigated, as shown by Figure 1. These are discussed as follows, with regard to engine monitoring implications.

Engine Derating Study

The objective here was to determine the relative benefits of derating a current technology engine, vs utilizing a non-derated advanced technology engine, to improve R&M characteristics.

"Derating", in this case, means to install an "oversized" engine, while maintaining the same aircraft performance. A "baseline" situation was assumed such that the "zero derate" case consisted of a Bell Model 206B "JetRanger III" with Allison 250-C20B engine, operating on a mission which requires "takeoff" (red-line) power for takeoff and landing, and "maximum continuous" power for cruise. Allison provided data for scaling engine size, weight, cost, specific fuel consumption (SFC), and maintenance cost to twice baseline rated power. Engines derated (i.e. oversized) in 10 percent increments were parametrically installed in the aircraft. Aircraft weight, cost, and fuel consumption were allowed to increase as engine size increased. The net result is shown by Figure 3. As expected, the cost of fuel goes steadily upward, due to higher power required (because of increased gross weight), and increased SFC due to part power operation. Insurance and depreciation increase due to increasing aircraft cost caused by increased weight empty and increased engine cost. Maintenance cost decreases due to operating at reduced gas temperatures. The net effect is negligable out to about 20 percent derating. Data for an advanced technology engine, rated at baseline power, were also provided by Allison and used in the same manner. In this case, aircraft weight and power required decreased, so that this engine was in effect derated also. The net effect is a significant saving in operating cost. To put these numbers in perspective, it should be noted that the current leasing rate for this aircraft is around $350/hour.

The engine monitoring implication is that the maintenance saving shown will not be realized without a monitoring system. This is because a large percentage of the saving is due to extended Time Between Overhaul (TBO) and component retirement times, as specified by the engine manufacturer. In order to grant these extensions, the manufacturer must be assured that the engine is in fact being operated in a derated mode. A fool proof, continuous on-board monitor will be required for this assurance.

Fuel Control

Fuel control improvements are centered around the Full Authority Digital Electronic Control (FADEC). This type of control is well suited to performing engine monitoring functions.

Configuration Changes

One of the configuration changes investigated is the effect of number of engines. One-Engine-Inoperative (OEI) requirements are increasing in importance, as evidenced by the number of new twin-engined designs entering the market. One problem, of course, is that the initial cost and operating cost of a twin-engined helicopter is considerably greater than that of a comparable single-engined design. This could be aleviated somewhat by special "contingency" ratings, to be used only in OEI situations. This would increase the power available from the remaining engine if one engine were lost, thus permitting a reduction in total installed power, weight, and improved SFC since the engines would normally operate at a higher percentage of rated power. This subject is thoroughly discussed in Reference 4.

Actual usage of the contingency rating must be closely controlled. Special inspections, parts replacements, or TBO reductions may be required if the rating is used. An engine monitor will be required to record the extent of usage of the contingency rating.

Another solution is to use three engines instead of two. Since loss of an engine would reduce the power available by 33 percent instead of 50 percent, total installed power can be reduced. Also, with three engines, one engine can be shut down for long range cruise, improving the power match and reducing SFC. The engine monitoring implication is that a system would be required for bookkeeping of engine operating time and cycles. Manual bookkeeping would be inordinately complex and error prone if the two-engine-cruise mode were frequently used.

Transmission System Improvements

Transmission and drive system monitoring could be incorporated into the engine monitoring system. One low-cost candidate function would be to record actuations of the various "chip detectors". These are magnetic devices in the lubrication system which collect magnetic debris from the oil. Collection of sufficient debris completes an electrical circuit and turns on a cockpit light. Detectors are now available which will self-clear normal wear particles. Frequent self-clearing, however, may indicate an incipient problem. Therefore, an indication of the frequency of clearing operations, whether automatic or pilot initiated, would provide useful diagnostic information. More sophisticated monitoring techniques based on vibration signal analysis are also available and could be incorporated into the monitoring system.

ENGINE MONITORING CONCLUSIONS

Engine monitoring implications of the study which have evolved thus far are summarized as follows:

Reciprocating Engine Monitoring

All aircraft engine monitoring programs of which we are aware are directed at turbine engines. However, most civil helicopters are powered by reciprocating engines. Although this percentage of the population is decreasing, it is estimated by the FAA that "recips' will still represent 40 percent of the population in 1992. Accident statistics indicate that reciprocating engines could benefit from a monitoring system. Automotive technology in this area is rapidly emerging, and should be considered for adaption to the general aviation fleet.

Extended TBO and Retirement Lives

In establishing TBO's and retirement schedules, engine manufacturers must assume "worst case" operation, since actual engine usage is unknown. Scheduled maintenance based on actual usage would be feasible if such usage were recorded by the monitoring system.

Engine Record Bookkeeping

Accurate logging of engine starts, run time, time above certain power levels, and engine cycles is a valuable monitoring function, even for a single-engine aircraft. This function becomes more useful as the number of engines increases, and is a practical necessity if engine run times are unequal, i.e. OEI cruise.

Contingency Rating Usage

If a "contingency" or "emergency" power rating is available for infrequent, short-term use, actual usage of this rating must be closely controlled. By recording such usage, the engine monitor can trigger any required maintenance actions.

Transmission and Drive System Monitoring

If the engine monitor is to be used in a helicopter, integration of transmission and drive system monitoring functions should be considered. Candidate functions are chip detector actuations, chip detector clearing operations, bearing temperatures, and vibration signal analysis.

Cost Effectiveness

Commercial operators are concerned that an engine monitor might cost more than it saves. The monitoring system must be tied to an extended or "on-condition" maintenance schedule so that cost effectiveness will be evident "up-front", not as a hopeful result.

REFERENCES

1. Dougherty, J.J. and Barrett, L.D.: Research Requirements to Improve Reliability of Civil Helicopters. NASA CR-145335, April 1978.

2. Census of U.S. Civil Aircraft, Calender Year 1978. U.S. Department of Transportation, Federal Aviation Agency, Office of Management Systems.

3. General Aviation Activity and Avionics Study, U.S. Department of Transportation, Federal Aviation Administration, Office of Management Systems.

4. Sample, R.D.: Emergency-Power Benefits to Multi-Engine Helicopters. Journal of the American Helicopter Society, July 1977, pg. 27.

U. S. CIVIL HELICOPTER POPULATION

CY 1978

MANUFACTURER	NO. OF MODELS	NO. REGISTERED
AEROSPATIALE	7	239
AGUSTA	1	12
BELL	7	3209
BRANTLY	2	153
ENSTROM	2	369
HILLER	2	85
HUGHES	2	1728
KAMAN	1	19
MBB	1	61
SIKORSKY	11	383
VERTOL	2	23
	TOTAL	6281
MISC		173
	TOTAL	6454

TABLE 1

HELICOPTER POPULATION SETS

CHARACTERISTIC	MODELS (NUMBER)	REGISTERED (NUMBER)	PERCENT POPULATION 1978	FLT. HOURS 1970-1978
RECIP	14	3779	60.2	46.4
TURBINE	24	2502	39.8	53.6
SINGLE ENGINE	27	5979	95.2	95.4
TWIN ENGINE	11	302	4.8	4.6
LIGHT	21	5549	88.3	91.0
MEDIUM	11	627	10.0	8.5
HEAVY	6	105	1.7	0.5
SINGLE ROTOR	35	6239	99.3	99.9
TANDEM ROTOR	3	42	0.7	0.1

TABLE 2

ACCIDENT CAUSE FACTORS BY ENGINE TYPE

TYPE	ACCIDENTS %	POPULATION %	FLT. HOURS %	% ACCIDENTS / % POPULATION	% ACCIDENTS / % FLT. HOURS
POWERPLANT RELATED					
RECIP	75.5	60.2	46.4	1.25	1.63
TURBINE	24.5	39.8	53.6	0.62	0.46
DRIVETRAIN RELATED					
RECIP	76.1	60.2	46.4	1.26	1.64
TURBINE	23.9	39.8	53.6	0.60	0.45
ALL CAUSES					
RECIP	75.9	60.2	46.4	1.26	1.64
TURBINE	24.1	39.8	53.6	0.60	0.45

TABLE 3

3M DATA ANALYSIS

TIME PERIOD: 1 YEAR 4/79 - 3/80
FLIGHT HOURS: 15000 - 27000

MODELS	SYSTEMS	CHARACTERISTICS
TH-57	ENGINE	MFHBF
UH-1E	MAIN DRIVE SHAFT	MFHBMA
UH-1N	MAIN TRANSMISSION	MMH / FH
SH-2F	TAIL ROTOR DRIVE SHAFT	MMH / MA
SH-3H	HANGER ASSEMBLY	EMT / MA
CH-53D	INTERMEDIATE GEARBOX	
CH-46D	TAIL ROTOR GEARBOX	

TABLE 4

3M HELICOPTER CHARACTERISTICS

MFGR	MODEL	NO. ENGINES	WT. CLASS	ROTOR SYSTEM	CIVIL EQUIV.	FLT. HOURS
BELL	TH-57	1	L	S	206	18079
BELL	UH-1E	1	M	S	205	14990
BELL	UH-1N	2	M	S	212	27536
KAMAN	SH-2F	2	M	S	NONE	16921
SIKORSKY	SH-3H	2	H	S	S-61	25440
SIKORSKY	CH-53D	2	H	S	NONE	15505
VERTOL	CH-46D	2	H	T	107	16890

TABLE 5

OPERATOR CONCERNS

- ELECTRONIC FUEL CONTROLS
- ENGINE MONITORING SYSTEMS
- ENGINE ACCELERATION DEVICES
- DRIVE SHAFT COUPLINGS
- OIL LEAKS
- ENGINE POWER CHECK PROCEDURES
- CORROSION
- REPAIRABILITY
- MODULAR INTERCHANGEABILITY
- TWIN ENGINES
- DERATING
- MILITARY TECHNOLOGY TRANSFER
- INSPECTION REQUIREMENTS

TABLE 6

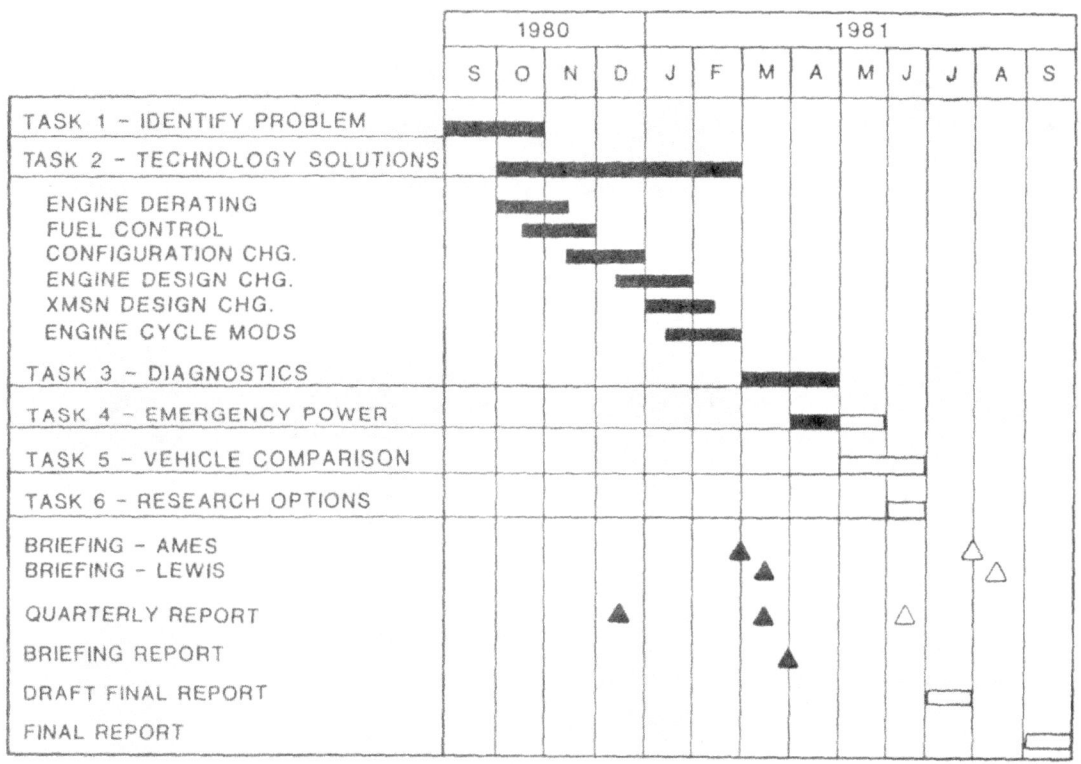

FIGURE 1

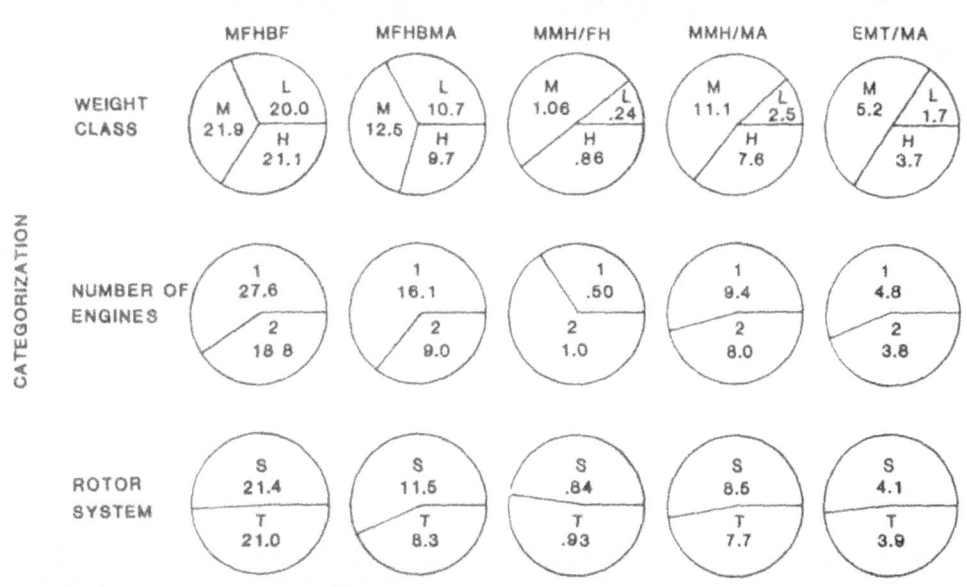

FIGURE 2

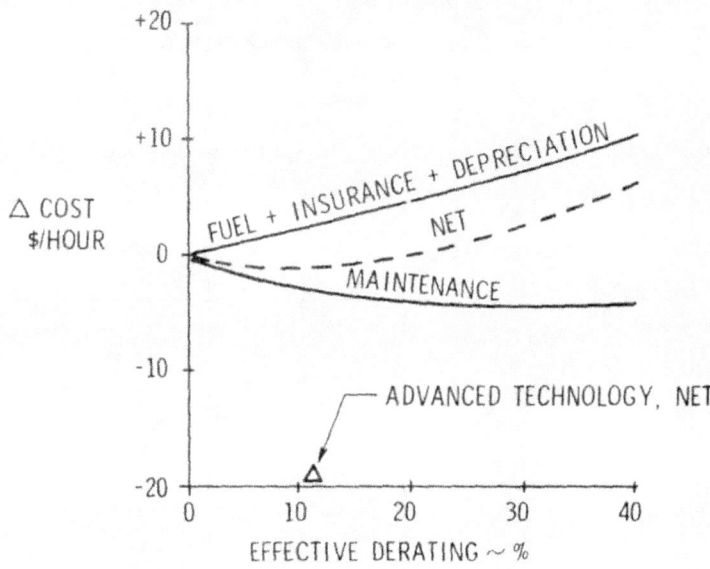

FIGURE 3

ENGINE HEALTH MONITORING SYSTEMS

TOOLS FOR IMPROVED MAINTENANCE MANAGEMENT IN THE 1980's

Jonathan C. Kimball
Pratt & Whitney Aircraft Group

ABSTRACT

The increased cost of fuel has placed an added importance on the "performance health" of commerical aircraft engines and greater emphasis on the performance-monitoring aspect of maintenance. This paper presents an overview of Engine Health Monitoring activities at Pratt & Whitney Aircraft. The development of Engine Health Monitoring, a description of systems currently used, and a summary of programs for improved monitoring in the 1980's are discussed.

INTRODUCTION

The first generation of commercial gas turbine engines (early JT3D and JT8D models) was largely maintained on a "hard time" or "as required" basis. Engine maintenance was performed when certain parts achieved a predetermined life limit (hard time) or when specific symptoms indicated maintenance was needed (as required).

Hard-time maintenance employs the same maintenance schedule for all similar engines. Since no two engines perform exactly alike, the schedule is based on an "average engine." Differences in engine performance may be the result of manufacturing variations and variations in engine mission and service experiences. Some engines may become fuel inefficient although safe and capable of continued operation.

There is, therefore, a need for scheduling maintenance on an on-condition, individual basis. Engine Health Monitoring satisfies this need by continually monitoring engine performance and providing the diagnostic tool for interpreting changes in performance in terms of maintenance requirements.

ENGINE HEALTH MONITORING AND FUEL COSTS

The repairing and refurbishing of engines at fixed intervals or when required has long been a standard practice in the airline industry and will probably persist at many airlines. These procedures have two advantages: 1) they are easily managed and 2) as-required maintenance tends to result in engines remaining on the wing for longer periods. However, these practices do not emphasize minimizing fuel costs.

Engine Health Monitoring can provide visibility into the performance levels of each engine, allowing better maintenance planning, as illustrated in figures 1, 2. Figure 1 shows a typical cycle of engine operation and repair, and figure 2 shows a repair schedule customized to the performance level of each engine. Fuel savings is the primary benefit. Since all engines do not deteriorate at equal rates, hard-time maintenance results in engines with high fuel consumption remaining in service too long and engines with low fuel consumption being repaired or refurbished too soon.

Using a system of Engine Health Monitoring to provide a customized maintenance schedule for each engine has benefits in addition to fuel savings. Lower overall maintenance costs are possible because engines with the highest deterioration levels will be repaired rather than left in service where they would be exposed to increasingly high turbine temperatures. Similarly, lower repair costs on engines with the lower deterioration levels can result by deferring the repair. An additional benefit of engine monitoring is greater reliability. Many developing engine problems can be foreseen with an engine monitoring system.

High fuel costs increase the need for determining repair requirements on an engine-by-engine basis with visibility from an Engine Health Monitoring system. The effect of fuel cost on repair interval is illustrated in figures 3, 4, 5. The total engine operating cost per flight cycle is the sum of the repair cost and fuel cost (fig. 3). As the time between repair increases within a reasonable range, repair costs per flight hour decrease and fuel costs increase, resulting in an interval of engine repair time over which total operating costs are minimized. Higher fuel costs shorten the optimum repair time and narrow the optimum band (fig. 4). Since the performance of each engine varies, the optimum repair intervals for individual engines are not the same (fig. 5).

Engine Health Monitoring can determine which engines require early or deferred maintenance to provide the lowest total operating cost.

DATA ACQUISITION SYSTEMS FOR AIRBORNE ENGINE HEALTH MONITORING

Systems used in flight to acquire data for Engine Health Monitoring vary considerably in both complexity and capability. The four systems most widely used by airlines today are illustrated in figure 6.

The most basic system (System 1) is also the system most commonly used today. The flight crew manually records the data, which can then be processed either manually or by computer.

Automated data acquisition systems expand on the capabilities of the mandatory flight data acquisition unit and digital flight data recorder. In System 2 (fig. 6) a flight data entry panel allows the crew to input documentary data (date, engine time, etc.) to the system. Data is recorded on cassettes (by means of the quick access recorder), which are easily removed for processing at a ground station and analyzed by computer.

An important addition to the data acquisition system is the on board computer, data management unit, which allows the system to perform many additional functions. With this unit the system can record data selectively, looking for appropriate parameter range and stability criteria. It can also make calculations and provide results to the crew by means of an on-board printer. If desired, the computer can scan the data and notify the crew of limits that have been exceeded. In system 3 (fig. 6), data acquisition is controlled by the data management unit, and the data is recorded with the on-board printer.

System 4 (fig. 6) is the most complex and capable of the in-flight data acquisition systems. Data recording is controlled by the data management unit. There is also an auxiliary data acquisition unit for more extensive analyses. Data can be stored on the quick access recorder (cassette) or on printouts from the on-board printer. The data is normally processed by computer, but can be analyzed manually if desired.

Automated data acquisition systems have, however, experienced a variety of problems. Problems have been experienced with unreliable instrumentation and inaccurate data, burdensome calibration and maintenance requirements, and difficulty with data management.

In order to improve the reliability and accuracy of the data acquired during flight, Pratt & Whitney Aircraft and Hamilton Standard jointly developed the Propulsion Multiplexer, an integral engine data acquisition system for acquiring the high quality data required for module performance analysis. The Propulsion Multiplexer (fig. 7) is a compact, durable system, housing pressure transducers, a microprocessor, and all required electronics for acquiring and sending multiplexed engine data to a recording device, such as the Airborne Integrated Data System.

INTERPRETATION OF ENGINE DATA ANALYSIS APPROACHES

Techniques for analyzing engine data obtained in flight or in test cells can be catagorized as follows:

- o Limit Exceedance Checking
- o Parameter Trend Monitoring
- o Module Performance Analysis
- o Turbine Life Accounting

Checks of specific engine parameters for operation in excess of limits is a fundamental requirement of post repair engine testing. In the test cell, compliance with engine limits indicates that an engine is suitable for service. Limits exceeded in flight can be monitored with an Airborne Integrated Data system. The information can provide cautionary warnings to the flight crew (similar to such warnings as low oil pressure) or help to define maintenance actions (as with Exhaust Gas Temperature exceedances).

Graphic display of key engine parameters is the most common method of monitoring the health trends of an engine in service (airborne data) or the performance trends of post-repair engines (test cell data). These plots provide indications of engine performance trends or instrumentation malfunctions.

The Engine Condition Monitoring Computer Program developed by Pratt & Whitney Aircraft has long been widely used to provide graphical parameter trends to improve visibility of the health of engines during service. The program can be used with all Pratt & Whitney Aircraft commercial engines and can operate with flight data acquired either manually or automatically. Manual data is recorded by the flight crew and then processed at a later date. Automatically recorded data is provided by an Airborne Integrated Data System.

The output of the Engine Condition Monitoring Program of primary importance to the user is the "plot report," which presents chronological trends of engine parameter shifts (fig. 8). Because engine parameter shifts are highly visible on the plot report, timely detection of developing engine problems is possible. The report also provides visibility into the long term deterioration trends of an engine or fleet and allows detection of large errors in measured parameters.

The primary advantage of the program is that a large amount of information about engine condition can be obtained without additional engine or airframe hardware, providing considerable benefits with little cost. A limitation to the program is that although it can recognize that a problem has occurred, it can not diagnose the cause. The user must apply judgment to determine the nature of the problem, and if necessary, request further investigation with other troubleshooting methods. For example, an experienced analyst would be required to distinguish between a bleed valve malfunction, a damaged engine module, or an error in measured engine pressure ratio.

Module Performance Analysis is a technique for using measured engine parameter shifts to determine specific engine module performance changes. This process can be illustrated using the example in figure 9. Measured parameter changes are first determined (shift in corrected high rotor speed ($\%\Delta N_2$) at a constant engine pressure ratio, for example). The analysis is used to calculate the most likely cause of these shifts, such as deterioration in high-pressure turbine efficiency. Finally, the shift in key parameters attributable to each module can be calculated. For example, exhaust gas temperature may have increased by 20°C relative to a new engine. The analysis will tell a user how this 20°C can be accounted for (e.g., 10°C due to high-pressure turbine deterioration, 5°C to fan performance losses, and 5°C to low-pressure compressor performance losses), thus indicating areas that may need maintenance.

Module Performance Analysis is currently most often used with test cell data as a tool for evaluating the effectiveness of a repair. The analysis is also used on prerepair data specifically acquired to help define shop work scope. The JT9D Test Cell Module Analysis Program, developed for analysis of JT9D engine data acquired in the test cell, combines a sea level data reduction system with module performance analysis and data validity screening. The program is very flexible, accomodating varied data input, analysis baselines, and test cell corrections. A sample output is shown on figure 10.

Module Performance Analysis systems have also been developed for use with data acquired in flight with an Airborne Integrated Data System. Data for inflight module performance analysis, including the additional parameters required for module performance analysis, must be recorded automatically by an airborne data system. The on-board computer of this system selects what data is to be recorded based on predetermined ranges of engine and aircraft parameters (data acquisition windows) and parameter stability criteria. Data from the airborne system may be manually input to the airborne module performance analysis program using data from an on-board printer or automatically using data transferred from an on-board recorder.

A typical program output, the module analysis plot report, is shown in figure 11. This report presents graphical trends of performance changes of each module in a highly visible format even if a module has been installed on a different engine.

The program can provide many benefits to a user. Knowledge of the performance of each module can be helpful in making maintenance decisions. For example, if an engine has a history of high exit gas temperature, the plot tells whether the high temperature is caused by the high-pressure compressor or the high-pressure turbine, or both. Appropriate maintenance can be planned. The In-flight Module Performance Analysis can be a useful tool for troubleshooting engine problems on the wing and can assist in improving shop scheduling.

The JT9D Airborne Integrated Data System/Module Performance Analysis Program is currently being used and evaluated at four major airlines. The program is emerging from the developmental stage and may soon be considered a developed engine monitoring tool.

Special attention must be paid to instrumentation in order to successfully perform module performance analysis. Parameters not normally measured are needed, as shown in Table I. In addition to the parameters normally acquired, temperatures at the discharge of the low-pressure compressor, high-pressure compressor, and low-pressure turbine are measured.

Special emphasis on data accuracy is also required for reliable Module Performance Analysis results. For example, if a fuel flow measurement is used only to determine engine suitability for service, a measurement error of 2% may go unnoticed. If the data is to be used for module performance analysis, a 2% error in fuel flow may be misinterpreted as an engine performance shift. Although the analysis systems now available have provisions for detection of erroneous data, a greater emphasis on data quality is necessary.

Module performance analysis capability, therefore, is an extremely useful engine maintenance tool. The use of module performance analysis with data from either a test cell or an Airborne Integrated Data System requires a commitment to additional instrumentation, closer instrumentation accuracy monitoring, and personnel trained in module performance analysis interpretation and use.

A Life Accounting Program can be used to calculate the fraction of life consumed for any set of critical high-pressure turbine airfoils. Since all routes are not equally severe on high-pressure turbine airfoils, large variations in part lives can exist. The life accounting program calculates the amount of life consumed for each critical airfoil, using analytical models of airfoil deterioration. The program can run as a subroutine of the Airborne Integrated Data System/Module Performance Analysis Program, using the accumulated time exposure of the parts to temperature, pressure, and rotor speed. The program can also be run by itself without airborne data, using the specific route structure and engine derate experience as input. A typical output from the life accounting program is shown in figure 12.

The primary purpose of the program is to maximize airfoil service life while minimizing the possibility of turbine damage. The program can also assist in efficiently scheduling hot-section maintenance, controlling inventories of airfoils, and in better planning of the hot-section assembly (e.g., a turbine could be assembled with airfoils having similar amounts of life remaining).

FUTURE TRENDS

Increasing airline fuel and maintenance costs have resulted in greater airline interest in Engine Health Monitoring. Since many aspects of this process can be addressed most efficiently by the engine manufacturer, Pratt & Whitney Aircraft is committed to providing superior Engine Health Monitoring system support.

A special emphasis will be placed upon data quality in future engine health monitoring systems. A common shortcoming of engine monitoring systems today is that analysis algorithms, although accurate, are unacceptably sensitive to sensor errors. The approach to data validity must be three fold:

1) encourage the development and proper use of accurate data measurement systems

2) develop software routines that recognize and report probable data errors for follow-up maintenance actions

3) design algorithms to be as insensitive to data errors as possible.

Present plans are to continue to develop and refine analysis software routines and to monitor current systems, making improvements where required. We will work with customers to define ways of monitoring different aspects of engine operation. Current systems now stress gaspath performance monitoring, but efforts are underway to increase the capability of monitoring the mechanical integrity of the engine and its subsystems, such as oil and bleed systems.

Table I
INSTRUMENTATION REQUIRED FOR JT9D TESTING
AND MODULE PERFORMANCE ANALYSIS

		Normally Measured On Test Cell	Required For MPA
P_{T2}	Engine Inlet Total Pressure	X	X
P_{T3}	LPC Discharge Total Pressure	X	X
P_{s3}	LPC Discharge Static Pressure	X	
P_{s4}	HPC Discharge Static Pressure	X	X
P_{s5i}	Turbine Cooling Air Static Pressure	X	
P_{T7}	LPT Discharge Total Pressure	X	X
T_{T2}	Engine Inlet Total Temperature	X	X
T_{T3}	LPC Discharge Total Temperature		X
T_{T4}	HPC Discharge Total Temperature		X
T_{T6}	HPT Discharge Total Temperature	X	X
T_{T7}	LPT Discharge Total Temperature		X
F_N	Engine Total Net Thrust	X	X
N_1	Low Rotor Spool Speed	X	X
N_2	High Rotor Spool Speed	X	X
W_F	Fuel Flow	X	X
β	HPC Variable Stator Vane Bellcrank Angle	X	X

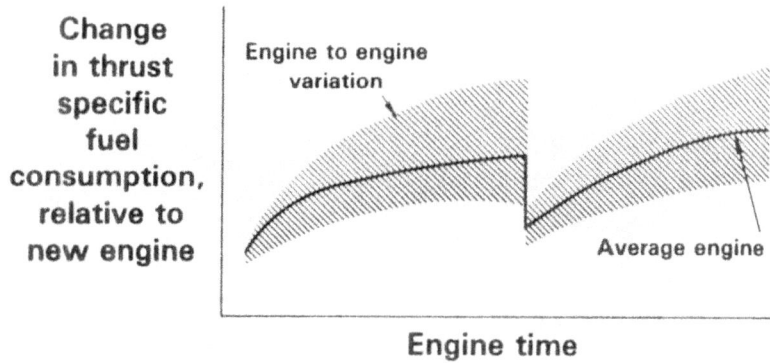

Figure 1   Typical Cycle of Engine Operation

FUEL SAVINGS RESULT IF ENGINE REPAIR IS SCHEDULED BASED ON ACTUAL ENGINE CONDITION

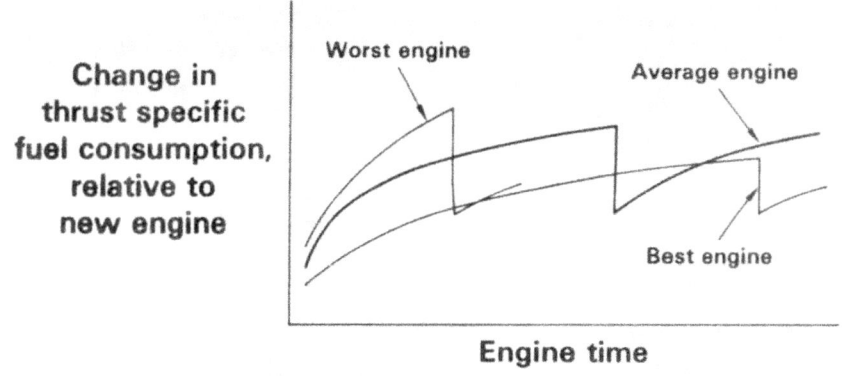

Figure 2   Repair Schedules Customized for Worst and Best Engines Compared With Schedule Based on Average Engine

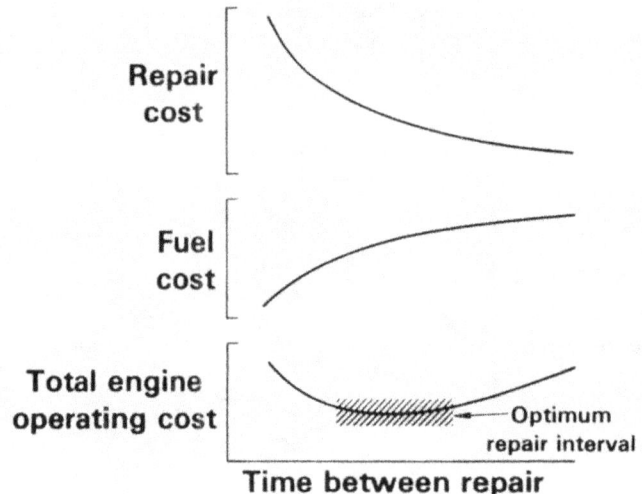

Figure 3   Effect of Time Between Repair on Repair Cost, Fuel Cost and Total Operating Cost

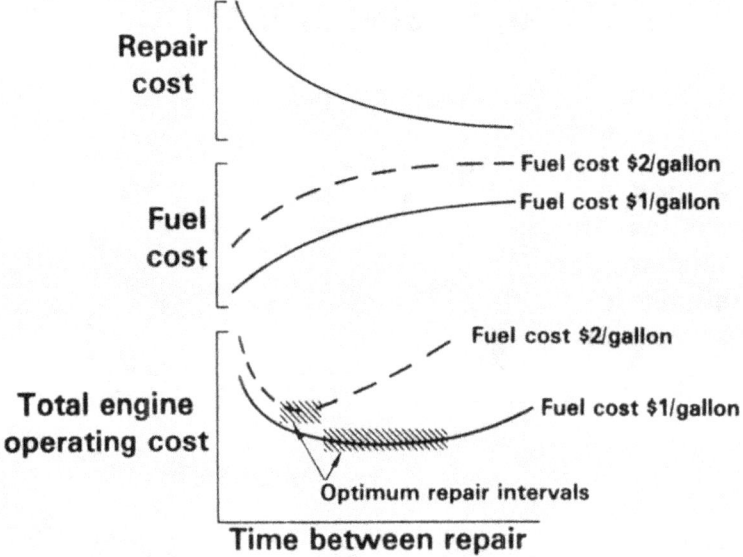

Figure 4   Effect of Fuel Cost on Optimum Repair Interval

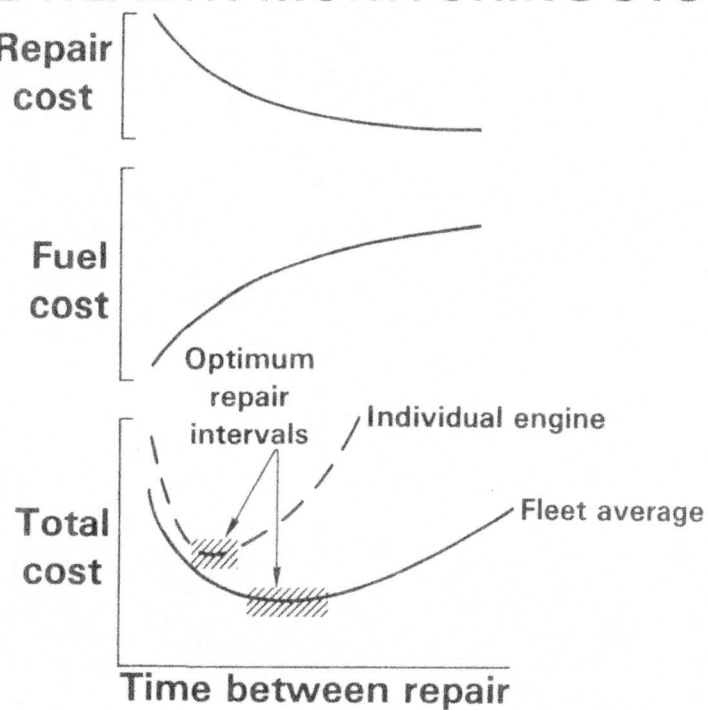

Figure 5   Comparison of Optimum Repair Intervals for Individual and Fleet Average Engine

AIRLINES USE VARIOUS DATA SYSTEMS

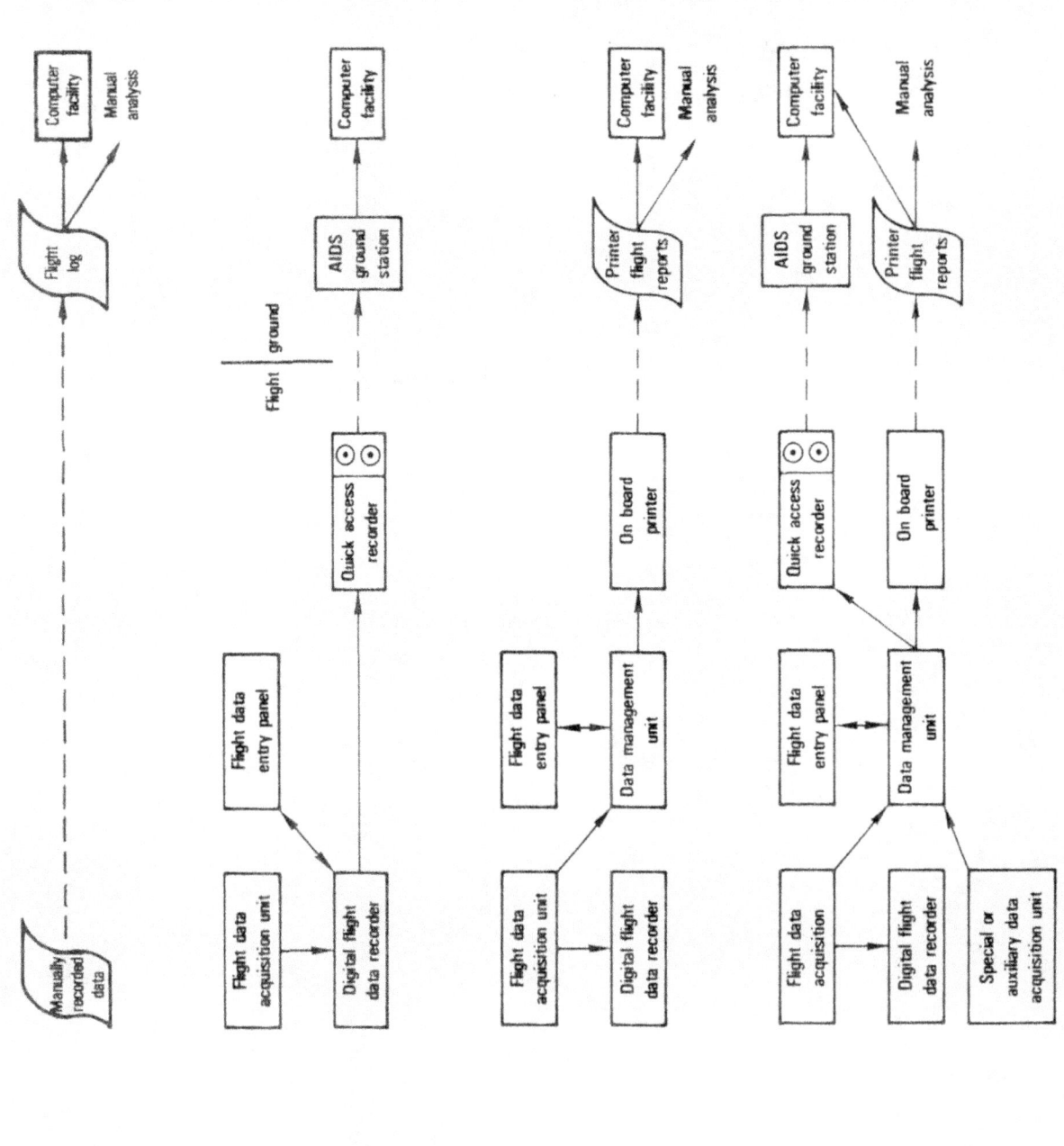

Figure 6   Various Data Acquisition Systems

THE JT9D MULTIPLEXER IS A COMPACT, DURABLE UNIT

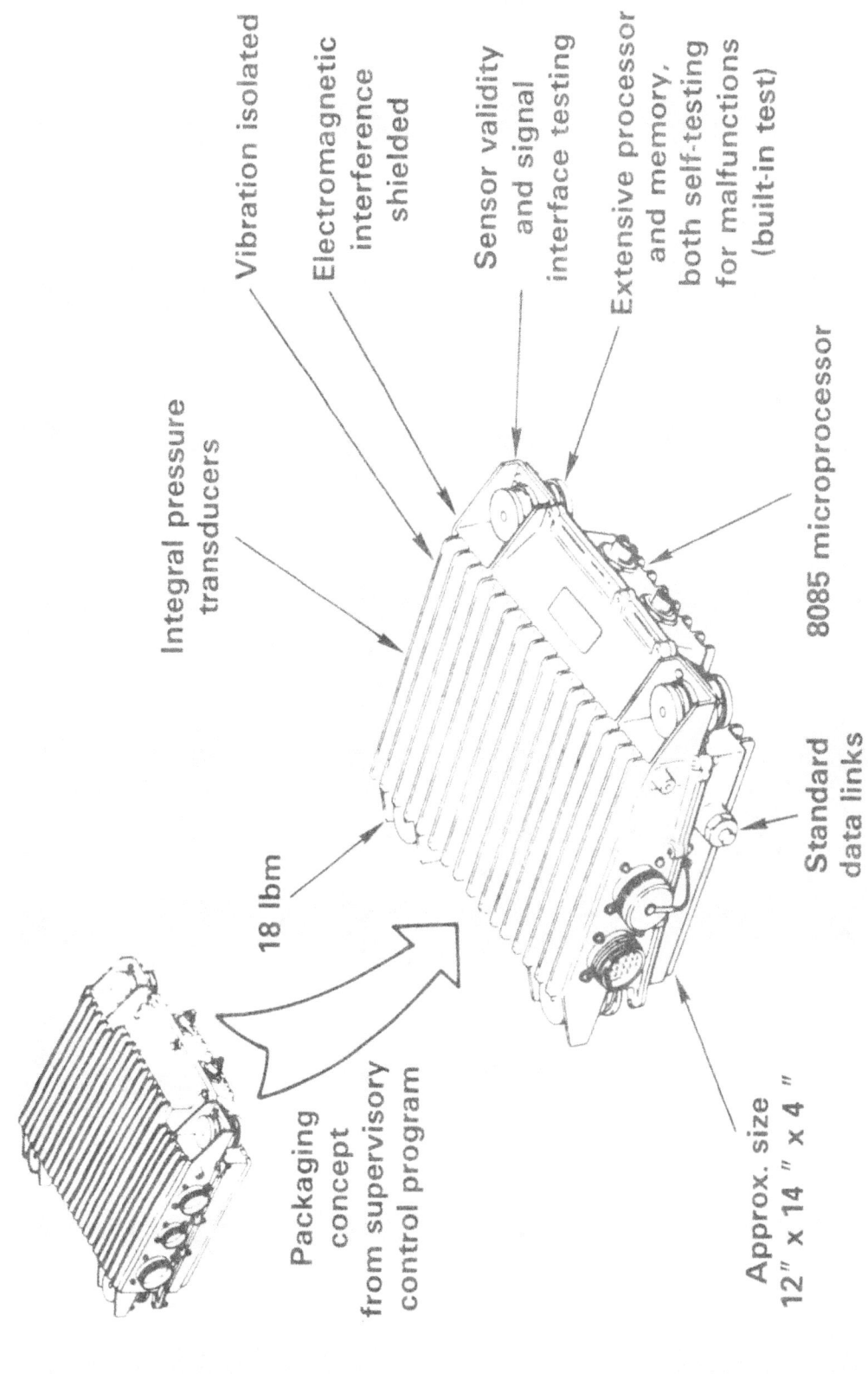

Figure 7 JT9D Compact, Durable Multiplexer

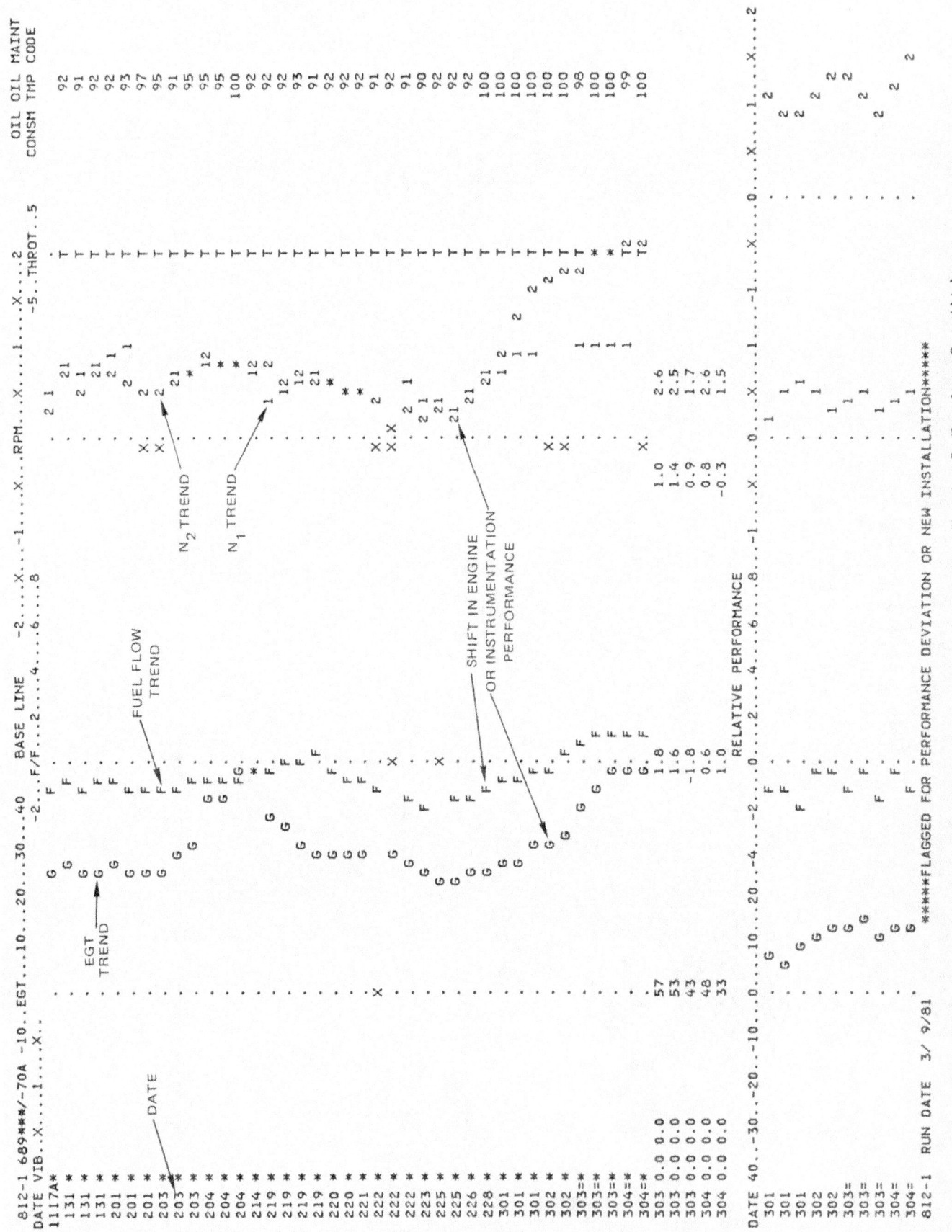

Figure 8. Sample Plot Report Showing Typical Engine Condition

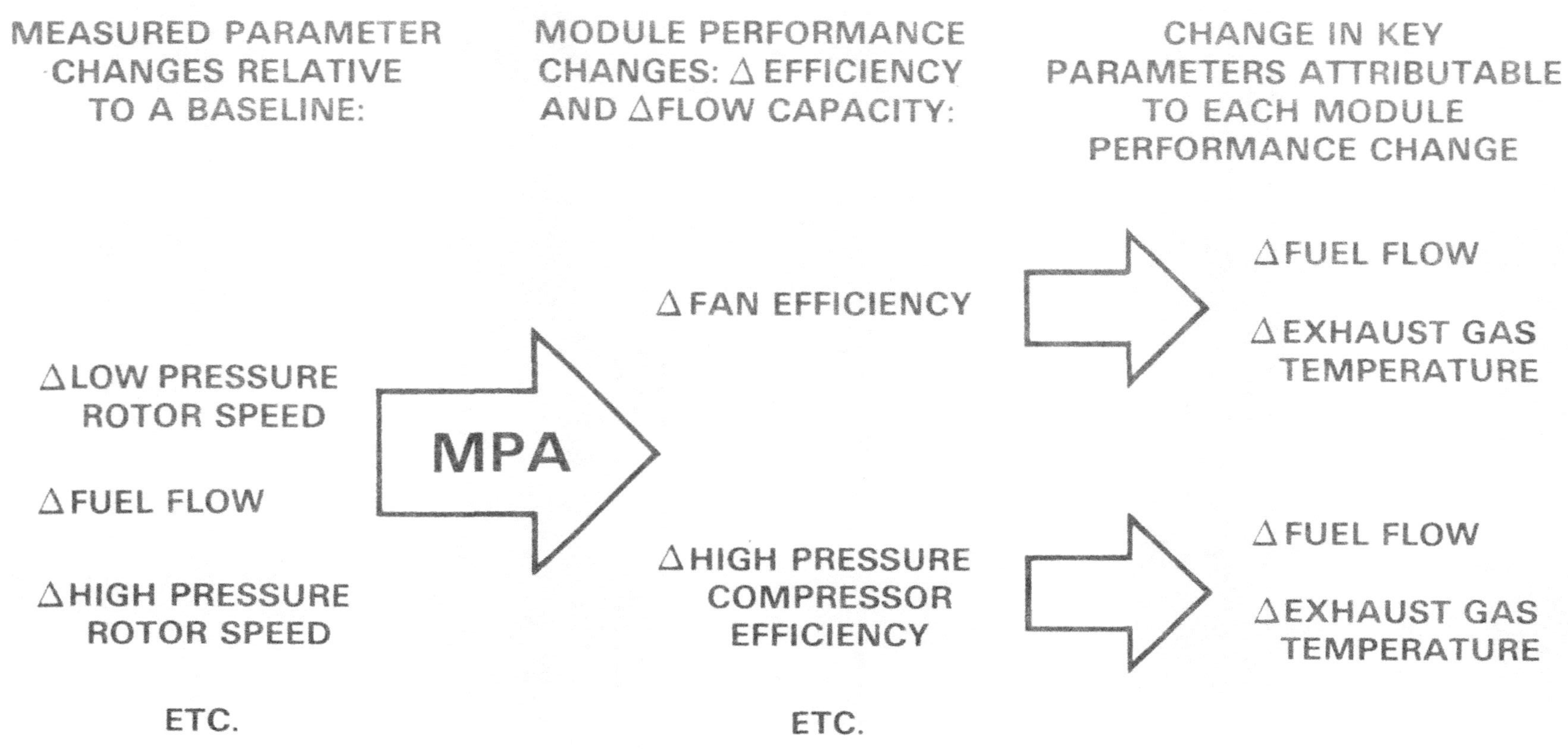

MEASURED PARAMETER
CHANGES RELATIVE
TO A BASELINE:

△ LOW PRESSURE
ROTOR SPEED

△ FUEL FLOW

△ HIGH PRESSURE
ROTOR SPEED

ETC.

⇒ MPA ⇒

MODULE PERFORMANCE
CHANGES: △ EFFICIENCY
AND △ FLOW CAPACITY:

△ FAN EFFICIENCY

△ HIGH PRESSURE
COMPRESSOR
EFFICIENCY

ETC.

⇒

CHANGE IN KEY
PARAMETERS ATTRIBUTABLE
TO EACH MODULE
PERFORMANCE CHANGE

△ FUEL FLOW

△ EXHAUST GAS
TEMPERATURE

△ FUEL FLOW

△ EXHAUST GAS
TEMPERATURE

Figure 9   Module Performance Analysis Example: Analysis Determines
Amount Each Engine Modules Contributes to Performance Change

POST REFURBISHMENT FOLLOWING 8 JULY 80 UER   13450   3933   A

OPERATOR: ENGINE MODEL: JT9D-3A   SERIAL NUMBER: P-******   DATE: 101980

```
*******************************************************************
******        ANALYZED MODULE ASSESSMENTS              *******
*******************************************************************

ETA    FAN    ETA    LPC    ETA    HPC    ETA    EFF    ETA
FAN    FCAP   LPC    FCAP   HPC    FCAP   HPT    A5     LPT
0.1%  -0.6%  -0.5%  -0.7%  -0.1%  -0.6%  -1.1%   1.6%  -0.0%
```

```
*******************************************************************
*******      RAW PARAMETER DIAGNOSTICS           *******
*******************************************************************
```

NO RAW PARAMETER ERRORS DETECTED.

```
*******************************************************************
*******           OUTLIER DIAGNOSTICS            *******
*******************************************************************
```

THE VALUE OF N1/ROT2   FOR POINT NUMBER   1 HAS BEEN REJECTED AS A PROBABLE OUTLIER.

THE VALUE OF WF/KCST2KH FOR POINT NUMBER   1 IS A POSSIBLE OUTLIER.

```
*******************************************************************
*******        CONFIGURATION DIAGNOSTICS         *******
*******************************************************************
```

ANALYSIS ENGINE DATA CONFIGURATION ADJUSTED TO BASELINE BMOD.

ANALYSIS ENGINE DATA CONFIGURATION ADJUSTED TO BASELINE A6CL.

```
*******************************************************************
*******       LARGE PARAMETER DIAGNOSTICS        *******
*******************************************************************
```

NO LARGE PARAMETER ERRORS DETECTED.

```
*******************************************************************
*******     PRIMARY NOZZLE AREA DIAGNOSTICS      *******
```

Figure 10   Typical Output From JT9D Module Analysis Program

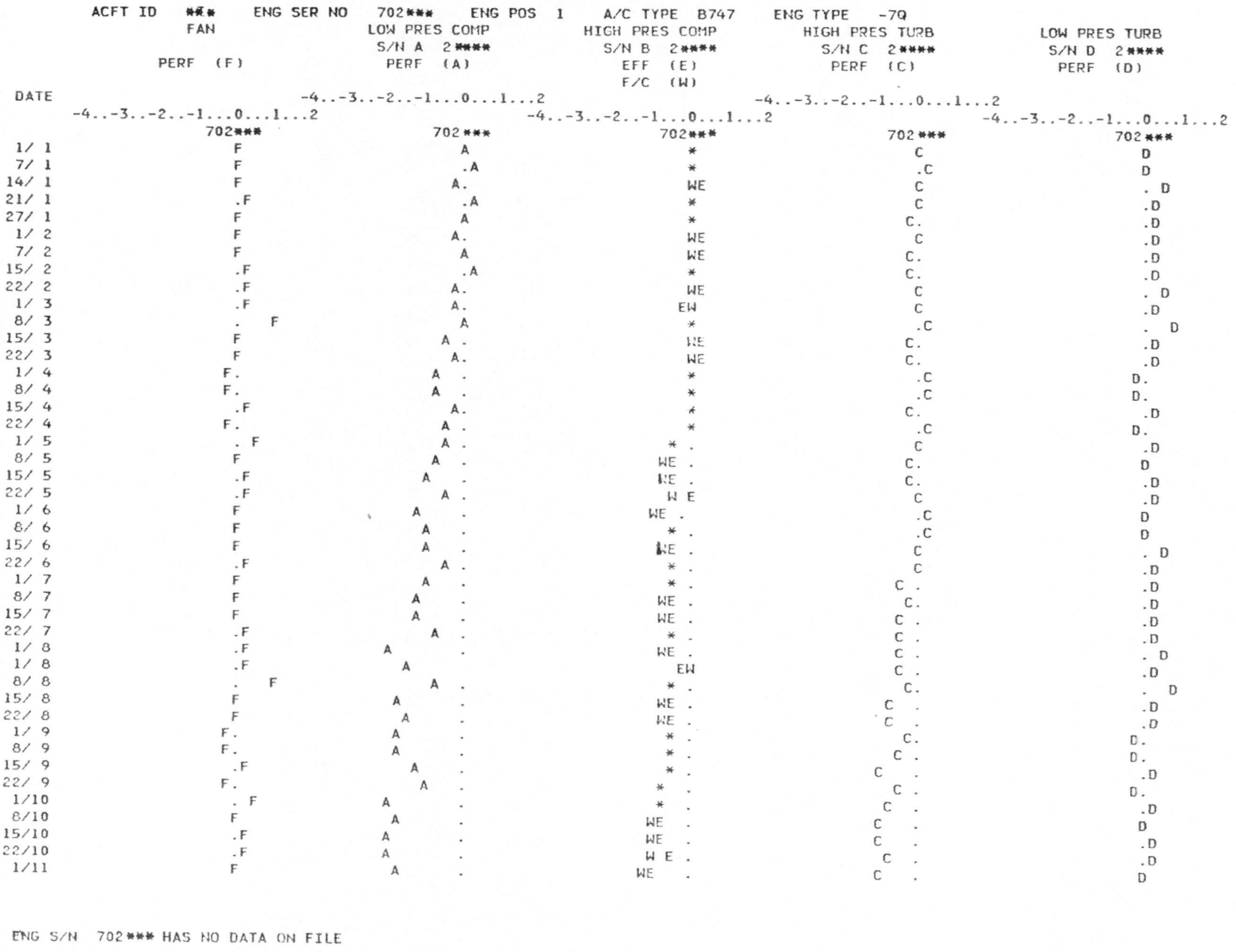

Figure 11 Typical Output From Airborne Integrated Data System/Module Performance Analysis System

LIFE ACCOUNTING PROGRAM REPORT ONE

AIR-CRAFT ID	ENGINE P NUMBER	FAIL MODE = IV PART- NUMBER	AMT	PCNT USED	LOT NO	FAIL MODE = 1BCF PART- NUMBER	AMT	PCNT USED	LOT NO	FAIL MODE = 1BOC PART- NUMBER	AMT	PCNT USED	LOT NO	FAIL MODE = 2V PART- NUMBER	AMT	PCNT USED	LOT NO	FAIL MODE = 2B PART- NUMBER	AMT	PCNT USED	LOT NO
8126	1 685709	750	66	30.1		773441	116	45.3		773441	116	55.3		735882	90	15.2		770102	66	10.1	
	2 685791	76027	66	22.1		773531	116	22.2		773531	116	33.3		772572	90	41.1		770102	138	15.0	
	3 686015	75278	15	10.1		773441	100	88.2	J	773441	100	92.3	J	735882	30	10.1		770102	138	60.1	
		75276	51	0.1		773441	16	22.2	I	773441	16	30.3	I	735882	30	20.1			0	0.0	
			0	0.0		773441	0	0.0	H	773441	0	0.0	H	772572	30	30.1			0	0.0	
	4 685706	755	5	0.1		773441	2	40.2	J	773441	2	5.3	J	735882	90	80.1		770102	138	0.0	
		75510	50	60.1		773441	5	90.2	I	773441	5	30.3	I		0	0.0			0	0.0	
			0	0.0		773441	5	80.2	H	773441	5	25.3	H		0	0.0			0	0.0	
			0	0.0		773441	10	76.2	G	773441	10	20.3	G		0	0.0			0	0.0	
			0	0.0		773441	20	60.2	F	773441	20	15.3	F		0	0.0			0	0.0	
			0	0.0		773441	3	50.2	A	773441	3	10.3E	A		0	0.0			0	0.0	
			0	0.0		773441	4	30.2	B	773441	4	0.3	B		0	0.0			0	0.0	
			0	0.0		773441	1	20.2	C	773441	1	0.3	C		0	0.0			0	0.0	
			0	0.0		773441	9	10.2E	F	773441	9	0.3	F		0	0.0			0	0.0	
			0	0.0		773441	7	5.2E	Q	773441	7	0.3	Q		0	0.0			0	0.0	

ID	P -7Q	FAIL MODE = IV				FAIL MODE = 1BOC			
8130	1 702043	77428	62	12.2		778741	100	54.3	
	2 702044	77428	62	12.2		778741	100	54.3	
	3 702045	77428	62	12.2		780441	100	54.3	
	4 702062	77428	62	12.1		780441	100	53.6	

Figure 12   Typical Output From Life Accounting Program

ENGINE "ON CONDITION" MONITORING - CF6 FAMILY

60's THRU THE 80's

H.J. Kent
General Electric Company

Dr. Gerwin Dienger
Lufthansa German Airlines

SUMMARY

The "On Condition" program which was introduced in the late 60's was immediately accepted by the industry. This program which provided the foundation for timely and economical maintenance procedures in fault detection and isolation has resulted in a significant reduction in material and labor cost. The in-flight shutdown rates (IFSD), the unscheduled engine removal rates (UER) and the departure reliability reported show that the airlines' "On Condition" monitoring programs are very effective in reaching a high level of reliability. In the near future, with the added emphasis on fuel conservation/economics and in conjunction with the advancements and refinements in electronics, it is anticipated that the on-board Engine Condition Monitoring Systems will become economically feasible through the expanded effectivenss in performance monitoring of the basic modules of the engine and the integration with the overall engine workscope. The CF6 Condition Monitoring experience beginning in the late 60's up through the future expectations of the 80's, is discussed in this paper.

INTRODUCTION

Monitoring of the overall engine condition has proved to be an effective maintenance tool both at the line station, as well as at the home base by the early detection of engine faults, erroneous instrumentation signals and by verification of engine health. It currently encompasses all known methods from the manual procedures to the fully automated Airborne Integrated Data Systems (AIDS). Future programs (Figure 1) will be built around the proven capabilities of today's systems, the continual growth in maintenance capability and effectiveness through improved data acquisition/analysis, and the projected module performance analysis program under development. Cost effectiveness, unquestionably, remains the prime criterion to an airline in the selection/definition of its monitoring system. Today's increased need for fuel conservation and to control the operating expense, coupled with recent advancements in the capabilities of on-board electronic equipment, has stimulated interest in the potential of the Expanded AIDS systems and, accordingly, the monitoring system architecture for future aircraft and engines.

Equally important in the establishment of future programs is the stressed need and acceptance of the team concept, a concentrated effort by members from the manufacturers of aircraft, engine, and AIDS equipment and the airlines. To date, the A310/CF6-80 AIDS team organized around the future application has proved to be most effective and is heartedly endorsed by the participants.

METHODS/SYSTEMS

The methods/systems (see Figure 2) utilized today and expected to be carried on into the future encompass the complete spectrum from fully manual (using only cockpit instrumentation) to completely automated with expanded instrumentation and data acquisition. The end objective of all of these methods/systems is, however, the same, and that is to afford the maintenance person the means to establish the required corrective action in a timely fashion.

Although the practices vary from airline to airline, the basic procedure is to routinely monitor the corrected trend data and the relative nominal data for variations. Whenever a shift is observed, the maintenance center analyzes the variation for validity, compares it to past experience and, thereby, establishes a level of severity, and then requests specific inspections and additional data signatures. These results are then compared with historical data and the maintenance manual to establish the corrective action. These tools (see Figure 3) have been proven and are being effectively utilized by airline maintenance today.

The integration of trend plots with non-destructive inspection results, together with the engine historical record, and with proven diagnostics affords maintenance the very means necessary to make effective decisions. It is recognized that individually, the separate techniques are somewhat inadequate, the secrets lie in the effective combination where they complement and support each other.

The parameters which are measured today on CF6 engines are shown in Figure 4. These parameters provide the basic information for trending an engine against itself for both short- and long-term diagnostics, the fuel consumption, mechanical integrity and the refined life cycle count programs. Initial effort toward Modular Performance Analysis has also begun with these same parameters. Additional parameters such as low pressure turbine discharge pressure and temperature are under study.

TREND PLOTS/EXPERIENCE

Manual

Manual trend plots are generated from data recorded from the cockpit instruments either on-board the aircraft by a member of the flight crew or on-ground by a maintenance person. Typical examples of manual trend plots are shown in Figure 5. These particular plots show the shift in engine parameters due to excessive bleed flow. Note the relative characteristics between the parameters differ substantially based on whether the bleed is recouped within the engine or dumped overboard. The excessive bleed from the manifold leak was dumped overboard. In this case, the trend plot shows increases in the exhaust gas temperature (EGT), fuel flow (FF), core speed (N_2) and a decrease in engine pressure ratio (EPR). A faulty HPT second stage seal is an example of an internal leak, that is, the air returns to the primary stream downstream of the high pressure turbine first and second stage nozzles, respectively. In this case, the on-board manual trend plots showed a significant increase in EGT over a short period of time without any apparent change in the level of core speed, fuel flow, engine pressure ratio and vibration. At first glance one would suspect faulty instrumentation; however, inspection of the high pressure turbine stator with a flexible borescope located the faulty seal and provided the basic information for the decision to remove the engine, replace the seal, rebuild the engine with original modules and place it on site as a ready spare.

Semi-Automatic

The semi-automatic method which is known in the industry as ADEPT, CEML, FML, TEMP, etc., provides computer printouts consisting of tabulations and trend plots. The input which is the same as that for the manual method consists of the engine parameters and flight conditions recorded from the cockpit instruments during steady-state cruise. The data is normally taken at least once during each flight, forwarded to the home base for processing through the on-ground computer. The output consisting of the tabulation of the flight conditions, engine parameters along with bleed conditions and the trend plots versus flights of vibration (V), exhaust gas temperature (G), fuel flow (F), core speed (2), etc., are reviewed daily by the maintenance person. In the future, it is anticipated that the input into the ground based computer will be automatic from magnetic tape recorded on-board the aircraft or by radio communication link.

Examples of the semi-automatic trending as obtained from a ground-based computer using the manually recorded data from cockpit instruments (see Figure 6) show typical changes due to instrumentation faults, compressor foreign object damage, and an on-wing fan trim balance. As shown, the shifts in the body of the trend plots appear to be gradual while in reality they are step changes. The reason for the gradual change is that the data versus flights is smoothed by averaging over several readings thereby damping a step change. The magnitude of the step change is reflected by the raw data which is shown at the bottom of the trend plot.

AIDS

AIDS is a fully automated system utilizing the output from the on-board printer and/or recorder as a direct input into the on-ground computer. As the computer is quite flexible, the output can be readily customized to satisfy the particular situation including transient as well as steady-state conditions.

Relative to manual methods, AIDS provides advantages in availability of data sequences during transients, recording additional parameters, greater accuracy and repeatability through established stability criteria. Some typical examples are shown in Figure 7. As seen, the high transient characteristics of the parameters during takeoff can be recorded and printed out in a comparison versus time. This is a most valuable tool in assessing variations in pressures, temperatures and rpm over short time intervals. Also shown, is the versatility in formating the trend plots in reverse chronological order against the latest, intermediate and the very early flights and, of course, the PLOT/LIST which tabulates the data versus time.

SOAP

Although some operators effectively monitor the lube system by checking only the screens and filters, others have found SOAP (Spectrographic Oil Analysis Program) to be a useful tool in monitoring specific wear problems, such as the number three bearing inner race hub wear (see Figure 8). It is to be recognized that the effective utilization of SOAP requires strict discipline in monitoring the lube system. As shown, any changes to the system, such as adding or draining oil, will completely distort the trend plot. Likewise, any contamination within the sampling and analytical equipment will distort the trend plot. And, of course, the sampling must be accomplished on a scheduled time interval.

Borescope and Radiograph

An area, in which great strides have been made and of which the industry can be very proud is that of borescope and radiographic non-destructive testing equipment and techniques. Equipment and techniques available today, not only afford the means to inspect deep within an engine, but also provide the means to readily visualize the actual condition. Figure 9 shows typical examples of the visibility these techniques provide and a measure of their value to the maintenance task. Many will undoubtedly remember when the borescope was hot enough to burn your hand and the visibility was completely inadequate.

Diagnostics

Not to be forgotten and, obviously, the most valuable link in this maintenance chain of events is the diagnostics/decision by the maintenance person or persons. One of the most valuable lessons learned is that the most

exotic tools and methods are most ineffective without dedicated people.
Humorously, but also quite seriously, Figure 10 tries to emphasize this point.
Our message, people and their vital communication links between the line
station and main base, are the key.

MEASUREMENTS

So how is the industry meeting the challenge. The trends (see Figure
11) on dispatch reliability, in-flight shutdowns, and shop visit rate
for the CF6-50 clearly show that they are doing very well and improving with
time. These measurements also show that the industries' dedication to the
"on condition" concept is fulfilling its objectives.

FUTURE

Now, let us as a member of the industry look at the future. In the late
70's, the airlines jointly provided specifications covering their future needs
for monitoring engine health. These specifications can be summarized into the
five basic programs (see Figure 1). The overall objective of these programs
is to provide an improved engineering/maintenance tool. As noted earlier,
the trending and mechanical integrity programs have been proven and are
essentially in place while the others are in various stages of development.
Several of the airlines are confident that with the team concept that the
Expanded AIDS will be a most effective tool. One of these operators is
Lufthansa German Airlines who, as a member of the ATLAS Group and an operator
of CF6-powered aircraft, today monitor engines by the semi-automatic method.
In the 80's, they will be operating the CF6-powered A310 with an Expanded
AIDS and plan to utilize it to the fullest.

Rational for DLH Expanded Aids/A310

Major elements in Lufthansa's decision to incorporate an Expanded AIDS
were the emphasis on the need for pre-shop identification of faulty engine
modules, for efficient corrective shop action and for engine operation at
minimum operating cost by making optimal usage of fuel and engine parts/
materials.

A simplified model (Figure 12) shows how fuel burn cost per engine
flight hour - caused by performance deterioration of a high spool module -
increases and how the module restoration cost per hour decreases versus the
flight cycles accumulated at time of restoration.

The summation of these two curves provides the accumulated module cost
versus flight cycles with the minimum indicating the performance life of the
module under consideration.

Superimposing appropriate curves of all the modules leads to the optimum point in time that the engine should be removed for restoration of the degraded module's performance.

Figure 13 shows the effect of increasing fuel cost on the minimum cost/performance life curve. The assumed fuel cost rise over the years to come will significantly reduce the usable performance life and also drastically steepen the overall cost curve which implies increasing losses in case of delayed performance restoration.

In view of the fact that the deterioration characteristics vary significantly from module to module of the same type, a typical cycle dependent deterioration rate cannot be assumed for identifying and scheduling the optimum restoration time. Hence, individual engine module performance analysis and monitoring is imperative.

The realization of such a concept centers around the additional instrumentation, an Expanded AIDS and the necessary ground based computer system with an effective analysis algorithm which is now under development by the engine manufacturer.

AIDS Configuration

The architecture (see Figure 14) of Lufthansa's A310 AIDS is built around an "On-Board" digital system incorporating an analog-to-digital multiplexer (PMUX), ARINC 429 data bus, Data Management Unit (DMU), Digital Flight Data Acquisition Unit (DFDAU), the Digital Flight Data Recorder (DFDR), a Control Display Unit (CDU), a printer and Quick Access Recorder (QAR). The output from the "on-board" printer and recorder will be input into the ground facility for further computation and documentation.

EXPECTATION

Now let us project the future potential from an industry standpoint (see Figure 15). The potential is great for further expansion and improvement of today's engineering/maintenance tools. The means and methods by which this will be achieved will certainly vary from operator to operator. A key approach is to build on proven experience and to extend the programs by: utilizing the increased memory/buffer size, increased sampling rates, improved stability criteria, optimizing the Modular Performance Analysis/Work Scope, improving diagnostics through effective means to detect stalls, isolating hot starts, establishing levels of EGT margin, incorporating/expanding divergence monitoring, enhancing short term monitoring through on-board exceedances/trends flags, and automatically documenting levels of reduced power.

In retrospect, "On-Condition" maintenance concept of the 60's and 70's have provided a sound foundation for an effective maintenance tool for the airline industry. While there is great promise and high potential for further expansion and improvement of the maintenance capability, any improvements and expansions must be economical and practical to become acceptable. The development and implementation of a cost effective engine diagnostic system will be the challenge of the 80's.

Engine Condition Monitoring

- **Programs Today/Future**
    - Trending
    - Module Performance Analysis
    - Fuel Consumption Survey
    - Mechanical Integrity
    - Refined Life Cycle Counting

FIGURE 1

Trending Methods/Systems

- Manual On-Board
- Manual On-Ground
- Semi-Manual (Computer)
- Aids (Airborne Integrated Data Systems)

FIGURE 2

Tools

- Trend Plots
    - Gas Path
    - Mechanical
    - Soap (Spectrographic Oil Analysis Program)

- Borescope Inspection

- Radiographic Inspection

- Diagnostics
  Fault Detection/Isolation

FIGURE 3

Condition Monitoring Sensors

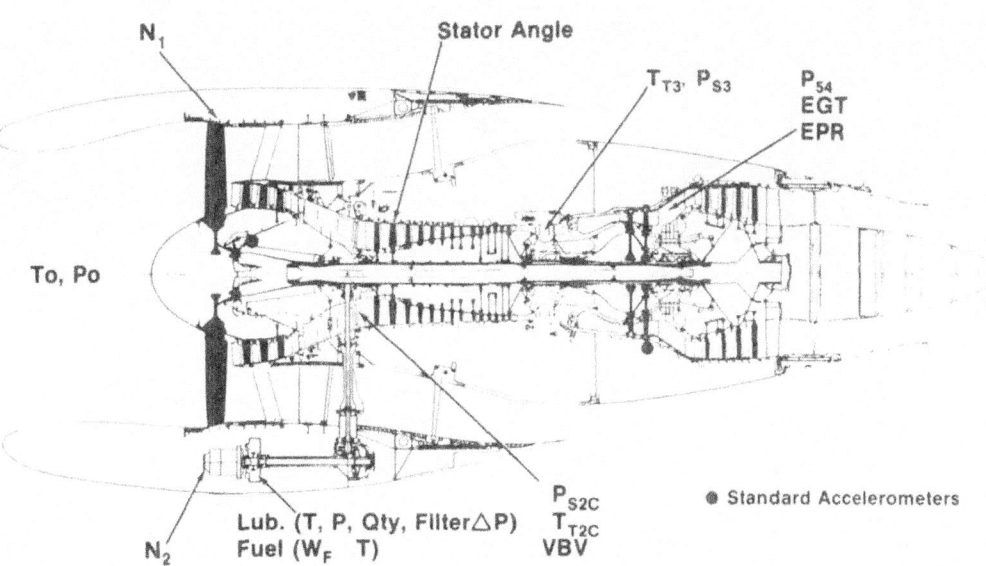

FIGURE 4

Trend Plots/Experience

- Manual
    - Typical Trend of Excess Bleed
    - On-Board Real Time Trend Plot
    - HPT Second Stage Seal

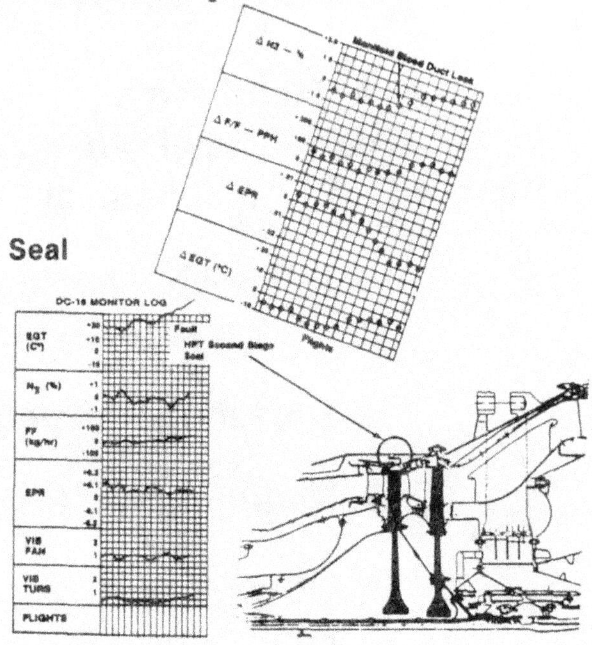

FIGURE 5

Trend Plots/Experience/Diagnostics

- Semi Automatic

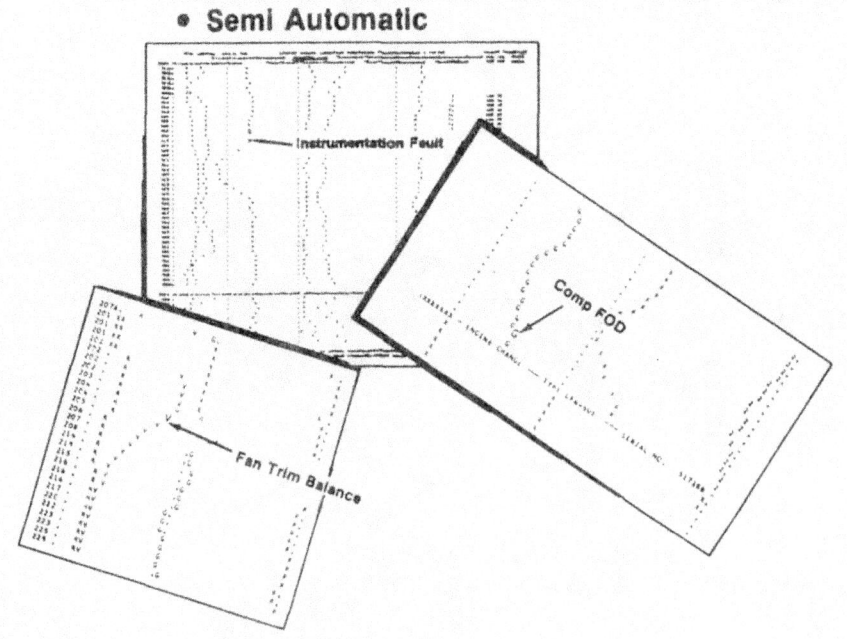

FIGURE 6

Trend Plots/Experience

- Aids

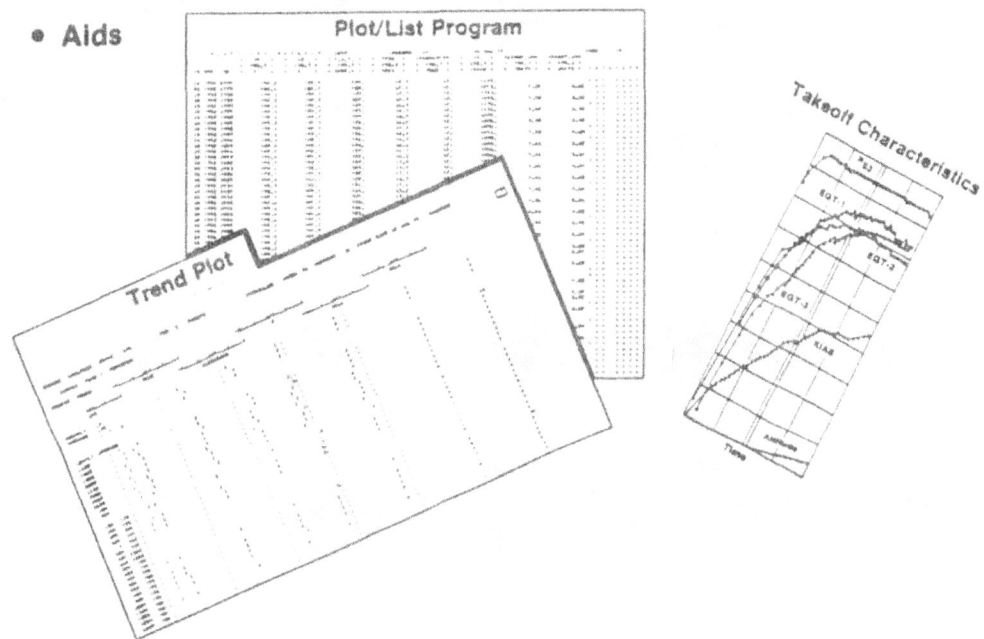

FIGURE 7

Trend Plot/Experience/Diagnostics

- SOAP

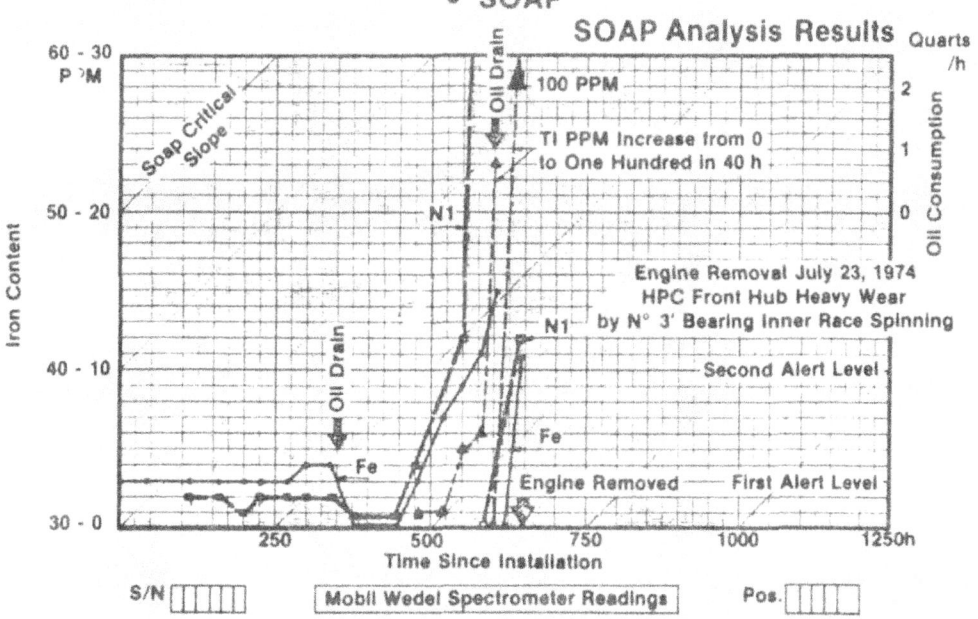

FIGURE 8

Borescope/Radiography

Borescope Viewing Detail

Radiographic Shot

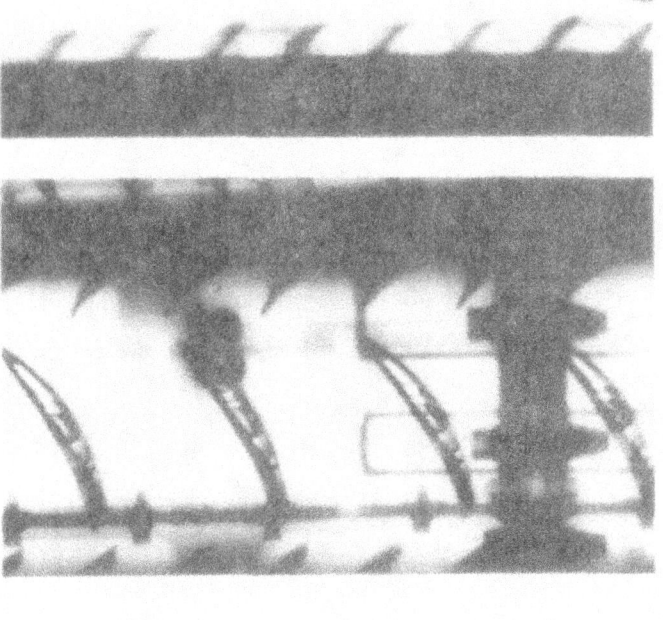

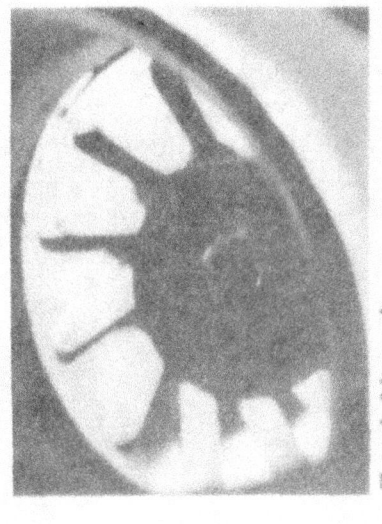

Fuel Nozzle

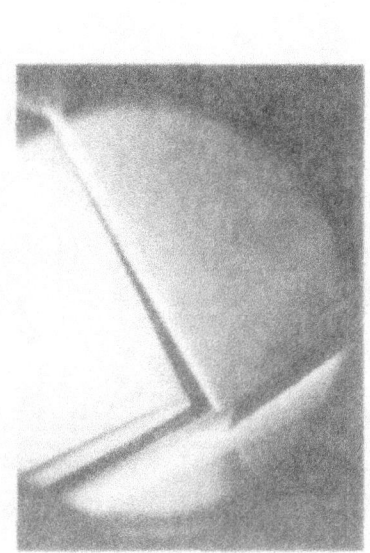

HP Compressor Blade

FIGURE 9

Diagnostics

- Key

    Personnel

    Understanding/Experience
    Engines and Systems

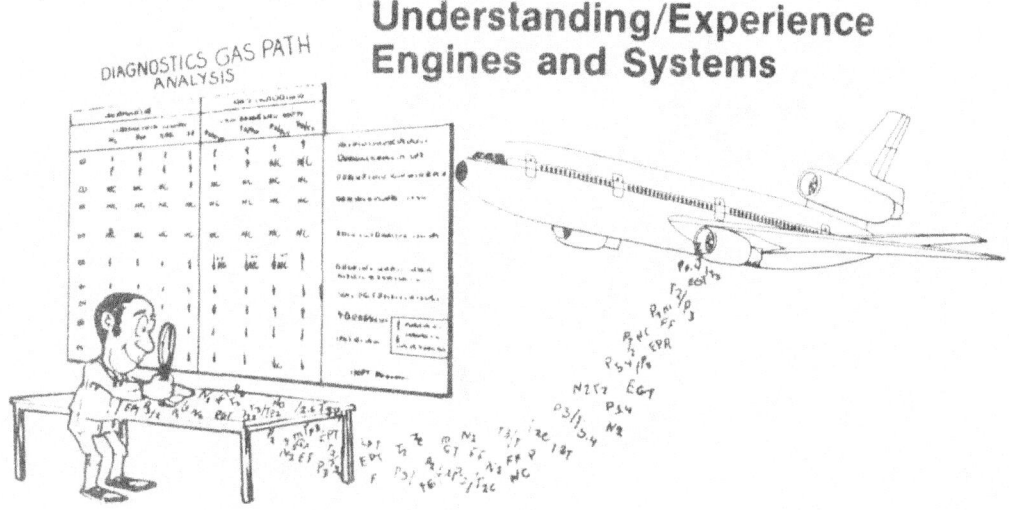

FIGURE 10

CF6-50 Engine Reliability

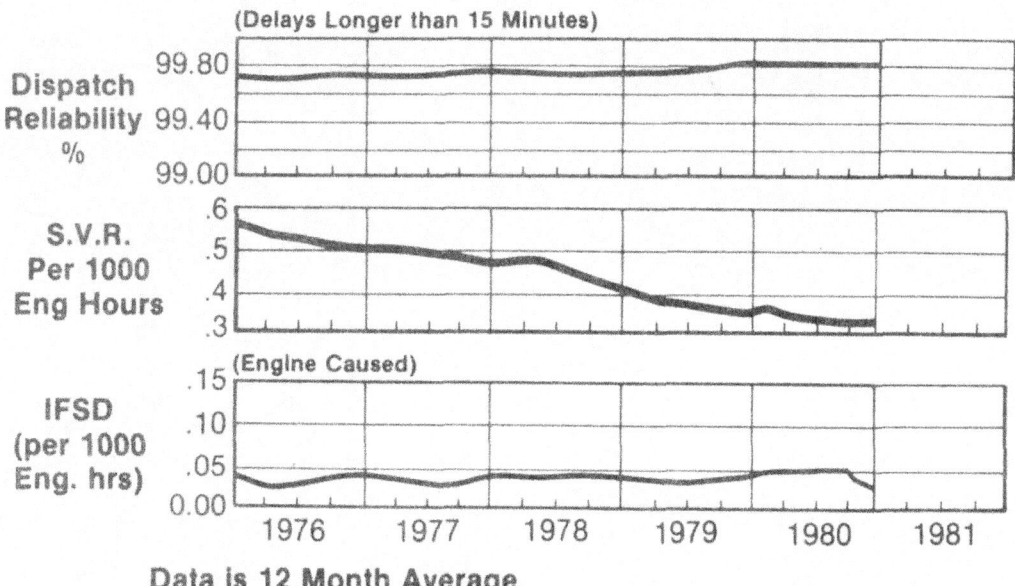

Data is 12 Month Average

FIGURE 11

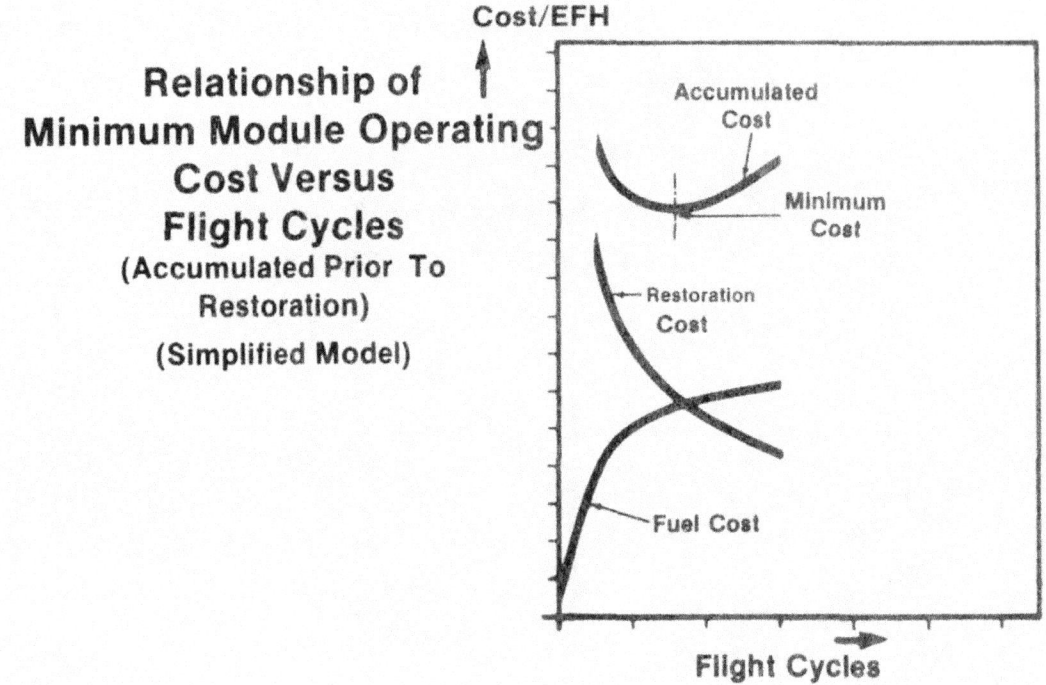

FIGURE 12

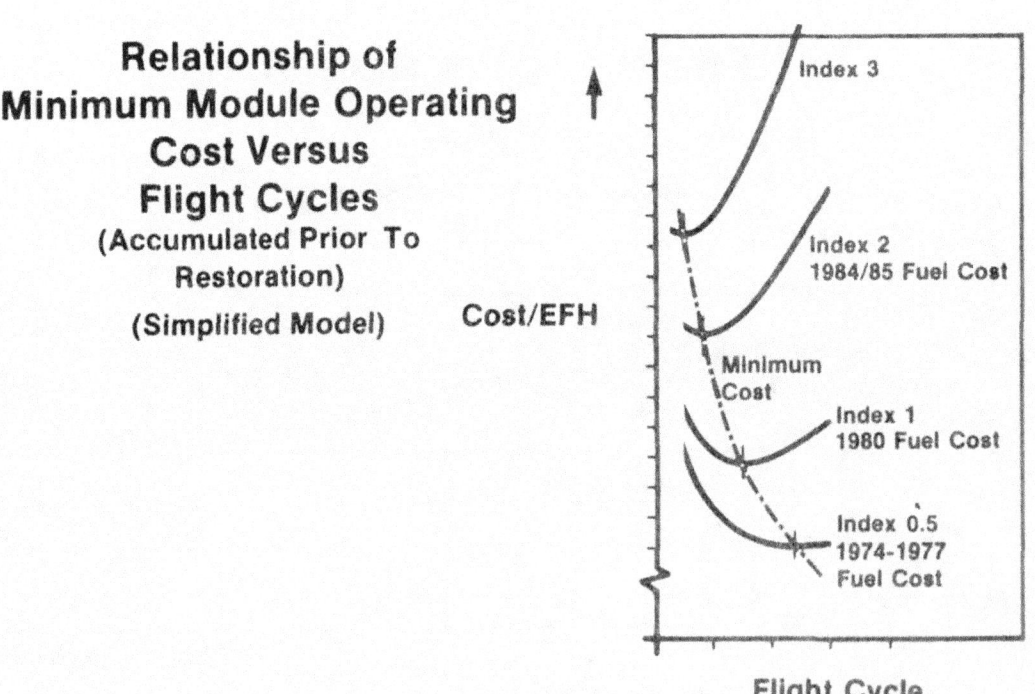

FIGURE 13

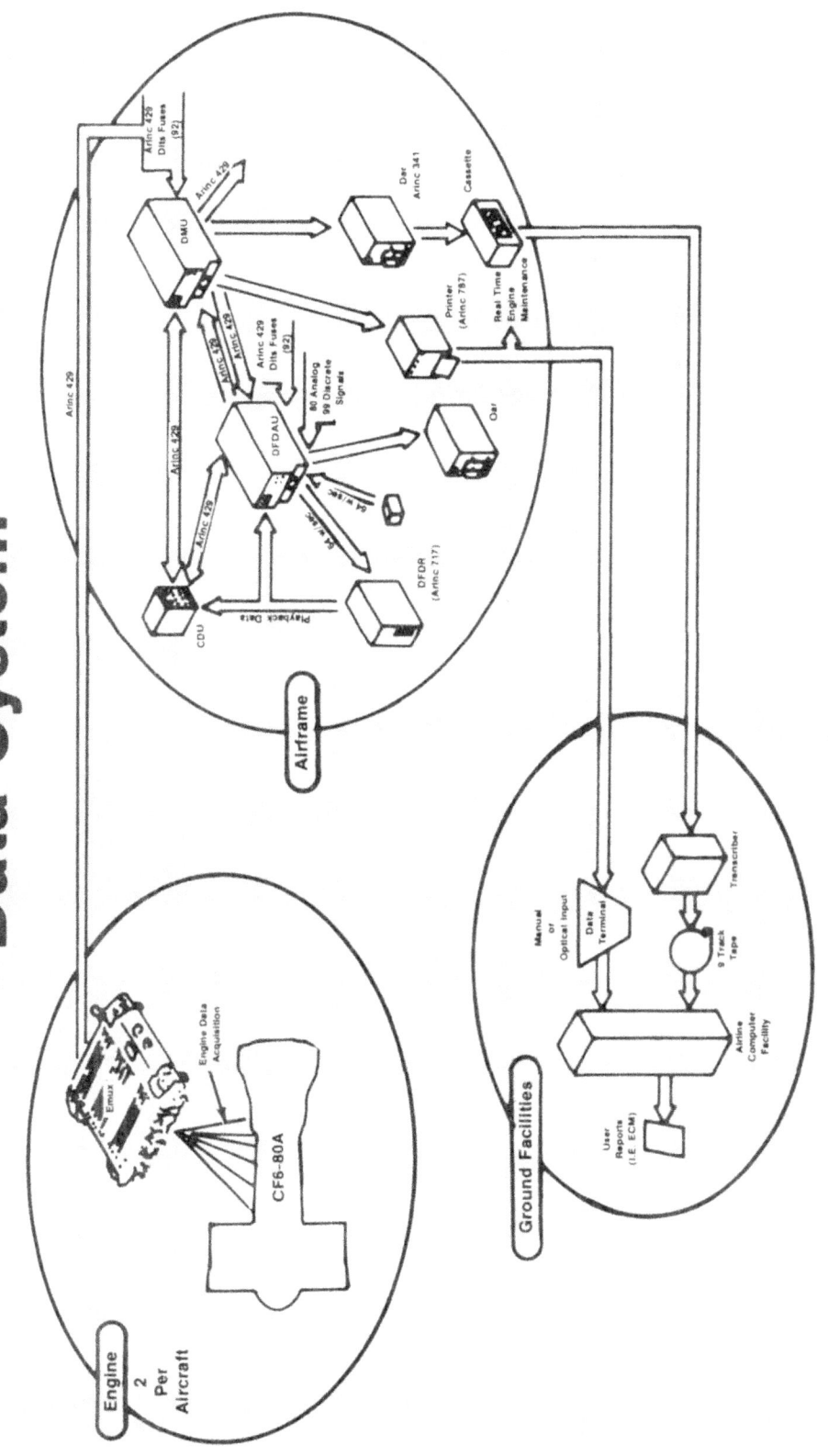

FIGURE 14

Future

Objective

- Expand on Effectiveness of Maintenance Tool

Approach

- Build on Proven CF6 Experience
- Extend Programs
    - Modular Performance Analysis
    - Improved Diagnostics
        Stall Detection
        Hot Starts
        EGT Margin
        Divergence Monitoring
        On-Board Trends
    - Benefits of Reduced Power

FIGURE 15

ENGINE HEALTH MONITORING - AN ADVANCED SYSTEM*

R.J.E. Dyson
General Electric Company

INTRODUCTION

The Advanced Propulsion Monitoring System (APMS) described in this paper fulfills a growing need for effective engine health monitoring. This need is generated by military requirements for increased performance and efficiency in more complex propulsion systems, while maintaining or improving the "cost to operate." This program represents a vital technological step in the advancement of the state-of-the-art for monitoring systems in terms of reliability, flexibility, accuracy, and provision of user-oriented results. It draws heavily on the technology and control theory developed for modern, complex, electronically controlled engines and utilizes engine information which is a by-product of such a system.

The General Electric Company has participated, and is participating, in a number of military and commercial engine health monitoring efforts which serve as a basis for this program. Most of the existing military systems would require a costly retrofit program of relatively mature vehicles with resultant cost, weight and space penalties. Therefore, this program is designed to progress concurrently with technologically advanced engines and electronic systems so that maximum advantage can be obtained from early development testing and then utilized in these engines and systems.

The intent of the APMS program is to demonstrate the usefulness of an efficient engine health monitoring system which ultimately could become part of a total aircraft data system. A concept for a future transport application is shown in Figure 1. One of the most important aspects is the development of MIL-STD-1553B data bus integration techniques which will be demonstrated in the APMS program. The system implementation will identify engine abnormalities, calculate and record engine life usage, and provide accurate and timely support information for flightline, intermediate and base maintenance personnel (see Figure 2 for a data flow schematic). With this equipment linked to a global data management network, as shown in Figure 3, an effective maintenance schedule and logistical support of an operational engine and weapons system can be realized. In short, APMS is being designed to support an On-Condition Reliability Centered Maintenance scheme.

---

* The work presented herein is being performed under a USAF-funded contract (No. F33615-79-C-2092) with Mr. K. R. Hamilton as the Project Engineer.

SYSTEM OVERVIEW

Several systems have adequately demonstrated the ability to acquire and record data but have suffered from shortcomings regarding the automatic analysis and presentation of useful data. Particular emphasis is being placed in the APMS program on three items:

- Improved result precision through data filtering, validation, and sensor degradation routines.

- Display simplicity and usefulness leading to increased system utilization.

- Life usage tracking information.

In the field of life usage tracking, the objective of this program is to collect and store sufficient data in the APMS processor to allow ground based determination of life consumption on specific life limited parts. This information could then ultimately be provided to the Comprehensive Engine Management System (CEMS) which would encompass the maintenance information, inventory accountability and technical information required by all levels of engine managers.

The specific purpose of the APMS program is to design, procure, and demonstrate an engine health monitoring system which utilizes, to the fullest extent, signals available from an electronic control. Other supplemental data required to provide useful information for all operational and maintenance levels will be acquired and integrated with the control information. In order to implement this, the following tasks will be performed:

- A comprehensive system operating analysis will be conducted in order to design a complete user-oriented system.

- The supplemental engine data will be handled by an on-engine signal conditioner and multiplexer (SCM) and transmitted, together with the control data, through a MIL-STD-1553B port or terminal.

- The flow of data on a MIL-STD-1553B data bus through a system of remote ports (terminals) will be demonstrated.

- An aircraft-type processor and memory will be provided for event detection, event storage, engine usage tracking, and acquisition of trend data.

- An off-engine support system will control the data bus, display flight-line data, store trend data, and obtain and process corroborating data. The plan for development and test of the APMS off-engine subsystem is shown in Figure 4.

- Software will be developed for the aircraft-type processor and the support system which will allow event, trend and engine usage data to be acquired, stored and made available for subsequent ground processing.

- The functional capability of the total hardware/software system will be demonstrated by testing on an advanced engine and by subsequent data display and analysis.

- Software for the WPAFB ASD CYBER 175 computer to provide analysis of long-term trends, life usage, modular fault isolation, and parts tracking is being developed under a parallel contract.

DESIGN OBJECTIVES AND IMPLEMENTATION

The objectives of the Advanced Propulsion Monitoring System (APMS) are to demonstrate a system which identifies aircraft turbine engine operating abnormalities early to minimize secondary damage, optimizes scheduling of engine repairs, improves effective use of maintenance facilities, increases aircraft availability, and reduces operating cost. This will be achieved through monitoring of signals from engine-mounted sensors through an on-aircraft computer which acquires and stores data and provides engine health status to the flight crew and maintenance personnel. Due to the limited processing capability of this on-aircraft computer, further processing of the accumulated data by a ground-based central computer will be performed to establish long-term engine health trends and computed cyclic life expenditure of life-limited engine parts. See Figures 5 and 6 for operational application. Figure 5 shows a typical fighter application with engine health monitoring data on a dedicated data bus. Figure 6 shows the associated flight line and central computer equipment.

The ultimate objective of the APMS is installation in an operational aircraft to demonstrate the feasibility of the system. The current program however will address only the installation on a demonstrator engine to be run in a test cell (see Figure 7). Under this condition, the APMS off-engine system will be housed in a room adjacent to the test cell instead of the electronics bay of the aircraft.

All of the signals required for engine health monitoring will be obtained from the SCM through a MIL-STD-1553B data bus. The SCM will obtain data from the engine control and from supplemental condition monitoring sensors. It will condition and digitize all signals necessary for engine health monitoring and provide a MIL-STD-1553B interface.

The APMS can be divided into two sybsystems for discussion purposes as follows:

On-Engine Data Acquisition Subsystem

This subsystem consists of the on-engine hardware, including sensors, cables, and the SCM with MIL-STD-1553B terminal. The SCM is capable of receiving digital data from a Full Authority Digital Electronic Control (FADEC) or analog data from an Augmentor/Fan Temperature (AFT) control. The following functions are performed on-engine:

a. FADEC

- Updates data every 10 ms.
- Provides limited sensor screening and self test.
- Furnishes digital output of control parameters.

b. AFT (IF USED)

- Provides analog output of control parameters.

c. SCM

- Interfaces with FADEC digital output or AFT control analog output.
- Interfaces with additional required engine health monitoring sensors.
- Provides signal conditioning where necessary.
- Provides sensor range checks and channel filtering.
- Interfaces with MIL-STD-1553B data bus, which is a military standard defining the requirements for digital, command/response time division multiplexing techniques on aircraft. It establishes uniform requirements and promotes standard digital interfaces.

Off-Engine Subsystem

This subsystem as shown in Figure 8 will acquire data through the digital data bus and will include the following:

a. APMS Processor

This represents the aircraft-mounted computer and memory, and performs the following prime functions:

- Acquires engine data from SCM via the data bus.
- Acquires airframe-related data via the data bus.

- Provides data-sensor processing validation and filtering.
- Recognizes conditions for taking trend and life usage data.
- Performs exceedance checks and transmits limit exceedances to display via data bus.
- Records event, trend and life usage data.
- Downloads to data computer.

b. Simulated Cockpit Display

This will provide real time indication in the cockpit of critical limit exceedances and events. This hardware will be in demonstrator form in order to validate a concept which could be included in a future application.

c. Air Data Signal Simulator Signals

These signals would normally be available from the air data computer and will be obtained from the test cell or simulated by the air data signal simulator and transmitted to the data bus.

d. Data Computer

This data computer and its peripherals are laboratory-type equipment which in total perform the following functions:

- MIL-STD-1553B Bus Controller (see Figure 9 - Bus System)
- Real Time Data Display
- Flight Data Retrieval and Storage
- Flight Line Data Display
  - Mission readiness from major cycles, minor cycles, time at temperature and total hours
  - Sensor failure
  - Manual event trigger
  - APMS loader and editor
  - Interface with the WPAFB CYBER 175

The most important task in the preliminary design is the System Requirement and Software Definition for data display, analysis and maintenance decisions. The definition of these requirements is an iterative process among General Electric, co-contractors, subcontractors and the Air Force and will integrate needs and existing hardware into an acceptable package within the cost/schedule constraints. A critical system requirements review was held eight months after contract go-ahead so that hardware and software specifications could still be influenced yet not delay procurement and programming. A final critical detail review will be held at the conclusion of the design phase.

SYSTEM FUNCTIONAL DESCRIPTION

System requirements will define the comprehensive system concepts which will form the basis for the final design. The system functions can be broken down to the following functions for discussion purposes.

Parameter Signal Sources

It is planned that twenty-three engine parameter signals will be acquired from either the control or from CM sensors. These conditioned, digitized, and multiplexed signals will be transmitted to the APMS processor via the MIL-STD-1553B data bus. The parameters are shown in Figure 10.

In addition to the above, a number of aircraft signals will also be transmitted to the APMS processor via the MIL-STD-1553B digital data bus. These are shown in Figure 11. Other aircraft signals could be added if required for a specific application.

System Operation

The Advanced Propulsion Monitoring System (APMS) consists of data acquisition and processing hardware and associated software logic. The purpose of the APMS is to monitor installed turbine engine behavior, to detect start-up and in-flight malfunction events, to track engine life usage, and to assess long-term performance degradation.

The APMS program will demonstrate a test system configuration representative of a flight-type installation. The major components are described below.

<u>Signal Conditioner/Multiplexer (SCM)</u> - The SCM acquires, conditions, and processes engine sensor data for transfer to the APMS. It is an engine-mounted, fuel-cooled unit, based on a SBP 9900 microprocessor which performs the following functions:

- Interfaces with FADEC or AFT control and CM sensor set

- Obtains data every 10 milliseconds

- Performs circuit built-in test

- Performs range checks on non-FADEC data
- Performs channel filtering on 10 samples
- Performs stall detection
- Outputs conditioned, digitized, serialized data in MIL-STD-1553B format every 100 milliseconds.

APMS Processor - The APMS processor receives data via the 1553B data bus from the SCM and the air data signal simulator (ADSS). It issues event messages to the cockpit display and status panel regarding detected conditions. It selectively retains data for subsequent transfer to a ground processing unit. It performs the following functions:

- Minor Loop Processing (every 0.1 sec.)
    - Data Acquired every 0.1 sec.
    - Engineering units conversion
    - Sensor validation
- Major Loop Processing (every 1 sec.)
    - Diagnostic Functions
    - Median Calculated (9 readings)
    - Data Mode
    - Event Storage of Exceedances
    - Trend Storage of Stabilized Data
    - Mission Profile Corner Points Calculated
    - Life Usage Updated

Data Computer - The data computer provides the bus controller function to effect data transfer between APMS components every 0.1 second during normal running. It controls off-line data transfer, it provides bulk data storage, engineering access to the system, and controls data transfer (via telephone MODEM) to the central computer, a CYBER 175 at Wright Patterson Air Force Base (WPAFB). The data computer operates in one of the following modes which are descriptive of its functions:

- Load - Load APMS processor through RS232
- Run - Normal APMS operation

- Display - Current APMS 1 second data scan
- Record - SCM 100 millisecond data to disc
- Data Transfer - by file
    - Engine History
    - Trend
    - Takeoff
    - Mission Profile
    - Flight Record
    - Exceedance
- Life Usage Calculation

DATA ACQUISITION/ANALYSIS EMPHASIS

In addition to the emphasis placed on system requirements, on-engine electronics and central data bus implementation, certain areas of data acquisition and analysis have been identified for receiving special attention. Two of these areas will be discussed below.

Data Filtering

Statistical approaches are presently in use in overhaul test cells to assess the performance of engine components and to estimate measurement errors. These approaches are being developed for "on-wing" applications, some of which will be demonstrated in this APMS program.

"Filtering" works by augmenting the measurement data with additional information which is available to the analyst but has not traditionally been incorporated into analysis programs. This additional information includes:

- typical values of the measured parameters;

- a list of possible engine problems which could cause a change in performance (reduced efficiency or pumping capacity, etc.);

- a signature or pattern which tells how each of the potential problems would be reflected in the measured parameters;

- a standard deviation for each of the potential engine problems which indicates what magnitude of change might reasonably be expected;

- a standard deviation for each of the measurements which indicates what level of measurement error might reasonably be expected.

This extra information is used to interpret the actual measurements. If an engine problem has occurred, it will normally be reflected in several of the measurements, and, hence, the "filter" can look for the pattern to recognize the problem. Deviations which do not fit an expected pattern are attributed to measurement error. To express the same thought in a more rigorous way, the "filter" finds the most probable combination of engine problems and measurement errors to explain the observed measurements given the additional information identified above. Examples of unfiltered and filtered data are shown in Figures 12 and 13.

The data in both figures has been normalized to account for variations in power setting and environmental conditions (i.e., speed, Mach number, altitude, etc.). In Figure 13, in addition to filtering the data as described above, the initial data point has been defined as "zero" to facilitate identifying subsequent changes in efficiency.

Implementation of Engine Parts Usage Tracking

Background - The On-Condition Maintenance Concept (OCM) depends heavily on the ability to change or replace parts only when necessary to preclude in-service malfunction or performance deterioration. This requires that an accurate evaluation of the operating history of the engine and its components be available to the maintenance and logistics specialists.

To achieve this, engine operating parameters, which significantly affect the potential lives of engine components, must be monitored. Also, the serial numbers and installation history of parts subject to wear-out, maximum operating time, low-cycle fatigue damage, or other limitations which are considered logistically important must be tracked.

Engines most recently introduced into the USAF inventory have been designed for the OCM concepts. Previous maintenance has been based on the requirement to perform teardown inspections and overhauls at prescribed engine flight-hour intervals (Hard Time). This normally requires the replacement of many items reaching predetermined time or cycle limits. This system does not provide for differentiation or discretion based on the severity of the exposure to life-consuming conditions. As an example, a simple time (hours) limit does not provide information on whether the time was accumulated at a high power setting or at a lower value which may have consumed less life. In the case of simple cycles (idle-max-idle), data is not normally available to identify the actual level of power extremes which may affect life consumption. In addition, other variations in power setting (80 percent to max, 90 percent to max, etc.) also consume life not accounted for in a gross measuring system.

To fully utilize the life potential of parts limited by low cycle fatigue, thermal fatigue, stress rupture, wear-out, operating time, etc., it is necessary to keep track of each part's operating exposure, location, and serial number. Assuming the life limiting parameters (cycles, hours, time at temperature, etc.) have been established for each life limited part, and a system devised to collect the data, a system must be implemented which provides maximum utilization of this data for logistics and maintenance purposes.

Engine Usage Tracking Description - It is the objective of this program to acquire appropriate data in the APMS processor to allow future ground-based determination of:

- Mission severity impact on maintenance and logistics indices
- Parts life consumption

Engine operation will be continuously monitored and data sets will be selectively saved to allow reconstruction of the mission profile defining changes in operating conditions which affect mission severity indices and parts life consumption, although neither of these computations are included in this program.

The parameters to be monitored and to be used to detect a meaningful change in operating condition and, thus, effect the saving of a data set, are:

- Fan Speed
- Power Lever Angle
- Compressor Inlet Temperature
- Ambient Air Temperature

The data set to be saved will include the above parameters, plus:

- Core Speed
- HPT Blade Temperature
- Altitude
- Mach Number
- Engine Run Time

The acquired data could be processed by a ground-based system to provide the following functions:

- Mission severity indices could be calculated to establish the relationships with "Normal" mission severity factors used to determine maintenance and logistic indices. Time-weighted composite severity indices could be used to adjust maintenance and logistics indices which reflect the way the fleet was "in fact" being operated.

- The data would be processed to count major and minor cycles accumulated during operation for those rotating parts with established LCF cyclic life limits.

- The data would be processed to count time at temperature (two levels) for assessment of hot section static parts life consumption.

- Total engine operating time could also be established.

CONCLUDING REMARKS

The hardware to implement these system requirements and functions is presently under development. The SCM is a flight-type SBP9900 microprocessor based, on-engine, electronic box. The APMS processor is a TI 9900 based avionics bay-type electronic box designed, using applicable provisions of MIL-E-5400R as a design guide. All other off-engine hardware is laboratory-type equipment intended to demonstrate the concepts described in this paper with a TI 990/10 mini-computer serving as the heart.

The APMS program will advance engine health monitoring concepts by offering an integrated and comprehensive approach. The purpose is to design, develop and demonstrate efficient methods of acquiring, processing and utilizing data and information obtained from an advanced engine system from engine inlet controls and from aircraft subsystems so that the monitoring system itself can be incorporated early in the design and development of aircraft weapons systems of the 1980's and 1990's.

The Advanced Prouplsion Monitoring System can provide the generic technology base for monitoring the health of sophisticated propulsion systems of high performance aircraft of the 1980's and 1990's.

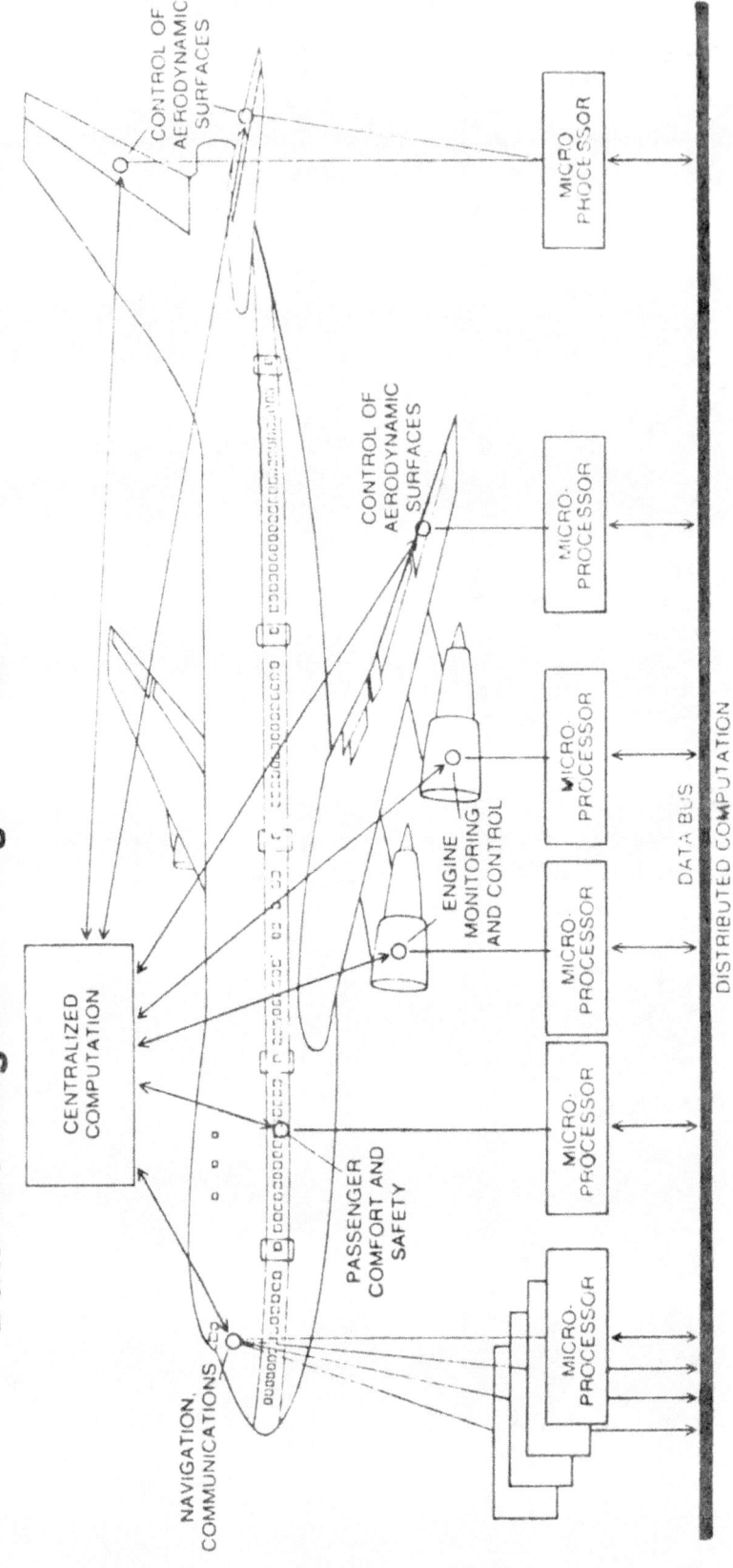

FIGURE 1

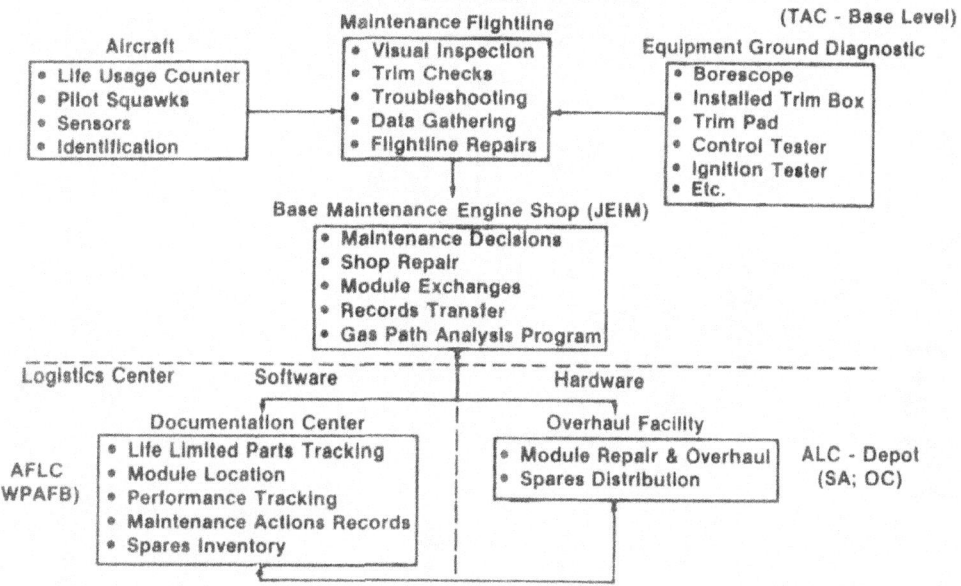

FIGURE 2

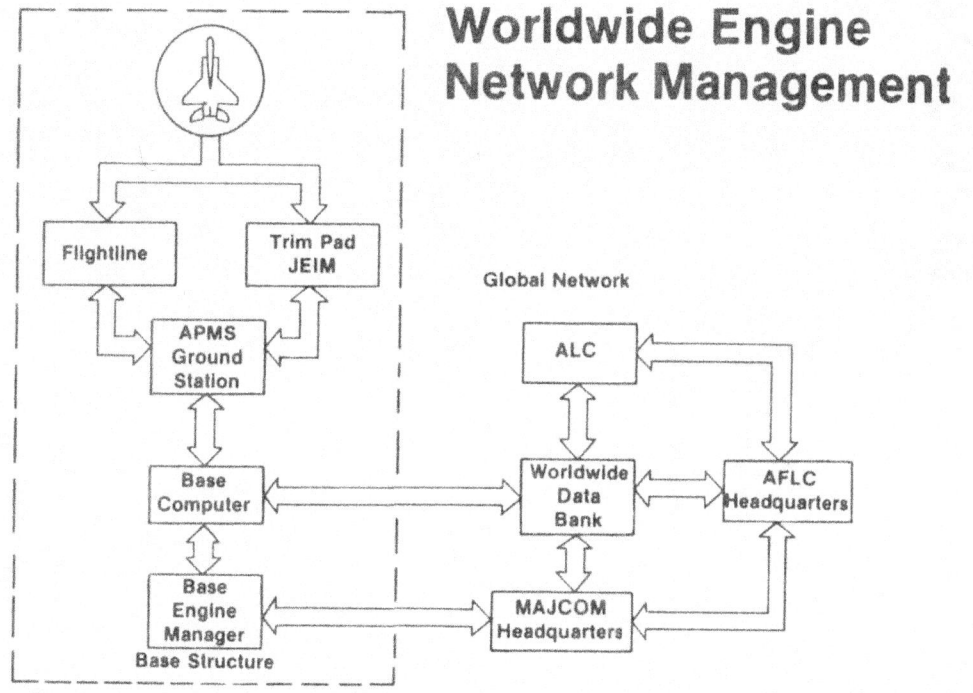

FIGURE 3

Off-Engine Subsystem Development Plan

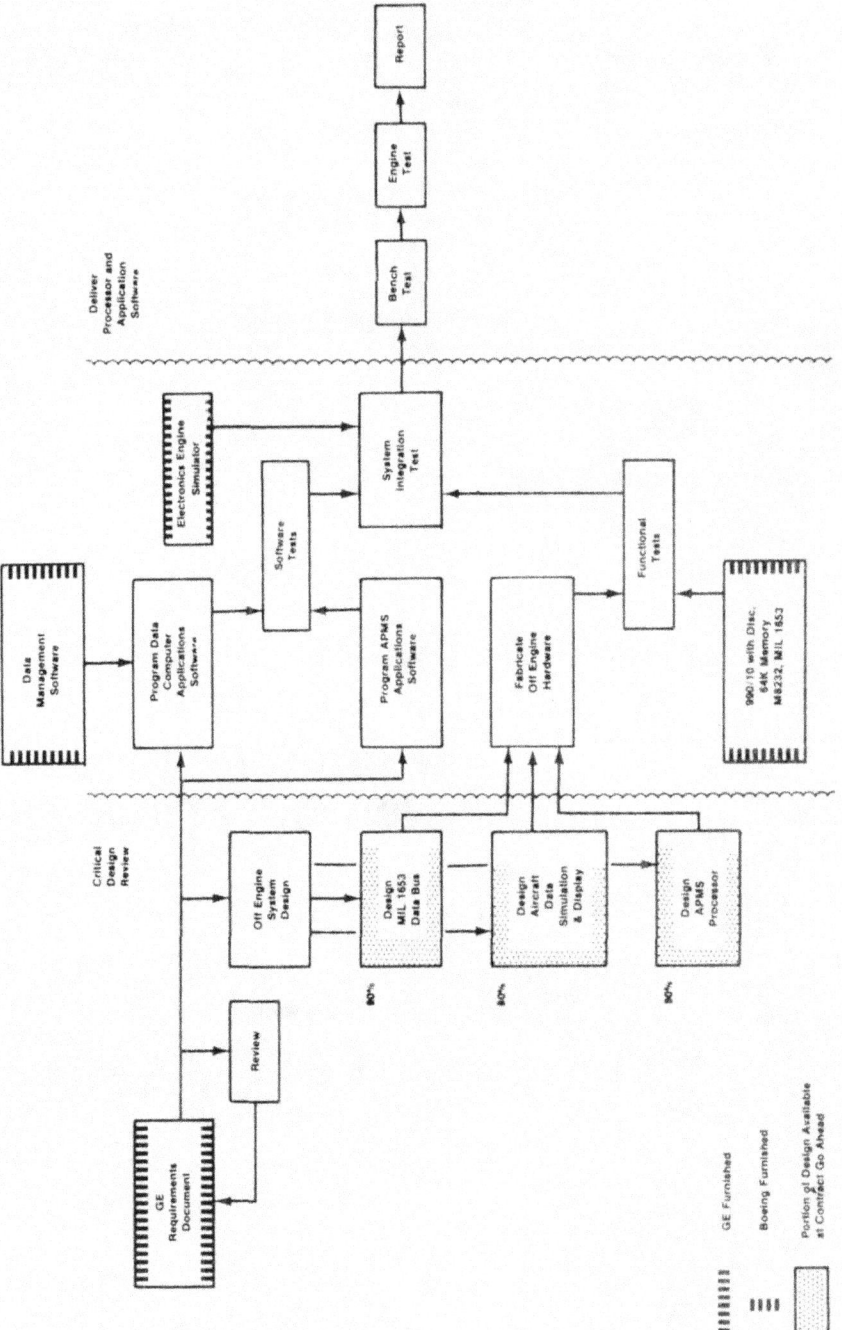

FIGURE 4

Advanced Propulsion Monitoring System
Proposed Follow-on Flight Configuration

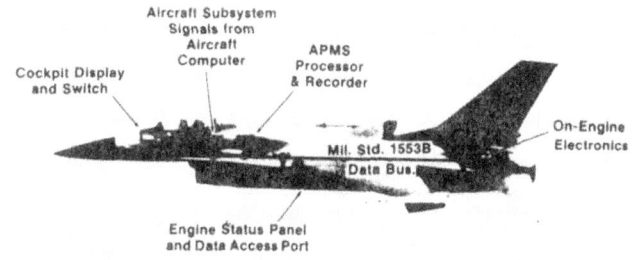

Typical Future Fighter Application with Engine Hours Monitoring Data on a Dedicated Data Bus.

FIGURE 5

Advanced Propulsion Monitoring System
Data Flow for Fighter Application

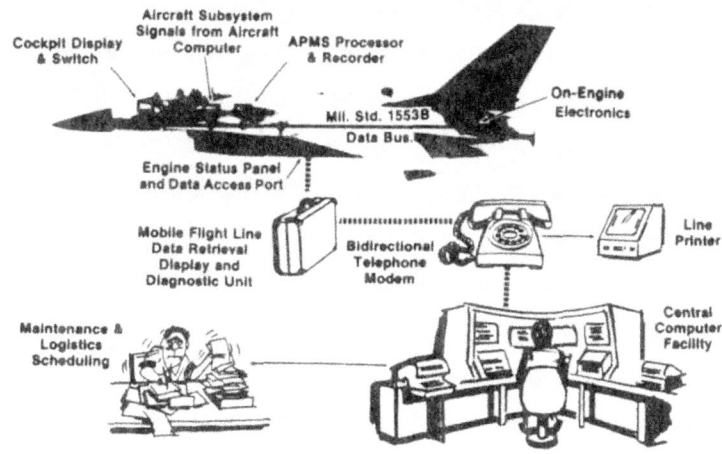

FIGURE 6

APMS Demonstration Configuration

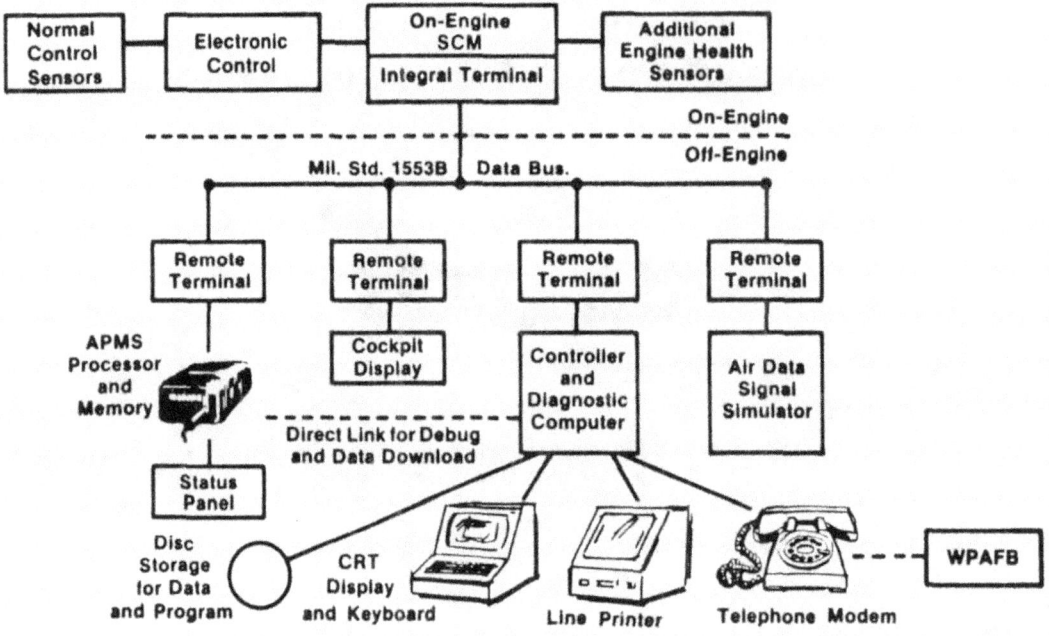

FIGURE 7

APMS Off-Engine Subsystem

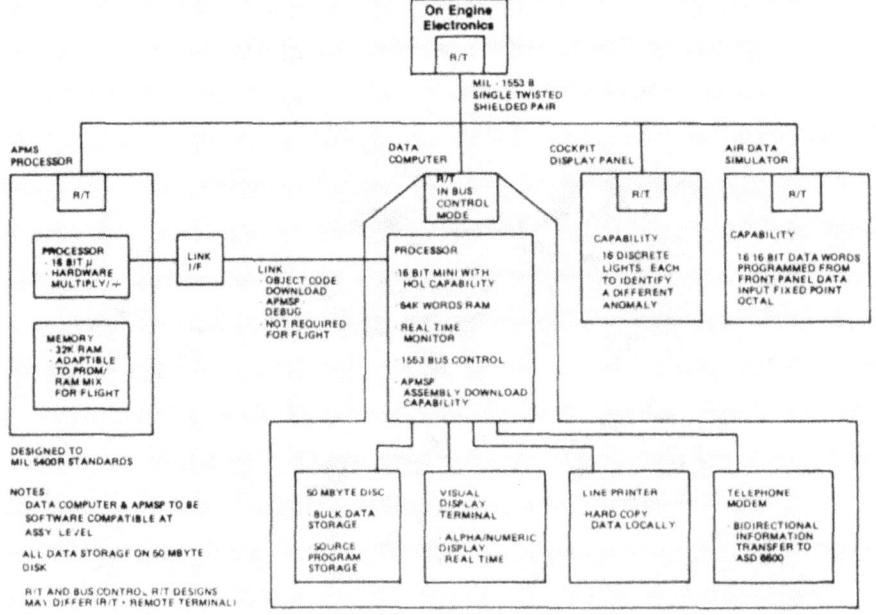

FIGURE 8

APMS Bus System

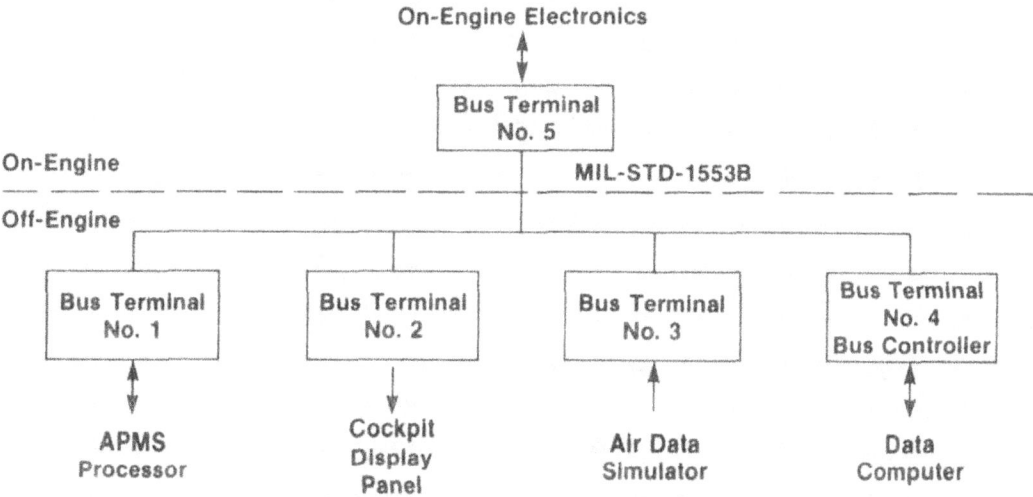

FIGURE 9

Condition Monitoring Parameters

Available from Electronic Control		Additional C/M Sensors	
A8	Exhaust Nozzle Area	T3	Compressor Discharge Temperature
BF	IGV Position	P49	Turbine Interstage Pressure
WFM	Main Fuel Flow	QL	Lube Quantity
WFR	Augmentor Fuel Flow	TL	Lube Temperature
T2	Inlet Temperature	PL	Lube Pressure
N1	Fan Speed	VF	Fan Vibration
N2	Core Speed	VC	Core Vibration
PS3	Compressor		
PLA	Power Lever Angle		
T4B	HPT Blade Temperature		
FDS	Flame Detector		
PAUG	Augmentor Switch		
DP14	Fan Pressure Ratio		
BC	Core Stator Position		
T25	Compressor Inlet Temperature		
PS14	Fan Discharge Pressure		

FIGURE 10

Air Data Simulator Signals

PAMB	Ambient Pressure	Real Cell Sensor
TAMB	Ambient Temperature	Real Cell Sensor
ALT	Altitude	Calculated in Signal Simulator
MN	Mach No.	Calculated in Signal Simulator
PLA	Power Lever Angle	Test Cell Sensor
WOW	Weight on Wheels	Manual Switch
POS	Pilot Option Switch	Manual Switch or Keyboard Entry

FIGURE 11

Component Efficiency vs Cycles

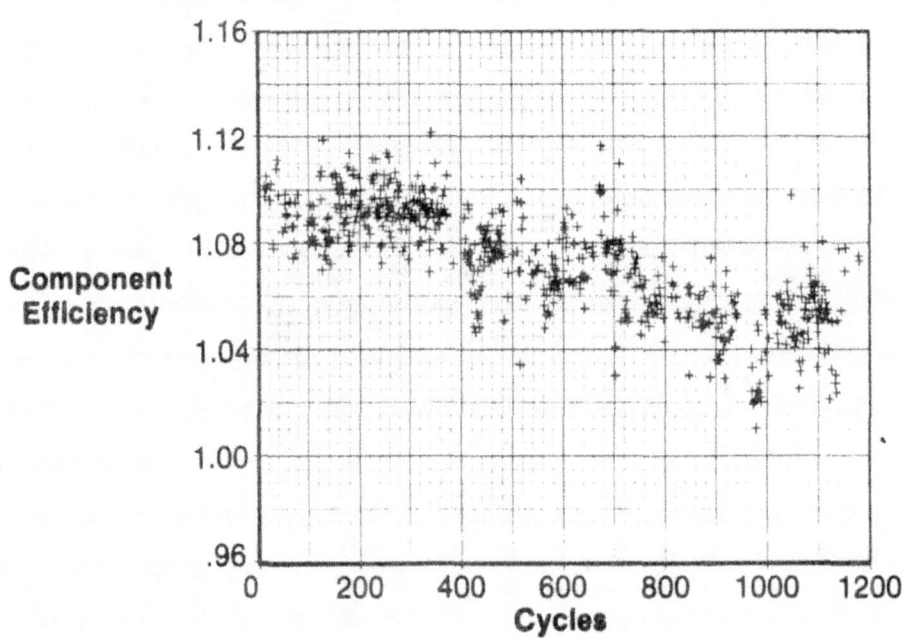

FIGURE 12

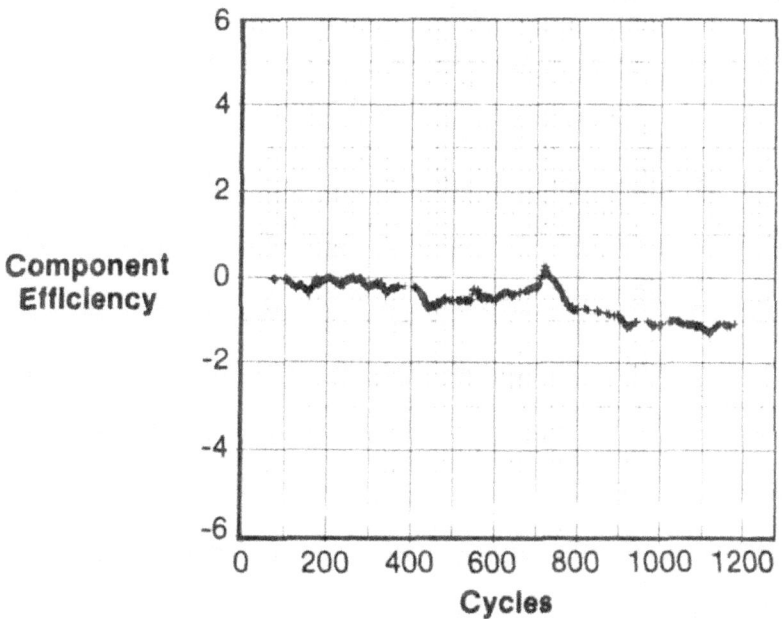

FIGURE 13

Page intentionally left blank

AN OVERVIEW OF SAE ARP 1587

"AIRCRAFT GAS TURBINE ENGINE MONITORING SYSTEM GUIDE"

John A. Murphy
Bell Helicopter Textron

The Society of Automotive Engineers (SAE) organized the E-32 Engine Monitoring Committee in November, 1976, with Tom Warwick of Pratt and Whitney as Chairman.

The committee is made up of members affiliated with:

- Engine manufacturers
- Equipment suppliers
- Airlines
- Government agencies
- Military services
- Aircraft manufacturers

Members serve as individuals, not as agents of any organization, and are expected to represent and vote their own opinions. Members are chosen because of their knowledge and expertise in the field, and need not belong to the SAE. There are currently 35 committee members. Countries represented, in addition to the USA, include Canada, France, Great Britian, Spain, and the Netherlands.

The primary task of the Committee thus far has been preparation of an SAE Aerospace Recommended Practice (ARP) document. This document has been completed, and is designated ARP 1587 "Aircraft Gas Turbine Engine Monitoring System Guide".

ARP 1587 outlines a systematic approach to developing an Engine Monitoring System (EMS). It presents an extensive shopping list of EMS capabilities and benefits. A team approach to developing an EMS is emphasized with a description of the responsibilities of each team member.

This is considered a keystone document, in that it lays the groundwork for future committee publications and activities. In addition to maintaining ARP 1587, the E-32 Committee is considering engine life usage methodology, "lessons learned" on past EMS programs, and hardware requirements including sensors amd "minimum system" definitions. These follow-on activities will be under the able direction of Bill Peters of General Electric, who is taking over as committee chairman. Tom Warwick will remain on the committee, but will assume a less active role. Tom has done a fantastic job over the past four years of guiding ARP 1587 from concept to reality.

It is planned that ARP 1587 will be available prior to the AIAA/SAE/ASME Joint Propulsion Conference to be held in Colorado Springs, July 27-29, 1981.

The ARP is currently priced at $6.00 per copy (SAE member or non-member). A companion Special Publication (SP 478), containing ARP 1587 and several related technical papers, will also be available at $20.00 per copy ($16.00 to SAE members).

For further information please contact:

David R. Bentley, Staff Engineer
Society of Automotive Engineers, Inc.
400 Commonwealth Drive
Warrendale, PA 15096

(412) 776-4841

ARP 1587 AIRCRAFT GAS TURBINE ENGINE MONITORING SYSTEM GUIDE

Recommends Systematic Approach Toward Engine Monitoring

FIGURE 1

ARP 1587 AIRCRAFT GAS TURBINE ENGINE MONITORING SYSTEM GUIDE

Presents Extensive List of Design Options

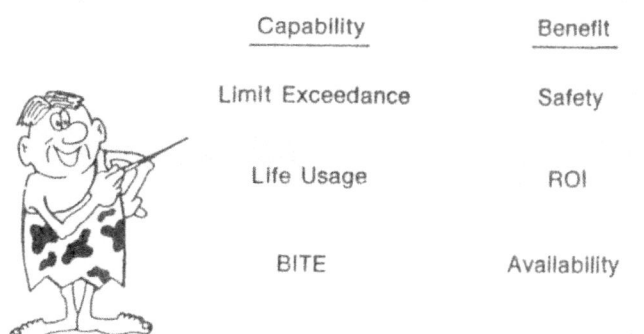

FIGURE 2

ARP 1587 AIRCRAFT GAS TURBINE ENGINE MONITORING SYSTEM GUIDE

Addresses User and Supplier Responsibilities

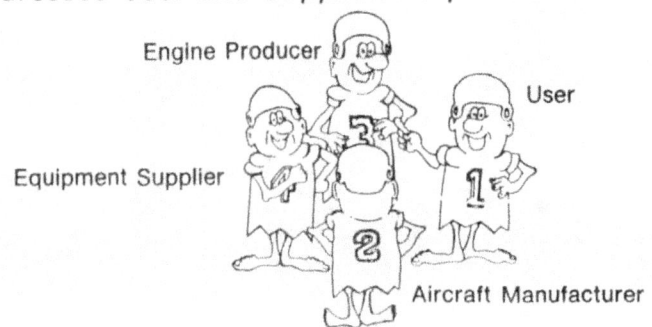

FIGURE 3

ARP 1587 AIRCRAFT GAS TURBINE ENGINE MONITORING SYSTEM GUIDE

Gets It All Together with Keystone Document

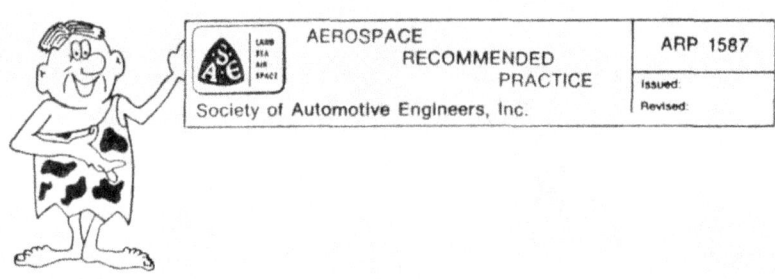

FIGURE 4

www.ingramcontent.com/pod-product-compliance
Lightning Source LLC
Chambersburg PA
CBHW081716170526
45167CB00009B/3602